The Private Science of
LOUIS PASTEUR

The Private Science of
LOUIS PASTEUR

Gerald L. Geison

PRINCETON UNIVERSITY PRESS

PRINCETON, NEW JERSEY

Published by Princeton University Press,
41 William Street, Princeton, New Jersey 08540
In the United Kingdom:
Princeton University Press, Chichester, West Sussex

Library of Congress Cataloging-in-Publication Data

Geison, Gerald L., 1943–
The private science of Louis Pasteur / Gerald L. Geison
p. cm.
Includes bibliographical references and index.
ISBN 0-691-03442-7
1. Pasteur, Louis, 1822–1895.
2. Scientists—France—Biography.
I. Title.
Q143.P2G35 1995
509.2—dc20
[B] 94-35338

This book has been composed in Berkeley with Benguiat Display

Princeton University Press books are printed on
acid-free paper, and meet the guidelines for permanence and
durability of the Committee on Production Guidelines
for Book Longevity of the Council on Library Resources

Printed in the United States of America

2 4 6 8 10 9 7 5 3

For Christopher and Andrew

MY FAVORITE SONS

[Pasteur] was the most perfect man who has ever entered the Kingdom of Science.

—Stephen Paget, *Spectator* 1910

. . .

Rarely . . . has history been so falsified—and with so much impudence.

—Philippe Decourt,
"Deuxième lettre à nos amis" 1975

. . .

In France, one can be an anarchist, a communist or a nihilist, but not an anti-Pastorian. A simple question of science has been made into a question of patriotism.

—Auguste Lutaud, *Pasteur et la rage*
1887

Contents

List of Illustrations and Tables ix

Preface xiii

PART I. BACKGROUND AND CONTEXT

Chapter One. Laboratory Notebooks and the Private Science of Louis Pasteur 3

Chapter Two. Pasteur in Brief 22

PART II. FROM CRYSTALS TO LIFE

Chapter Three. The Emergence of a Scientist: The Discovery of Optical Isomers in the Tartrates 53

Chapter Four. From Crystals to Life: Optical Activity, Fermentation, and Life 90

Chapter Five. Creating Life in Nineteenth-Century France: Science, Politics, and Religion in the Pasteur-Pouchet Debate over Spontaneous Generation 110

PART III. VACCINES, ETHICS, AND SCIENTIFIC VS. MEDICAL MENTALITIES: ANTHRAX AND RABIES

Chapter Six. The Secret of Pouilly-le-Fort: Competition and Deception in the Race for the Anthrax Vaccine 145

Chapter Seven. From Boyhood Encounter to "Private Patients": Pasteur and Rabies before the Vaccine 177

Chapter Eight. Public Triumphs and Forgotten Critics: The Debate over Pasteur's Early Use of Rabies Vaccines in Human Cases 206

Chapter Nine. Private Doubts and Ethical Dilemmas: Pasteur, Roux, and the Early Human Trials of Pasteur's Rabies Vaccine 234

PART IV. THE PASTORIAN MYTH

Chapter Ten. The Myth of Pasteur 259

Appendixes 279

Author's Note on the Notes and Sources 305

Notes to the Chapters 309

Acknowledgments 343

Bibliography 345

Index 367

List of Illustrations and Tables

PLATES

1. Pasteur's mother. From a portrait drawn in pastel by Pasteur at the age of thirteen.
2. Pasteur's father. From a portrait drawn in pastel by Pasteur at the age of fifteen.
3. Pasteur's birthplace in Dole.
4. Pasteur in 1846, while a student at the Ecole Normale Supérieure. From a drawing by Lebayle, based on a daguerrotype.
5. Pasteur in 1857, while dean of the Faculté des sciences at Lille.
6. Pasteur's first laboratory at the Ecole Normale Supérieure.
7. Pasteur and Madame Pasteur in 1884.
8. Emile Roux.
9. Pasteur in 1884.
10. Pasteur observing rabbits injected with the rabies virus. From *La Science illustré*, 15 September 1888.
11. Joseph Meister in 1885.
12. Jean-Baptiste Jupille in 1885.
13. From the famous painting by Rixens of Pasteur's jubilee at the Sorbonne.
14. Pasteur, in 1892, with his grandson.
15. The original building of the Institut Pasteur, inaugurated in November 1888.
16. Pasteur in 1895, the last photograph taken of him in the gardens of the Institut Pasteur.
17. Pasteur's funeral procession through the streets of Paris, 5 October 1895.
18. Pasteur's mausoleum at the Institut Pasteur.
19. "The Death of Pasteur. Exhibition of the Body at the Institut Pasteur."

20. "La mort du Pasteur," *Le journal illustré*, 6 October 1895.
21. "Pasteur est eternal."
22. Pasteur as "Benefactor of Humanity." Frontispiece from Fr. Bournard, *Un bienfaiteur de l'Humanité: Pasteur, sa vie, son oeuvre*.
23. "National Homage: From France to Louis Pasteur."
24. "Pasteur Destroys the Theory of Spontaneous Generation." Advertising card for La Chocolaterie d'Aiguebelle.
25. "Pasteur Discovers the Rabies Vaccine." Advertising card for La Chocolaterie d'Aiguebelle.
26. Pasteur seated in his laboratory. Advertising card for the Urodonal Company in honor of the centenary of Pasteur's Birth.
27. "Wine Is the Healthiest and Most Hygenic of Beverages." Advertisement on the official map of the Métro subway system.

FIGURES

3.1. Hemihedral crystals of sodium ammonium tartrate. 55
3.2. The path to Pasteur's discovery of optical isomers—the standard story. 57
3.3. From Pasteur's first laboratory notebook, "Notes divers," 2/5/2 on the microfiche owned by Seymour Mauskopf. 63
3.4. Pasteur, "Notes divers," 1/3/1. 65
3.5. Pasteur, "Notes divers," 1/3/2. 67
3.6. Pasteur, "Notes divers," 1/3/3. 68
3.7. Pasteur, "Notes divers," 1/3/7. 74
3.8. Pasteur, "Notes divers," 1/3/8. 75
3.9. Pasteur, "Notes divers," 1/3/9. 76
3.10. Pasteur, "Notes divers," 1/3/10. 77
3.11. The path to Pasteur's discovery of optical isomers—as reconstructed from his laboratory notes. 82
5.1. Experiment against spontaneous generation: Pasteur's apparatus for collecting solid particles from atmospheric air and then introducing them into a previously sterile flask. 115

5.2. The "swan-necked" flasks used in Pasteur's most elegant experiments against spontaneous generation. 117

5.3. Pasteur on the correlations between internal structure, external form, optical activity, and life. 136

6.1 (a,b,c). Pasteur's handwritten record of the agreed upon protocol for the trial of the anthrax vaccine at Pouilly-le-Fort. 152

6.2. This page, from the same laboratory notebook as figure 6.1 (a,b,c), establishing that Pasteur did in fact use the potassium bichromate vaccine at Pouilly-le-Fort. 155

6.3. Schematic diagram of "The Secret of Pouilly-le-Fort." 157

7.1. On the relativity of immune responses. 186

7.2. Pasteur's path to his rabies vaccine: the published papers. 194

7.3. Pasteur's laboratory notes on the presumably rabid M. Girard, his first "private patient." 196

7.4. Pasteur's laboratory notes on Julie-Antoinette Poughon, his second "private patient." 199

8.1 (a,b). Pasteur's laboratory notes on the treatment of Joseph Meister, beginning on 6 July 1885. 208

8.2 (a,b). Pasteur's laboratory notes on the treatment of Jean-Baptiste Jupille, beginning on 20 October 1885. 210

8.3. The Roux-Pasteur technique for preserving spinal marrow from a rabid rabbit. 214

9.1. Pasteur's path to his rabies vaccine, 13 April 1885 through 6 July 1885: Animal experiments and human trials with dried spinal cords. 244

9.2. The results of Pasteur's experiments on dogs treated by the "Meister Method," 28 May 1885 through 6 July 1885. 252

TABLES

2.1. Outline of Pasteur's career. 28

2.2. List of Pasteur's major prizes and honors. 34

2.3. Chronological outline of Pasteur's major research interests. 37

3.1. Pasteur's list of eight tartrates. 66

3.2. "Tartrates (Questions to resolve)." Pasteur's four anomalies in the relation between crystalline isomorphism and waters of crystallization. 69

9.1. Results of Pasteur's "post-exposure" experimental trials on dogs after they had been bitten by a rabid dog, August 1884 through May 1885. 241

Preface

FROM THE DAY I began this project, I have been asked why we needed yet another study of Louis Pasteur. His career had already been fully described so many times, beginning with the standard two-volume biography by his son-in-law, René Vallery-Radot, published in French in 1900 and translated into English a year later.

My response was and is fourfold. First, René Vallery-Radot's standard biography, for all of its detail and other virtues, is hagiographic through and through, and much of the subsequent biographical literature is derivative and similar in tone. Second, the last major—in fact still the best—scientific biography of Pasteur, René Dubos's *Louis Pasteur: Free Lance of Science*, was published almost half a century ago, in 1950. Third, none of the book-length biographies of Pasteur meet current scholarly standards; even Dubos's widely admired book lacks footnotes or other scholarly apparatus, so the sources of his insights are often obscure. Fourth, and most important by far, students of the Pastorian saga can now draw on a vast collection of manuscript materials deposited at the Bibliothèque Nationale in Paris. This stunning archival collection became generally available to scholars as long ago as the mid-1970s, but surprisingly little use has been made of it thus far. Particularly revealing, I think, are the one hundred or so unpublished laboratory notebooks Pasteur left behind, and they serve as crucial sources for the reassessment of his life and career that this book represents.

Long ago, I decided not to publish the results of my archival research in isolated bits and pieces. Of the chapters that follow, none has appeared in precisely this form. Parts of the book, especially Chapter Two, do make liberal use of my essay on Pasteur in the *Dictionary of Scientific Biography*, published in 1974, before I had begun my archival research. Chapter Three, on Pasteur's discovery of optical isomers, is a slightly revised version of an article published by James Secord and me in *Isis* in 1988. In fact, that article was based largely on Jim's research on Pasteur's very first (and still unpublished) laboratory notebook, and I am deeply grateful to him for allowing me to repeat so much of that article here. Parts of Chapter Five, on the spontaneous generation debate, are adapted from an article that John Farley and I published in 1974 in the *Bulletin of the History of Medicine*, and I am most grateful to John for permission to make use of our collective effort here. The other chapters, except for scattered passages, are wholly new, published here for the first time.

In the course of producing this book, I have accumulated a heavy burden of debt to a host of people and institutions. So long is the list that I have saved it for a separate entry on Acknowledgments at the end of the book. By then, I hope my creditors will still be glad to be mentioned there; they are of course absolved of any responsibility for defects in the book. There is, however, one debt so large and so overdue that I must acknowledge it here. For the plain fact is that this book would never have seen the light of day without the inspiring scholarly example and patient support of my mentor, Larry Holmes.

Princeton, New Jersey
August 1994

Background and Context

Laboratory Notebooks and the Private Science of Louis Pasteur

IN 1878, WHEN he was fifty-five years old and already a French national hero, Louis Pasteur told his family never to show anyone his private laboratory notebooks.[1] For most of a century those instructions were honored. Pasteur's notebooks—like the rest of the manuscripts he left behind at his death in 1895—remained in the hands of his immediate family and descendants until 1964. In that year, Pasteur's grandson and last surviving direct male descendant, Dr. Pasteur Vallery-Radot, donated the vast majority of the family's collection to the Bibliothèque Nationale in Paris.[2] But access to this material was generally restricted until Vallery-Radot's death in 1971, and there was no printed catalog of the collection until 1985.[3]

The Pasteur Collection at the Bibliothèque Nationale is stunning in its size and significance. It is a tribute not only to Pasteur's own awesome productivity as scientist and correspondent, but also to the tireless efforts of Pasteur Vallery-Radot, who greatly increased the size of the initial family collection by gathering additional correspondence and manuscripts by and about his grandfather from every conceivable source. There are, to be sure, other significant collections of manuscript materials by or relating to Pasteur—at the Académie des sciences and the Archives Nationales in Paris, for example, or at the Wellcome Institute for the History of Medicine in London, and at the National Library of Medicine in Bethesda, Maryland, in the United States. But the collection at the Bibliothèque Nationale is the largest and most important by far.

As now deposited in the Salle de Manuscrits at the Bibliothèque Nationale, the Papiers Pasteur includes fifteen large bound volumes of correspondence by, to, or about Pasteur. Another fifteen volumes contain lecture notes, drafts of published or unpublished manuscripts, speeches, and related documents. Most important, the Papiers Pasteur includes a meticulously preserved collection of more than 140 notebooks in Pasteur's own hand, of which more than one hundred are laboratory notebooks recording his day-to-day scientific activities over the full sweep of his forty years in research. Until these manuscripts are deciphered, edited for publication, and subjected to critical scrutiny, our understanding of Pasteur and his work will remain incomplete. There is no prospect that this monumental task will be accomplished anytime soon, not even with the stimulus of the centenary of Pasteur's death in 1995. Indeed, the task has not even begun in any systematic way, and a full and proper edition of Pasteur's papers and manuscripts will require a massive investment of time and resources.

For the foreseeable future, we shall have to contend with a vast reservoir of unedited and unpublished manuscripts. True, Pasteur Vallery-Radot long ago published a small but significant sample of the collection, including notably a four-volume selection of Pasteur's correspondence.[4] Some of these letters, when read critically in the light of other sources, already reveal a Pasteur who was more complex and interesting than he has been seen, or indeed wished to be seen. Yet even these published letters have been surprisingly under-utilized by students of Pasteur's career. They have done little to add nuance or depth to the standard Pastorian legend. In the popular imagination, Pasteur remains the great and selfless "benefactor of humanity" who single-handedly slashed through the prejudices of his time to discover a set of scientific principles unmatched in their impact upon the daily lives and well-being of humankind.

But as the centenary of Pasteur's death approached, his oft-examined career attracted still greater attention, some of it more critical than the usual celebratory accounts. Much of the revaluation now underway has focused on Pasteur the man, whose human foibles and difficult personality have never been entirely absent from the published record but are now gaining wider publicity. But Pasteur the scientist is also being subjected to the more systematic critical scrutiny that his importance and influence deserve. That is not to suggest that Pasteur's life can be neatly divided into its scientific and nonscientific aspects. In some ways, his scientific style seems a virtual extension of his personality, and one theme of this book will be that his scientific beliefs and modus operandi were sometimes profoundly shaped by his personal concerns, including his political, philosophical, and religious instincts.

As this book unfolds, it will become clear how much the standard Pastorian legend needs to be qualified, even transformed. That point will be made most explicitly in the last chapter, "The Myth of Pasteur," which will also serve as a bibliographical essay of sorts. Long before that last chapter, however, the standard Pastorian saga will begin to unravel. For now, I want only to emphasize that the most important revelations in this book are the result of focusing on what I have chosen to call "the private science of Louis Pasteur."

PRIVATE SCIENCE AND LOUIS PASTEUR

The choice of this phrase for the very title of this book deserves a preliminary discussion and justification, if only because some readers may consider it a contradiction in terms. If, as many assume, the very definition of science implies a public (usually published) product—if, as Charles Gillispie has written, "science is nothing until reported," or if, in Gerard Piel's words, "without publication, science is dead"[5]—whatever can "private science" mean?

The notion of private science is indeed problematic, and not only in the sense that these commentators probably have in mind. Strictly speaking, there may be no such thing as purely private science or knowledge—or even a purely private thought. Even the most solitary scientist is heir to a tradition of thought, practices, techniques, training, and social experiences. Perhaps this was part of what the Victorian physicist John Tyndall had in mind when he wrote in 1885, in his introduction to the English translation of the first biography of Pasteur, that "[t]he days when angels whispered into the hearkening human ear, secrets which had no root in man's previous knowledge or experience, are gone for ever."[6] Tyndall's immediate purpose was to convey his inductivist skepticism toward the alleged role of "preconceived ideas" in Pasteur's research, but his general point can be extended to the realm of seemingly private thoughts or practices of any sort.

For, in fact, there is always a continuum between private thought or practices and public knowledge, whatever the field. The thoughts of the individual scientist alone in his or her study or laboratory will perforce be filtered not only through an inherited tradition, but also through the scientist's *anticipations* of audience response to the communication of those ideas. The scientist will always be aware that the anticipated audience may be large or small, friendly and receptive, or skeptical or hostile. According to the Russian cultural critic Mikhail Bakhtin (1895–1978), thought itself is nothing but "'inner speech,' or social conversations we have learned to perform in our heads." On this view, "when we think, we organize possible 'dialogues' with other people, whose voices and implicit social values live within us."[7] One might even say that something like a "sociology of the mind" is always at work. As we shall see in the case of Pasteur, and as the famous example of Darwin amply reveals, this sociology of the mind can temper, modify, repress, or forever silence a "passing thought."[8]

Similarly, the "private" correspondence of a scientist (or anyone else) is obviously written with at least one recipient in mind. In the case of famous correspondents, including the mature Pasteur, some presumably private letters are clearly also being addressed to that larger audience known as "posterity." More generally, as Stephen Jay Gould has suggested, there is little reason to suppose that "private letters somehow reveal the 'real' person underneath his public veneer." This common notion, says Gould, is a "misplaced, romantic Platonism":

> People have no hidden inner essence that is more real than their overt selves. If [a scientist] reacted one way to most people in public life, and another to his sister in letters, then the public man *is* most of the whole. We meet a different [scientist] in these letters, not the truer core of an essential personality. These letters do not show us the real man. They simply remind us once again that people have the damnedest ability to compartmentalize their lives; one can be a fine statesman and a cad at home, a financial genius and an insensitive lout, a lover of dogs and a murderer of people.[9]

Gould's point can be extended to private documents of any sort, including even laboratory notebooks. They may provide revealing insights into a scientist and his or her work, but they do not offer uniquely privileged access to the "real" story as opposed to the public "myth." In the case at hand, Pasteur's public performances must also be incorporated into our understanding of him and his science, as with any other social actors and their work.

"Private science" becomes a still more problematic category when the research involves assistants and collaborators, as it did throughout much of Pasteur's career (and as it does in most modern laboratory research). Even Pasteur, despite his secrecy and "Olympian silence" about the direction of his research, could not always conceal his work or thoughts from his closest collaborators.[10] And a few of them did not always and forever honor Pasteur's stricture that the research carried out in his laboratory should remain a totally private affair within the Pastorian circle unless and until he chose to disclose the results himself or specifically authorized others to do so. True, Pasteur's collaborators did honor this demand to a degree that may seem astonishing in our less discreet world, and nearly all of them continued to do so even after the master's death. But there is evidence to suggest that these severe restrictions on public disclosure did not always sit well with some of Pasteur's assistants and co-workers. By 1880, for example, Emile Roux, his major collaborator in research on anthrax, rabies, and other diseases, was warning Pasteur that outsiders had begun to regard his laboratory at the Ecole Normale as a "mysterious sanctuary."[11] Eventually, the veil of secrecy was pulled back in part, most notably in the anecdotal reminiscences of Pasteur's own nephew and sometime personal research assistant, Adrien Loir, who did, however, wait half a century to publish his revelations in a widely ignored series of essays that carried the apt title, "In the Shadow of Pasteur."[12]

One could raise still other objections to the whole notion of "private science," but I will proceed as if the term embodies a meaningful distinction. Throughout this book, I will use the term "private science" in the informal sense of those scientific activities, techniques, practices, and thoughts that take place more or less "behind the scenes." That definition might be less appropriate in the case of a scientist whose activities and career were less theatrical than Pasteur's, but his carefully orchestrated public performances invite a close examination of the private dress rehearsals. Finally, I should stress that my notion of "behind the scenes" is not restricted to activities and thoughts that were literally kept out of public view, but will occasionally be extended to matters that can be found in the published record if one looks hard enough, but have been lost from that collective public memory represented by the standard Pastorian legend.

This approach means, among other things, that I will sometimes highlight relatively obscure features of Pasteur's published papers or correspondence, and will pay much closer attention than usual to some of the supporting cast, including a few of the once public but now mostly forgotten critics of the star. Nonetheless, the most striking revelations come when one brings to center stage some of the activities and ideas recorded only in Pasteur's unpublished manuscripts. This book makes selective use of the full range of the manuscript materials that Pasteur left behind. In the most dramatic cases, however—including Pasteur's crowning work on vaccines against anthrax and rabies—the crucial evidence will come from his laboratory notebooks. It is therefore worth saying something now about my attitude toward these very special documents.

PASTEUR AND HIS LABORATORY NOTEBOOKS

The most private of the manuscript materials Pasteur left behind are the 144 holographic notebooks that his grandson donated to the Bibliothèque Nationale in 1964. Of these 144 notebooks, 42 fall outside the category of laboratory notebooks, consisting instead of collections of newspaper clippings, draft sketches of projected books that never appeared, lecture outlines, and reading and lecture notes. The remaining 102 notebooks represent the most precious documents in the Papiers Pasteur. They consist of careful and detailed records of experiments carried out by Pasteur and his collaborators during forty years of active, almost daily research. They are the central repository for the private science of Louis Pasteur, the documents he once asked his family to keep forever out of public view. During his lifetime, he carefully guarded them from others, including his closest collaborators. Even when he left Paris for trips or holidays, Pasteur took the most current of the laboratory notebooks with him. His co-workers sometimes experienced inconvenience or worse because of his insistence on total control of the notebooks.

In late November 1886, for example, while Pasteur was resting at a villa on the Italian Riviera for the sake of his fading health, his collaborators in Paris were suddenly faced with a legal problem connected with the death of a boy who had undergone the Pastorian rabies treatment (a story to which we shall return in Chapter Nine). As we know from his retrospective personal testimony, Pasteur's nephew-assistant Adrien Loir had to be dispatched quickly to Italy in order to retrieve important details about the boy's treatment—information that was recorded only in a laboratory notebook the master had taken with him to the Italian villa.[13] Earlier, in July 1883, when Emile Roux wanted to gather together some of the results of his important work on rabies for his doctoral thesis, he had to seek Pasteur's permission to use information recorded in the laboratory notebooks. To ensure the master's assent, Roux promised to expose only those results already made known in a general way in Pasteur's published papers, submitted a draft version to the master for his corrections and revisions, and "inscribed your [i.e., Pasteur's] name on the first page of this exposition of studies that belong to you."[14]

In 1896, a year after Pasteur's death, Roux gave a revealing, if surprisingly restrained, account of the master's proprietary attitude toward his laboratory notebooks. Roux's account also deserves attention because it reveals the extent to which the work in Pasteur's laboratory had become a collaborative affair by the time Roux participated in it:

> In order to be nearer the work, master and disciples lived in l'Ecole Normale. Pasteur was always the first to arrive; every morning, at 8 o'clock, I heard his hasty step . . . over the loose pavement in front of the room which I occupied at the extremity of the laboratory. As soon as he had entered, a bit of paper and pencil in his hand, he went to the thermostat to take note of the state of the [microbial] cultures and descended to the basement to see the experimental animals. Then we made autopsies, cultures and the microscopic examinations. . . . Then Pasteur wrote out what had just been observed. He left to no one the care of keeping the experimental records; he set down most of the data which we gave him in all its details. How many pages he has thus covered, with

> his little irregular, close-pressed handwriting, with drawings on the margin and references, all mixed up, difficult to read for those not accustomed to it, but kept nevertheless with extreme care! Nothing was set down which had not been established; once things were written, they became for Pasteur incontestable verities. When in our discussions, this argument resounded, "It is in the record book," none of us dared to reply. The notes being taken, we agreed upon the experiments to be made; Pasteur stood at his desk ready to write what should be decided upon. . . .
>
> Then we spent the afternoon in making the experiments agreed upon. . . . Pasteur returned toward five o'clock. He informed himself immediately of all that had been done and took notes; his notebook in hand, he went to verify the tickets fastened on the cages, then he told us of the interesting communications heard at the [Académie des sciences earlier in the afternoon] and talked of the experiments in progress.[15]

As Roux reports, Pasteur did indeed keep a detailed and meticulous record of the experiments carried out in his laboratory. I have never counted the pages that Pasteur filled with experimental data in his sometimes crabbed and microscopic hand, but they probably exceed ten thousand. As some of the illustrations in this book suggest, the task of deciphering and interpreting Pasteur's entries is often daunting. Like most laboratory notebooks, Pasteur's usually consist of bare records of experiments, with only occasional hints as to their aim or theoretical significance. The meaning of such documents cannot begin to be grasped without an intimate familiarity with the scientist's published work. Beyond that, their would-be interpreter should ideally possess a combination of skills akin to those of the paleographer, cryptographer, and mind-reader. It is a species of detective work in which tantalizing clues too often lead to dead ends.

But the effort is exhilarating as well as exhausting. Words cannot fully convey the sense of excitement that comes from turning the pages of any one of Pasteur's laboratory notebooks. It is as if one were looking over his shoulder as he designed and carried out experiments ranging from the trivial to the profound. The laboratory notebooks form a virtually unbroken chain of documents that record Pasteur's day-to-day dialogue with a sometimes recalcitrant nature. They are, I think, the most revealing of all the manuscript materials he left behind. Perhaps that is to be expected, since Pasteur did after all spend most of his waking hours at work in the laboratory.

To produce a detailed account of all of Pasteur's one hundred laboratory notebooks, several decades of work will surely be required. I have therefore focused attention instead on a few episodes in Pasteur's career where there are distinct—and sometimes astonishing—discrepancies between the results reported in his published papers and those recorded in his private manuscripts. This approach is open to several objections. It is one thing to be selective in order to reduce the task to manageable limits. But why choose such special and possibly misleading criteria? If most of Pasteur's published accounts are consonant with his laboratory records, why focus on the exceptions? Can such an approach give us a balanced assessment of Pasteur's usual scientific practice? Will not the full range of his achievement be lost through such an episodic treatment of his career? And is this not an especially suspect approach at a time when so much public attention is being drawn to a few spectacular examples of real or alleged fraud in science?[16] Is even Pasteur to be swept up in the current fashion for muckraking exposés of science and its legendary heroes?

Only as this book unfolds can the reader begin to judge whether or how far these objections have been met. But it may be useful to address them in a preliminary way even now. In doing so, I will be able to clarify my aims and to insist on some of the virtues of my approach. Let me emphasize at once that I have no intention of denying Pasteur's greatness as a scientist. To be sure, my definition or conception of a "great scientist" may differ somewhat from the conventional. For me, there is no reason to suppose that a great scientist must also display personal humility, selfless behavior, ethical superiority, or political and religious neutrality. The historical record often enough reveals the opposite. For me, past scientists are not great insofar as they were the "first" to advance concepts that look "right" in the light of current knowledge, nor insofar as they adhered to the precepts of an allegedly clear-cut Scientific Method that their lessers and rivals presumably violated. For me, rather, past scientists are great insofar as they persuaded their peers to adopt their ideas and techniques and insofar as those ideas and techniques were fertile in the investigation and resolution of important research problems. Pasteur was no exemplar of modesty, selflessness, ethically superior conduct, or political and religious neutrality. Nor was he always "first," "right," or a rigorous practitioner of the Scientific Method as usually conceived. But he was a remarkably effective and persuasive advocate for his views, and his concepts and techniques were immensely fertile in the pursuit of a wide range of important scientific and technical problems. By these criteria, he deserves his reputation as one of the greatest scientists who ever lived.

But let me turn, at greater length, to the more specific objection that it is misleading and unfair to adopt an episodic approach that emphasizes the "exceptional" discrepancies between Pasteur's published writings and his "private science." To begin with, the episodes on which I focus are far from trivial: each concerns a major phase or turning point in Pasteur's research.

Nor are they concentrated in any narrow field or period of his career. They span his active career and concern fields as varied as crystallography, molecular asymmetry, fermentation, spontaneous generation, vaccination and immunization, and veterinary and human medicine. The three episodes examined most closely here through the use of Pasteur's laboratory notebooks concern his first great discovery (of optical isomers in the tartrates), his most famous public experiment (the anthrax vaccination experiment at Pouilly-le-Fort), and his most famous achievement of all (the application of a rabies vaccine to human subjects). With the admittedly significant exception of his investigation of the silkworm diseases, the only major topics of Pasteur's research that receive no focused attention here are his conceptually undistinguished studies on the manufacture and preservation of vinegar, wine, and beer.

Nor is it likely that the discrepancies on which I focus are really exceptional. My sample is far from complete. Many additional examples will surely emerge as the entire corpus of Pasteur's notebooks is subjected to systematic analysis. On the other hand, it is crucial to emphasize that the discrepancies between Pasteur's public and private science do fall into two very different categories of very different significance.

LABORATORY NOTEBOOKS, SCIENTIFIC FRAUD, AND THE RHETORICAL CONSTRUCTION OF SCIENTIFIC KNOWLEDGE

Most of the discrepancies between Pasteur's public and private science are of a sort that will come as no great surprise to working scientists, or to anyone who has been attentive to recent historical scholarship on laboratory notebooks. To these audiences, it will be obvious that such discrepancies are part and parcel of the process by which "raw data" are transformed into published "results." In the interests of brevity, clarity, logical coherence, and rhetorical power, the published record always projects a more or less distorted image of what the scientist "really" did.

For some reason, laboratory notebooks were long overlooked by historians of science, but their virtues as a strategic site of inquiry have become evident in recent years. The recognition of their special value owes much to the pioneering work of M. D. Grmek and F. L. Holmes, both of whom used the laboratory notebooks of Pasteur's friend and contemporary, the great French physiologist Claude Bernard (1813–1878), to produce two brilliant and complementary books published twenty years ago. Grmek's book of 1973 focused on Bernard's work on poisons (notably curare and carbon monoxide), while Holmes's book of 1974 gave an exhaustive account of Bernard's early research in digestive physiology.[17] In the wake of these path-breaking works, other valuable analyses of laboratory notebooks have already appeared—two striking examples being David Gooding's work on the notebooks of Michael Faraday (1791–1867) and Gerald Holton's investigation of the laboratory notes of the American Nobel laureate in physics, Robert Millikan (1868–1953).[18] But it is Holmes who has become the leading advocate and practitioner of the study of laboratory notebooks. In the years since his book on Bernard, Holmes has produced comparably detailed and insightful analyses of the laboratory notebooks of the great eighteenth-century French chemist Antoine Lavoisier (1746–1794) and Nobel laureate biochemist Hans Krebs (1900–1981) of "Krebs Cycle" fame.[19] We can surely expect other significant studies of this sort as historians uncover more examples of scientists who have earned our gratitude by preserving these traces of their daily work in the very special literary genre known as the laboratory notebook.

Much remains to be done in this line of research. But in every case thus far in which records of "private science" have been closely investigated, one can detect discrepancies of one sort or another between these records and published accounts. Even the best scientists routinely dismiss uncongenial data as aberrations, arising from "bad runs," and therefore omit or "suppress" them from the published record. Equivocal experiments are sometimes transformed into decisive results. The order in which experiments were performed is sometimes reversed. And the actual nature or direction of research is otherwise simplified, telescoped, and generally "tidied up." There is rarely anything sinister about such practices, rarely any intention to deceive, and their existence has long been recognized. As long ago as the seventeenth century, Francis Bacon noted that "never any knowledge was delivered in the same order it was invented," while Leibniz expressed his wish that "authors would give us the history of their discoveries and the steps by which they have arrived at them."[20] From time to time ever since, scientists and others, including the influential American sociologist of science Robert K. Merton, have drawn renewed attention to this "failure of the public record to record the actual course of scientific inquiry."[21]

More recently, analysts of the scientific enterprise have moved from expressions of regret about the discrepancies between private and public science to a recognition of their rhetorical import in the construction of scientific knowledge through the literary genre of the scientific paper. In the case of Millikan, for example, Holton shows us a country bumpkin from rural Illinois who was initially so naive about the genre that he included *all* of his experimental data about the quantity of charge on the electron, supporting his view of its unitary charge by publicly assigning more or fewer "stars" to what he considered good or bad runs. Millikan was quickly enlightened by his experience and the advice of others; never again did he resort to public displays of his less persuasive data. And Holton insists that Millikan's later published papers can actually be seen as "better" (i.e., more persuasive) science than that represented in his first paper, with its needlessly candid full disclosures.[22]

More recently still, Holmes has extended his approach beyond the analysis of laboratory notebooks to ask broader questions about the history of the practice of laboratory record keeping and its relation to the published record of science. In the case of Lavoisier, Holmes has shown the extent to which a scientist's ideas can be altered in the very process of "writing up" the results from laboratory notebooks for publication, and in the case of Krebs he has had the rare opportunity of comparing his historical reconstructions of events from laboratory notebooks with Krebs's own recollections of his investigative trail. In neither of these cases, nor in the case of Bernard, does Holmes suggest that his historical actors engaged in deliberately deceptive practices. Instead, he maintains that Lavoisier, Bernard, and Krebs simply and wisely adopted the standard practices and rhetorical strategies that always intervene between private laboratory records and their effective and persuasive presentation in the public domain.[23]

Against this background, it should be clear that Pasteur was not committing "scientific fraud" whenever his laboratory notebooks reveal a course of research different from that recorded in his published works. Long before his day, and perhaps especially in France, the institutionalization of the scientific paper—its progressive codification into a formulaic literary genre—had reached a point that discouraged instructive disclosures of the sort Bacon and Leibniz once thought might emerge from a closer fit between private research and its public presentation.[24] On Holmes's account, the institutionalized scientific paper did not (and does not) deliberately "suppress" uncongenial private data, but rather seeks to provide an efficient and authoritative public presentation of the most pertinent results to an expert audience with little need of elaborate additional detail.[25] By Pasteur's day, a pattern of formulaic discrepancies between public and private science was already long-standing and widespread, if not overtly sanctioned.

But the existence of this practice does not make such discrepancies insignificant or uninteresting, in Pasteur's case or any other. Precisely because they were and are so common, these formulaic discrepancies deserve much closer attention. To ignore or trivialize them is to miss the force of Peter Medawar's now-hackneyed warning that "scientific 'papers' [do] not merely conceal but actively misrepresent the reasoning that goes into the work they describe."[26] As Medawar suggests, to rely solely on the published record is to distort our understanding and appreciation of science as it actually gets done. The effect is impoverishing in several respects. By making the results of scientific inquiry look more decisive and straightforward than they really are, the published record tends to conceal the pliability of nature. It eviscerates science of its most creative features by conveying the impression that imagination, passion, and artistry have no place in scientific research. It makes it seem as if scientific achievement and innovation result not from the impassioned activity of committed hands and minds, but rather from passive acquiescence in the sterile precepts of the so-called Scientific Method. More specifically, as Medawar emphasizes, the published record tacitly endorses a naive and long-outmoded "inductivist" or "empiricist" philosophy of science, according to which scientific truth emerges from the innocent and unprejudiced observation of raw facts. The superficially objective and dispassionate image of science thus conveyed is bought at the price of much of its zest and human appeal. The construction of scientific knowledge is a much more interesting process than its published record suggests.

There are, of course, those who insist that "genuine" scientific knowledge is independent of the process by which any particular scientist arrives at his or her conclusions. In very different ways, philosophers and sociologists of science tend to be suspicious of historical studies of individual "scientific creativity." For philosophers in the tradition of Karl Popper, such studies seem to be pursuing a will-o'-the-wisp, an elusive "psychology of discovery," at the expense of a clear-cut "logic of justification." For them, the object of study is the published text, and the "scientificity" of a given text is to be assessed in terms of logical and methodological criteria that transcend particular individuals, particular social groups, or any contingent historical circumstances.

For sociologists of knowledge, by contrast, studies of individual scientific activity run the risk of ignoring the extent to which scientific knowledge is a community affair—the outcome of a complex process of social negotiation. On this view, scientific knowledge is constructed within a culturally limited space. For some, the boundaries of that space are set by the broadly cultural "interests" of participants. More recently, attention seems to have shifted to more sharply localized, "internal" material and technical constraints—a trend that may invite the risk (or opportunity) of a return to positivist or inductivist epistemologies.

Often lost from sight in such theoretical discussions is the real individual scientist who tries to navigate a safe passage between the constraints of empirical evidence on the one hand and personal or social interests on the other. To chart such individual passages is certainly to leave aside some important general issues about the nature and construction of scientific knowledge. Yet there remains a place for studies of individual scientists and their creative activity. To proceed as if scientific knowledge were somehow achieved all apart from the activity of individual scientists is itself a distortion of reality. For the historian, one way to reduce such distortions is to explore the process of scientific research as recorded day to day in surviving laboratory notebooks.

That is not to say—to repeat a point already made—that these private documents somehow permit direct access to the "real" work of the scientist. Even laboratory notebooks are incomplete *traces* of activity, much of which remains tacit, none of which can be observed directly, and all of which must be deduced from recorded inscriptions that are often difficult to decipher and interpret. Sociologists and anthropologists of knowledge have the advantage of being able to interview and observe participants *in the very process* of doing science, and some important results have already emerged from recent research along these lines. Responding—sometimes explicitly—to Medawar's challenge to subject science to "an ethological enquiry," to study what scientists actually *do* by "listening at the keyhole," some sociologists and anthropologists of science, notably Harry Collins and Bruno Latour, have uncovered important elements of what is variously called the "private," "personal," "tacit," or "craft" knowledge that is fundamental to the actual practice of science but finds few echoes in the published literature—or, for that matter, in unpublished laboratory notebooks. These sociologists or anthropologists can watch the scientist go about his or her "craftsman's work" and thus observe the nonverbal activity that accompanies and gives rise to verbal and other symbolic accounts. In short, they can go much further toward recovering the actual *activity* of science before it becomes encoded in fading and incomplete verbal or graphic "inscriptions," including laboratory notebooks.[27]

But if historians lack these advantages, they can be relatively sure that the episodes they choose to study are already known to be of special interest. Anthropologists of science may hang around a laboratory for a year or more and witness no obvious peaks of productivity. Historians, by contrast, can be selective in their choice of notebooks, which nonetheless bring them closer in time and place to the creative work of scientists than do any published results. At a minimum, laboratory notebooks give the historian another set of "texts" to read, and the work of Grmek, Holmes, Holton, and others has already provided ample evidence that a comparison of these "private" texts with the published literature can yield important insights of general significance.

In the spectacular case of Pasteur, we are fortunate to have a complete set of his unpublished laboratory notebooks—those one-hundred-odd tidy and meticulously preserved records of his day-to-day research. By exploring his laboratory notebooks in the full context of his life, work, and social setting, we can gain unusual insight into the construction of scientific knowledge at the concrete level of an extraordinarily creative individual scientist.

This book can only begin the task, and for the most part these more general concerns will only emerge implicitly. Yet it should gradually become clear that some of Pasteur's most important work often failed to conform to ordinary notions of proper Scientific Method. In particular, it will become clear that Pasteur sometimes clung tenaciously to "preconceived ideas" even in the face of powerful evidence against them. And it should also eventually become clear just how far the direction of his research and his published accounts of it were shaped by personal ambition and political and religious concerns. We will become aware of his ingenious capacity for producing empirical evidence in support of positions he held a priori. In other words, one aim of this book is to show the extent to which nature can be rendered pliable in the hands of a scientist of Pasteur's skill, artistry, and ingenuity. But it will also suggest that not even Pasteur's prodigious talent always sufficed to twist the lion's tail in the direction he sought. Nature is open to a rich diversity of interpretations, but it will not yield to all.

PASTEUR AND THE ETHICS OF BIOMEDICAL RESEARCH

These themes and issues continue to appear in the second part of the book, which concerns episodes in Pasteur's veterinary and biomedical research. But now an additional focus begins to take center stage, and it relates directly to the second and very different category of discrepancies between Pasteur's public and private science. Here we deal not with mere acquiescence in the formulaic genre of scientific papers and the associated "inductivist" image of science, but with discrepancies between Pasteur's public and private science in cases where the word "deception" no longer seems so inappropriate, and even "fraud" does not seem entirely out of line in the case of one or two major episodes. These are serious allegations, and they will be treated with the care they deserve.

Only a very few episodes are in question here, and two of them are so close in time and so similar in nature that it is better to conflate them into one. Moreover, as we shall see, this "double episode" is relatively easy to explain and excuse, since it concerns "therapeutic experiments" on seemingly doomed victims of rabies and is at worst an example of deception by omission. Instead of informing the public and the scientific community of the dramatic results of these two human trials, Pasteur chose to remain completely silent.

The other episodes concern the two most celebrated achievements in Pasteur's career: his bold public demonstration of a vaccine against anthrax in sheep at Pouilly-le-Fort in 1881, and the first known application of his rabies vaccine to a human subject, young Joseph Meister, in July 1885. In the first case, as we shall see, Pasteur deliberately deceived the public and the scientific community about the nature of the vaccine used in the experiments at Pouilly-le-Fort. In the second case, the nature of Pasteur's deception is less clear-cut, but here too we will find some striking discrepancies between the public and private versions of the famous story of Joseph Meister.

Let it be clear at the outset that I am less concerned to *expose* Pasteur's public deceptions than to *explain* them. True, the ascription of motives to historical actors is a notoriously risky business, and this is very definitely the case here. In every case, it is possible to offer exculpatory explanations for Pasteur's behavior—though credulity is sometimes strained, especially in the case of the sheep-vaccination experiments at Pouilly-le-Fort and certain aspects of his work on rabies. But the effort to analyze Pasteur's ethically dubious deceptions is justified by the importance of the larger questions these few episodes raise. In what circumstances, and under what pressures, is a scientist of Pasteur's stature tempted to deceive? To what extent is such conduct explicable in terms of personal circumstances or character, and to what extent in terms of a competitive ethos or other more general cultural forces? Are the presumed norms of scientific conduct always reconcilable? Do scientific advance and the public welfare sometimes require scientists to tell "white lies"? How can the public or even other scientists be expected to appreciate the intuitive basis for actions that cannot be fully justified in strictly "scientific" terms? Is there a difference between "scientific ethics" and "medical ethics"? Especially in the face of dread disease and terrified people, how much prior evidence from animal experiments is required before preventive measures are applied to human cases? At least implicitly, Pasteur's deceptions raise these and other equally important questions about the ethics of research in general and of biomedical research in particular.

But in the midst of these absorbing and more or less timeless issues, it should not be forgotten that our subject is a particular individual in a specific historical context. We must not wrench Pasteur from his historical circumstances for the sake of facile insights into our current concerns. There are profound differences between the intellectual, social, and ethical climate of his day and our own. His ethical conduct, like his scientific achievements and practices, should and will be assessed by applying criteria and standards that were recognized by his contemporaries and, indeed, by Pasteur himself.

WHAT DO WE DO WHEN PRIVATE SCIENCE BECOMES PUBLIC KNOWLEDGE?

At this point, it will prove useful to circle back to the beginning of this chapter and to disclose the context in which Pasteur instructed his family to keep his laboratory notebooks forever out of the public eye. Pasteur did not fear the exposure of some deep and dark secret recorded only in his notebooks. Instead, his directive was a plausible response to a specific wrenching experience he had just gone through.

In February 1878, Pasteur mourned the death of his friend and compatriot, the great experimental physiologist Claude Bernard. About six months later, one of Bernard's disciples instigated the publication of some fragmentary laboratory notes he had left behind. The contents of Bernard's hitherto private notes surprised Pasteur and their publication placed him in an awkward position. In essence, these private notes disputed Pasteur's "germ theory" of fermentation. While alive, Bernard had never challenged that theory in public nor even in conversation with Pasteur. Pasteur felt obliged to respond to these now public manuscript notes, lest his deeply held theory of fermentation be undermined by appeal to the authority of the revered Bernard. If he felt uncomfortable about attacking the private work of his late friend and frequent public supporter, who could no longer disavow or defend the experiments in question, Pasteur did nonetheless publish a full-length critique of Bernard's manuscript notes. By carefully repeating Bernard's experiments and comparing them with his own, Pasteur went a long way toward establishing his claim that Bernard's results were mistaken, dubious, or misinterpreted. Both in tone and substance, the critique was devastating.[28]

Pasteur's conduct in this affair was by no means universally approved. Half a century later, Paul de Kruif, whose best-selling book *The Microbe Hunters* did so much to popularize Pasteur's work in the United States, fulminated against Pasteur's behavior in this case. For de Kruif, Pasteur's conduct when faced with the publication of Bernard's private notes served as the most striking example of his inability to accept criticism of any sort. Worse yet, it displayed Pasteur's willingness to stomp on the grave of a revered and recently deceased colleague solely for the sake of his own reputation.[29]

Pasteur was himself concerned that this tirade against Bernard would be unpopular among important segments of the French scientific community and larger public. To justify his assault against the work of one of France's scientific heroes, Pasteur adopted a two-pronged strategy. On the one hand, he impugned the motives of the man who had arranged for the publication of Bernard's private notes, the distinguished French chemist Marcellin Berthelot (1827–1907), a long-standing advocate of a modified "chemical" theory of fermentation as opposed to Pasteur's strictly "biological" theory. Pasteur accused Berthelot of misusing and debasing Bernard's reputation by publishing these crude preliminary experiments. If his critique tarnished Bernard's memory, Pasteur insisted, then Berthelot must accept much of the responsibility. For it was he who had tried to bolster his own misguided and doomed campaign against the germ theory of fermentation by bringing unauthorized public attention to bear on Bernard's private and preliminary experiments on fermentation.[30]

But Pasteur also justified his critique on methodological grounds. For him, Bernard's manuscript notes represented an instructive example of the danger of "systems" and "preconceived ideas." Bernard himself had done much to expose this danger in his famous *Introduction to the Study of Experimental Medicine* (1865), a masterful discussion of Scientific Method by one of its leading practitioners. Yet somehow, Pasteur insisted, Bernard had forgotten his own wise precepts in these private notes on fermentation. Bernard had been led astray, Pasteur continued, by his a priori conviction of a fundamental opposition between organic syntheses and organic decompositions. He supposed that organic syntheses were peculiarly vital phenomena, while organic decompositions—including fermentation, combustion, and putrefaction—were physicochemical rather than vital processes. For Bernard, in effect, organic syntheses were associated with life, while fermentation and other organic decompositions were associated with death. Because Pasteur's theory linked fermentation with life, Bernard privately rejected it and undertook experiments in hopes of refuting it. In Pasteur's eyes, Bernard was secretly opposed to the biological or germ theory of fermentation because it clashed with his general conception of organic processes—with his "system" of "preconceived ideas" about such phenomena.[31]

It is less important here to assess the validity of Pasteur's charges against Berthelot and Bernard than to recall that they arose in response to the posthumous and unauthorized publication of Bernard's laboratory notes. For it was also in response to this event that Pasteur instructed his family to protect the privacy of his own notebooks.[32] He clearly feared that the publication of some of his laboratory notes might do similar damage to his reputation. At that point, he was presumably concerned only about his reputation for experimental probity and methodological propriety, for none of the ethically dubious episodes discussed in this book had yet occurred.

Pasteur criticized Bernard's posthumously published notes in large part to defend his own theory of fermentation. But he also seized the opportunity to draw methodological lessons from Bernard's once-private laboratory notes. In doing so, Pasteur supplied an inadvertent precedent and justification for exposing his own manuscripts to critical scrutiny. And the results, as we shall see, bear no resemblance to the lesson that Pasteur professed to find in Bernard's manuscript notes.

In presenting Bernard's private experiments as an example of the "tyranny of preconceived ideas," Pasteur wrote as if he were surprised to discover that a scientist of Bernard's stature and methodological self-consciousness could sometimes stray from the path of objectivity. He expressed dismay that even Bernard could sometimes be seduced by that "greatest derangement of the mind . . . believing things because one wants to believe them."[33] In the context of this polemic, Pasteur presented himself as a practitioner of the "inductive scientific method, working outside of theories."[34] Yet elsewhere he spoke of the fertility of his own "preconceived ideas,"[35] and he sometimes seemed to advocate something like the hypothetico-deductive method now favored by many philosophers of science.

In truth, Pasteur did not think very deeply about questions of Scientific Method, and he presented conflicting accounts of his own methodology depending on the audience and purpose at hand. To understand and appreciate Pasteur's scientific modus operandi, it is essential to examine what he actually did in his laboratory rather than to read his scattered and inconsistent remarks about Scientific Method. The crucial source for penetrating the ways in which Pasteur produced scientific knowledge is the extensive set of laboratory notebooks he left behind. Unlike Bernard's notebooks, moreover, Pasteur's manuscripts also bring us face-to-face with important questions about the ethics of biomedical research. To that extent, we may hope to learn even more from them.

Given Pasteur's concern about exposing his laboratory notebooks to public scrutiny, it may seem surprising that they survived him at all, let alone that he should have preserved them so meticulously. Perhaps his concern passed with time, but there is no reason to suppose that he would have welcomed the prospect of a future inquiry of the sort embodied in this book or other recent scholarship. It may be doubted, in short, that Pasteur saved his laboratory notebooks with future historians in mind. True, he did profess great interest in the history of science, even suggesting that it should be taught as part of the regular science curriculum at the Ecole Normale Supérieure.[36] He often sprinkled his memoirs and lectures with historical allusions and wrote a substantial historical article on the life and work of Lavoisier.[37] As a working scientist, however, Pasteur valued the history of science only insofar as he thought it could advance the cause of science and scientists. He held a heroic conception of the history of science according to which great men bring us ever closer to absolute truths about nature. And when he proposed that the history of science be incorporated into the science curriculum at the Ecole Normale, he did so in the belief that it might inspire students to respect and honor their elders and forebears by revealing how difficult it was to produce original scientific work.[38]

If Pasteur believed that a future study of his own laboratory notebooks or other manuscripts might contribute toward these or other worthy goals, he did not say so. The pains he took to preserve his notebooks can almost surely be traced instead to two very different considerations: (1) he repeatedly returned to his records of old experiments to inspire or test new ideas, and in that sense his laboratory notebooks were of direct and continuing utility to him; and (2) like a pack rat, he saved absolutely *everything* anyway, as many an archivist would attest after trying to make sense of the mounds of isolated and sometimes trivial slips of paper he left behind.

We are, in any case, fortunate that Pasteur left us these detailed records of his ongoing research. Indeed, one's sense of gratitude is so great that one might feel almost churlish about using them in any way that their author did not intend or foresee. But Pasteur's notebooks are now public property, available to anyone who gains access to the manuscript room of the Bibliothèque Nationale in Paris. In an important sense, it is no longer possible to invade Pasteur's privacy, for his "private science" has now become part of the public domain. We are thus, in some ways, placed in a situation like the one facing Pasteur upon the publication of Bernard's laboratory notes on fermentation. And it is precisely for that reason that we can insist that the standard Pastorian legend requires revision and even transformation. As the contents of these once private documents find their way into public view, a fuller, deeper, and quite different version of the Pasteur story will perforce emerge. There is, in effect, a new "history of Pasteur" to be written.

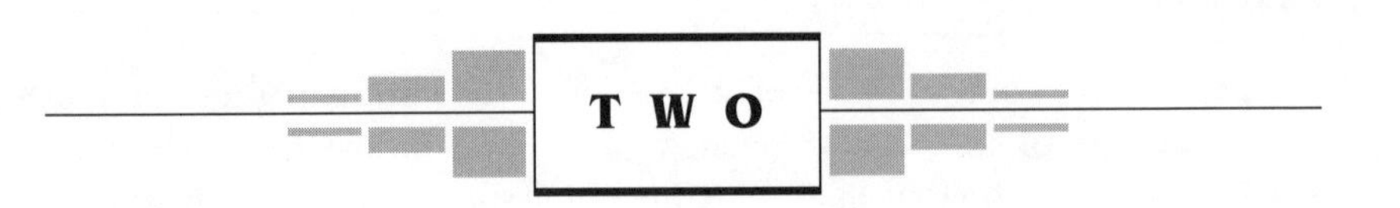

Pasteur in Brief

PASTEUR sprang from humble roots. For centuries his ancestors lived and worked as agricultural laborers, tenant farmers, and then modest tradesmen in the Franche-Comté, on the eastern border of France. The shift from agriculture to trade came five generations before Louis was born. For two generations, in the early eighteenth century, the Pasteurs were millers in service to the Count of Udressier. Pasteur's three immediate male ancestors, including his father, were small-scale tanners. His father, Jean-Joseph Pasteur (1791–1865), was drafted into the French army at the age of twenty. Assigned to the celebrated Third Regiment of Napoleon's army, he served with distinction in the Peninsular War. By 1814, when he was discharged, he had attained the rank of sergeant major and had been awarded the cross of the Legion of Honor. Jean-Joseph Pasteur often looked back proudly to his brief military service, and he instilled in his only son a yearning for those glorious days when Napoleon and France seemed on top of the world.[1]

Upon his return to civilian life, Jean-Joseph settled into his work as a tanner, initially at Besançon, where his father had plied the same trade. In 1816, he married Jeanne Etiennette Roqui, daughter of a gardener from an old proletarian family of the Franche-Comté. They moved to Dole, where the first four of their five children were born. Louis, their third child, was born two days after Christmas in 1822. He was preceded by a son who died in infancy and a daughter born in 1818. Two more daughters came later. Pasteur thus grew up as the only brother of three sisters. The family moved twice before Louis was five, first to Marnoz, the native village of the Roqui family, and then in 1827 to the neighboring town of Arbois, on the Cuisance River, where a tannery had become available for lease. As at his birthplace in Dole, the tannery was also home, the family being lodged above the half-dozen tanning tubs that provided its modest income. It was Arbois, a picturesque town of eight thousand inhabitants in the foothills of the Jura mountains, that Pasteur came to think of as home and to which he later returned for extended summer vacations and at moments of family tragedy.[2]

As one might expect in a family whose men had long worked as modest tradesmen, Louis absorbed at the hearth the traditional values of the petit bourgeoisie—familial loyalty, moral earnestness, respect for hard work, and concern for financial security. His father, who had received little formal education, had no greater ambition for his son than that he should become a teacher in a local lycée, an elite upper-level secondary school. This modest aspiration seems entirely in keeping with Louis's early performance at school. Until quite near the end of his secondary schooling, he was considered just a cut above the average student. Only his genuine, if immature, artistic talent seemed to promise anything at all exceptional. Several of Pasteur's early portraits of family, friends, and teachers have been preserved. Two sensitive character sketches of his parents, done when he was a teenager, reveal a talent quite beyond the ordinary. His powerful visual imagination and aesthetic sense come through in some of his later scientific work, especially that in crystallography.

ACADEMIC CAREER

If Pasteur ever seriously considered a career as an artist, he was dissuaded by his pragmatic father and by his mentors at the Collège d'Arbois, who gradually came to appreciate his scholastic talents. During the academic year 1837–1838, when he was fifteen, Louis swept the school prizes. He was now encouraged to prepare for the Ecole Normale Supérieure in Paris, the institution of choice for those seeking a career in French secondary and higher education. With admission to the Ecole Normale as the eventual goal, it was arranged that he enter a preparatory boarding school in Paris. Within a month, however, Louis returned to Arbois, overwhelmed by homesickness. His superb performance again that year at the Collège d'Arbois kept alive his ambition to enter the Ecole Normale.

To secure his baccalaureate in letters, the standard entrée to professional careers in France, Pasteur had to pursue his studies beyond the offerings of the Collège d'Arbois, which lacked the requisite class in philosophy. He therefore matriculated at the Collège Royale de Besançon, forty kilometers from Arbois, where he was awarded the degree in August 1840, three months shy of his eighteenth birthday. He received a mark of "good" in all subjects except elementary science, in which he was considered "very good." Now determined to seek entrance to the science section of the Ecole Normale, Louis stayed at the collège in Besançon to prepare for a second baccalaureate degree, this one in science. His family's financial burdens were eased by his appointment there as "preparation master" or tutor, which paid room and board as well as a small annual salary. After two years of study in the class of special mathematics, Pasteur received his baccalaureate in science in August 1842, though in physics he was considered merely "passable" and in chemistry "mediocre." Two weeks later he was declared admissible to the Ecole Normale, but he was dissatisfied with his rank of fifteenth among twenty-two candidates and declined admission for the time being.

In September 1842, having also considered a career as an engineer, Pasteur took, but failed, the entrance examination of the famous Ecole Polytechnique in Paris.[3] He then decided to spend another year preparing for the Ecole Normale. To do so, he returned to Paris and a boarding school run by one M. Barbet, himself a Franc-Comtois. This time, unlike four years before, he overcame his homesickness and stayed at the school, whose students attended the classes of the Lycée Saint-Louis, one of the leading preparatory schools for the Ecole Normale. By now Pasteur's discipline and diligence were beginning to be matched by his achievements. At the end of his first year in Paris, he took first prize in physics at the Lycée Saint-Louis and was admitted fourth on the list of candidates to the science section of the Ecole Normale, which he entered at the start of the next academic year.

For the next five years, from his twenty-first through his twenty-sixth year, Pasteur studied and worked at the Ecole Normale. To qualify for a position in secondary education, he competed in the two national certifying examinations, the *license* and the *agrégation*. He placed seventh in the *license* competition of 1845 and third in the physical sciences in the *agrégation* of 1846. In October 1846 he was appointed *préparateur* in chemistry at the Ecole Normale, a position that allowed him to continue working toward his doctorate. In August 1847 Pasteur became *docteur-ès-sciences* on the basis of theses in both physics and chemistry. While awaiting appointment elsewhere, he continued to serve as *préparateur* in chemistry at the Ecole Normale and quickly began to win a reputation in scientific circles for his work on the relation between chemical composition, crystalline structure, and optical activity in organic compounds.

Certainly by this point, if not long before, Pasteur had far outgrown his father's early aspirations for him. The prospect of a teaching career in a provincial lycée no longer satisfied him. Like other candidates for positions in the state educational system, Louis did still expect to begin his career in the French provinces. But he now hoped to be spared the heavy lycée teaching load and to be appointed instead to a university-level faculty of science, where he might be able to continue his research. And he already had his sights firmly fixed on an eventual career among the scientific élite in Paris.[4]

When revolution rocked Paris in February 1848, young Louis at first took no part. But in April, after the Second Republic had been declared, he briefly joined the National Guard, a municipal militia charged with the maintenance of civil order, and contributed his savings of 150 francs to the republican cause.[5] At the end of May 1848, when his immediate future was yet to be settled, his mother suddenly fell sick and died, apparently the victim of a cerebral hemorrhage. Pasteur blamed her death partly on her anxiety about his living in strife-torn Paris. His father, who shared this concern, now also had sole responsibility for Louis's three sisters, all of whom were still at home in Arbois and one of whom had been severely retarded since being struck by a cerebral fever at the age of three. Louis knew that some of his father's anxieties would be reduced if he left Paris. He therefore asked the Ministry of Public Instruction to release him from his position at the Ecole Normale and to appoint him instead to some provincial post, even if that meant that he would be forced to go to a lycée.

On 16 September 1848, Pasteur was named professor of physics at the lycée in Dijon, though he was allowed to remain in Paris through the first days of November in order to complete some exciting new research on optical activity and crystalline asymmetry in tartaric and racemic acid. When his duties at the lycée could no longer be postponed, he took consolation in the relative proximity of Dijon to his father and sisters and in his expectation that he would not be there for long.[6] Pasteur's prediction was confirmed even sooner than he expected. By late December 1848, just a few weeks after he started teaching at Dijon, he had applied for and won appointment as *professeur suppléant* (acting professor) of chemistry at the Faculty of Sciences in Strasbourg. After a fleeting concern about the possible effects of this distant move on his family, he eagerly looked forward to his transfer of duties, finally arriving in Strasbourg toward the end of January 1849.[7]

A whirlwind courtship must have begun right away, for in less than a month he proposed marriage to Marie Laurent, daughter of the rector of the Strasbourg Academy. In a formal letter of proposal to her father, dated 10 February 1849, Pasteur spoke of his family's solvent but modest financial circumstances, putting the value of its total assets at no more than 50,000 francs, which he had already decided should go to his sisters. All that he had to offer, he wrote, was "good health, a good nature, and my position in the University."[8] At the age of twenty-six, he married Marie Laurent on 29 May 1849.

At Strasbourg, where he spent nearly six years, Pasteur continued and greatly extended his work on optical activity and crystalline asymmetry in spite of expanding teaching duties. From 1850 on, his letters reveal an increasing impatience with his position as acting professor. While pressing his claims upon his friends and the Ministry of Public Instruction, he followed closely the rumors and intrigues of French academic life in hopes of securing a more satisfactory position. In November 1852, immediately after a well-publicized voyage to Germany and Austria in search of racemic acid, Pasteur was promoted to titular professor of chemistry at Strasbourg. In 1853, for his work on racemic acid and crystallography, he received a prize of 1,500 francs from the Société de Pharmacie and membership in the Legion of Honor. His reputation was already such as to bring his name into consideration for membership in the Académie des sciences in Paris, though in fact nearly a decade was to pass before that long-standing ambition was finally realized.

By September 1854, it was clear that Pasteur was going to be named professor of chemistry and dean of the newly established Faculty of Sciences at Lille, though the appointment did not become official until 2 December 1854. Located at the center of the most flourishing industrial region in France, the Faculty at Lille was designed in part to bring science to the service of local industry. In his inaugural address at Lille, Pasteur strongly supported this goal as well as two innovations brought to the French faculties of science by imperial decree of 22 August 1854: the opportunity for students to do their own laboratory work; and the creation of a new diploma, the "certificate of capacity in the applied sciences," designed for students who wished to become factory managers and to be awarded at the end of two years of theoretical and practical studies at the faculties of science.[9]

In his three years as dean of the Faculty of Sciences at Lille, Pasteur displayed considerable administrative and organizational talent. Under his leadership, laboratory teaching was soon established in all scientific subjects there. With regard to the teaching of "applied" subjects, however, Pasteur moved more cautiously, emphasizing that "theory is the mother of practice" and that without theory, "practice is mere routine born of habit."[10] Despite some pressure from the Ministry of Public Instruction, he resisted any emphasis on applied subjects at the expense of basic science and opposed suggestions that the Lille Faculty should train secondary teachers. He also consistently emphasized that professors at the Faculty owed allegiance to scientific research as well as to teaching, and complained that too many of the auditors were idle amateurs who sought mere entertainment or immediately "useful" information. Equally frustrating to Pasteur was the conservatism of Lille industrialists, their lack of attention to basic science, and their aversion to scientifically trained employees.[11]

For his part, Pasteur believed he was fulfilling his duty to forge bonds between industry and the Faculty at Lille. Among other things, he led his students on excursions to metallurgical factories in Belgium and undertook to test manures for the department of the Nord. In his own courses, he taught the principles and techniques of bleaching, of sugar making and refining, and especially of fermentation and the manufacture of beetroot alcohol, an important local industry. During part of 1856, by which time his research interests had turned to fermentation, Pasteur went regularly to the beetroot alcohol factory of M. Bigo, where he sought to discover the cause of and remedies for recent disappointments in the quality of that product. Such efforts had just begun to yield results when, in September 1857, the directorship of scientific studies at the Ecole Normale fell vacant. Pasteur immediately announced his intention of seeking the position at his alma mater, insisting that the Ecole Normale had become "but a shadow of its former self," beset with apathy and in need of vigorous new leadership.[12]

On 22 October 1857, at the age of thirty-four, Pasteur was named director of scientific studies at the Ecole Normale as well as administrator, which made him responsible for "the surveillance of the economic and hygienic management, the care of general discipline, intercourse with the families of the pupils and the literary or scientific establishments frequented by them."[13] These positions carried with them neither laboratory nor allowance for research expenses, and in order to continue his scientific work, Pasteur was obliged to evade bureaucratic regulations and to rely on his own ingenuity. He managed at once to secure the use of two tiny unoccupied rooms in an attic of the Ecole Normale, where he pursued his research on fermentation despite being unable to stand at full height. With the tacit collusion of colleagues in the bureaucracy, he covered the small costs of essential equipment and supplies by diverting funds from the household budget of the Ecole Normale.[14]

In December 1859 Pasteur gained possession of a small pavilion at the Ecole Normale, which was considerably expanded in 1862. For this expansion, he clearly depended on the support of Emperor Louis Napoleon, whom he had approached by way of the imperial aide-de-camp and to whom he revealed his intention of working on the diseases of wine and infectious diseases in general. Within a few years, through constant appeals to governmental officials, Pasteur had also secured the services of a series of research assistants, funds to cover the expenses of field trips in connection with his studies of fermentation, and an annual laboratory allowance of 2,000 francs.

In his new laboratory at the Ecole Normale, Pasteur continually expanded his research interests and achievements. His well-publicized efforts on behalf of the germ theory of fermentation and against the doctrine of spontaneous generation brought him new honors and recognition. On

Table 2.1 Outline of Pasteur's Career

1829–1831	Student at Ecole Primaire, Arbois
1831–1839	Student at Collège d'Arbois
1839–1842	Student at Collège Royal de Besançon
1842–1843	Student at Barbet's School and Lycée St.-Louis, Paris
1843–1846	Student at Ecole Normale Supérieure (Paris)
1846–1848	Préparateur in chemistry, Ecole Normale
1849–1854	Professor of chemistry, Faculty of Sciences, Strasbourg *suppléant*, 1849–1852 *titulaire*, 1852–1854
1854–1857	Professor of chemistry and dean of the Faculty of Sciences, Lille
1857–1867	Administrator and director of scientific studies, Ecole Normale
1867–1874	Professor of chemistry, Sorbonne
1867–1888	Director of the laboratory of physiological chemistry, Ecole Normale
1888–1895	Director of the Institut Pasteur (Paris)
	In addition:
Sept.–Dec. 1848	Professor of physics, Lycée de Dijon
1863–1867	Professor of geology, physics, and chemistry in their application to the fine arts, Ecole des Beaux-Arts (Paris)

Source: Geison 1974, pp. 350–351.

8 December 1862, a few weeks before his fortieth birthday, Pasteur was elected to membership in the mineralogy section of the Académie des sciences, thus realizing an old dream and succeeding in his third formal campaign for the honor.[15] Thereafter, the weekly meetings of the Académie regularly took him away from his laboratory. So did his lectures at the Ecole des Beaux-Arts, where from November 1863 to October 1867 he was the first professor of geology, physics, and chemistry in their application to the fine arts, and where he introduced laboratory procedures oriented toward the problems of art and the preparation of its materials.[16] After 1865 he faced a much larger demand on his time in the form of the French silkworm blight, which he agreed to study at the government's request, and which took him away from Paris to a field laboratory in the south of France every summer through 1870. Even at the Ecole Normale, Pasteur's activities were scarcely confined to the laboratory. An innovative administrator and fastidious organizer, he displayed a remarkable devotion to detail. He invested great time

and energy in his administrative duties, proposing and carrying through a series of important institutional reforms.

Notable among these reforms were those having to do with the *agrégé-préparateurs*, laboratory assistants who were graduates of the Ecole Normale. Although he did not create these positions, as is sometimes supposed, Pasteur did propose an expansion in their number from three to five and a decrease in the period of their appointment from seven or eight years to two, his goal being to encourage more *normaliens* to seek doctorates and a career in research.[17] With the success of his proposal, Pasteur himself became a major beneficiary of the reform, beginning in 1863 when the first of a series of *agrégé-préparateurs* was assigned to his own laboratory. Less successfully, Pasteur urged that the Ecole Normale should overcome its excessive dependence on the Faculty of Sciences in Paris by developing its own integrated two-year science curriculum, including instruction in the history of science.[18] Finally, in the most tangible and enduring of his innovations, Pasteur founded a new journal, *Annales scientifiques de l'Ecole Normale Supérieure*, devoted to the publication of original papers by *normaliens*. He directed the *Annales* himself from its first issue in 1864 until ill health forced his resignation in 1871.

It is one index of Pasteur's administrative success as well as his scientific reputation that the number of candidates for the scientific section of the Ecole Normale increased enormously during the decade he served as director of scientific studies. By the end of his directorship, the number of candidates, for an average of fifteen places annually, had reached 200–230, compared to only 50–70 before his appointment. Also during his directorship, every student admitted simultaneously to the Ecole Normale and the famous Ecole Polytechnique chose the former. Twice, in 1861 and then again in 1864, the top-ranking candidate at the Ecole Polytechnique resigned this title in order to study at the Ecole Normale—an event entirely without precedent before Pasteur arrived on the scene.[19] He took great pride in these institutional achievements and especially in the challenge the Ecole Normale now posed to the Ecole Polytechnique, where Pasteur himself had failed to gain entrance two decades before.

Pasteur's talents as an administrator did not extend to the handling of student discipline, to which task he brought the full measure of his respect for order, his moral earnestness, and his inflexible and authoritarian manner. His interaction with students at the Ecole Normale has been described as "hardly frequent" but "often disagreeable."[20] He dealt summarily and unsympathetically with student complaints about the food and strict rules, and by 1863 was openly appalled by the insubordination of the students, especially those in the humanities. In March of that year, Pasteur expelled two students who had left the school for a few hours without permission, "not for what they have done, but because of the detestable spirit which reigns at the school."[21] He also announced that anyone caught smoking in the future would be expelled, again not because of the offense itself, but rather because they would have ignored his injunction. Disturbed by Pasteur's severity and rigidity, three-fourths of the students at the Ecole Normale signed a protest petition offering their resignation. Eventually, through the intervention of less rigid officials, peace was restored after the imposition of minor punishments.[22]

From this point on, Pasteur's letters reveal an increasing dissatisfaction with his position and with the general direction of the Ecole Normale. He was particularly irritated by the absence of a single clear line of authority in the school and felt that he was being denied de facto the influence and status that was his de jure. He therefore proposed a thorough institutional reorganization and registered his protest against increases in salary given other administrative officers of the school while his remained static. Without some reorganization or at least an increase in his salary, Pasteur threatened to resign as administrator and retain only his position as director of scientific studies.[23] In the event, however, he retained both offices until 1867, when a more serious student disturbance ended with the closing of the Ecole Normale and the replacement of its three major administrative officers, including Pasteur.

This time the student disturbance was bound up with external political events. In July 1867 a student at the Ecole Normale wrote a letter in support of a celebrated speech by Senator Sainte Beuve defending free thought and deploring an attempt to remove allegedly subversive books from a provincial library. The letter, which claimed to express the views of three-fourths of the students, found its way into print in a newspaper to which it had been transmitted by two of the author's schoolmates. Besides violating a university bylaw forbidding any collective political activity by students, the published letter referred ironically and with seeming approval to a recent attempt to assassinate Emperor Louis Napoleon. When the author of the letter was provisionally expelled, the students of the Ecole Normale protested nearly en masse, and even the Ministry of Public Instruction seemed to disapprove of the action.

Pasteur, however, remained rigidly in support of a decision for which he seems to have been chiefly responsible. Both to the Ministry of Public Instruction and to the students of the Ecole Normale, Pasteur demanded the provisional expulsion of the author of "this ridiculous and culpable address,"[24] as well as of the two students who had taken it to the newspaper. Unless this were done and agreed to by the students, Pasteur said that he would resign immediately and that the school ought to be closed. Apparently spearheaded by the humanities students, virtually the entire student body of the Ecole Normale was soon marching in protest in the streets of Paris. At this point the school was closed, and when it reopened in October, Pasteur and the other administrative officers had been replaced.[25]

These events left the Ministry of Public Instruction with the problem of finding an appropriate position for so distinguished a scientist as Pasteur, who was now on the verge of his forty-fifth birthday. He was at first offered the inspector generalship of higher education, but withdrew when it became known that one of his former mentors wished to assume this office himself. The Ministry then offered Pasteur a professorship of chemistry at the Sorbonne and a position as *maître de conférences* in organic chemistry at the Ecole Normale, with the right to retain his apartment and laboratory there. In place of this offer, Pasteur submitted an alternative proposal of his own, addressed simultaneously (on 5 September 1867) to the minister of public instruction and to Emperor Louis Napoleon. Pasteur agreed fully with his appointment as professor of chemistry at the Sorbonne, but objected to the proposed position at the Ecole Normale on several grounds, including his concern that two teaching positions would leave him too little time for his own research. Instead, Pasteur proposed the construction at the Ecole Normale of a new, spacious, and well-endowed laboratory of physiological chemistry in which he would continue his own research. He supported his proposal by referring to "the necessity of maintaining the scientific superiority of France against the efforts of rival nations," and by projecting studies of immense practical importance, particularly on infectious diseases.[26]

Emperor Louis Napoleon immediately expressed his support for Pasteur's project in a letter to the minister of public instruction. Construction of the laboratory began in August 1868, the cost of 60,000 francs being shared equally by the Ministry of Public Instruction and the Ministry of the House of the Emperor. The large new laboratory was to be linked by a gallery with the pavilion Pasteur had occupied since 1859. Largely because of the Franco-Prussian War, however, the laboratory remained incomplete and unoccupied as late as 1871. During the war, Pasteur withdrew to the provinces and launched a study of beer, his explicit aim being to serve France in "a branch of industry in which Germany is clearly superior to us."[27]

In September 1871, following the departure from Paris of the Prussian troops and the crushing of the Communard uprising, Pasteur returned to his nearly finished new laboratory and immediately asked to be relieved of his teaching duties at the Sorbonne. While seeking to retain the directorship of his laboratory, he declared himself unfit to teach any longer on account of his health. Although not yet fifty, he argued that his resignation from the

Sorbonne chair ought to bring him a retirement pension, since he had already spent thirty years in university service (including his days as a tutor at the collège in Besançon). He further requested a national recompense in recognition of the contributions he had made to his country through his research. In repeating his requests at least twice to the president of the new Third Republic, Pasteur pointed out that the abdication of Emperor Louis Napoleon had unluckily deprived him of a Senate seat and a national recompense that were to have been his by imperial decree.[28] Within three years, officials of the Third Republic met all of Pasteur's requests, despite his close and long-standing ties with Louis Napoleon. In a landslide vote of July 1874, the National Assembly awarded him a national recompense of 12,000 francs annually, roughly equal to the salary he lost by resigning his professorship at the Sorbonne.

By this point, in his early fifties, Pasteur had reached a watershed in his career. He had achieved a set of opportunities and facilities for research that almost matched his expansive needs and wants. He had a large new laboratory and had been relieved of all teaching duties. He had solved to his satisfaction the problem of the silkworm diseases, thus discharging a duty that had cost him dearly in time, energy, and health between 1865 and 1870. He had access to research assistants from the Ecole Normale and an annual research budget of 6,000 francs. And he had already declared his intention to mount a focused attack on a new and potentially vast area of research: the study of infectious diseases. In effect, he had a governmental mandate to do just that, for the construction of his new laboratory had been approved by Emperor Louis Napoleon in 1867 on the understanding that it would be devoted mainly to the investigation of the infectious diseases.

A decade passed before Pasteur redeemed this pledge. Nearly five years were lost to construction, to his work on the silkworm diseases, and to the disruptive effects of the Franco-Prussian War. The next five years of delay are less easily explained. For a man of his bold readiness to tackle major problems in virtually any area of science, and for a man who had long been an influential indirect participant in medical debate through his work on fermentation, spontaneous generation, and silkworm diseases, Pasteur hesitated a surprisingly long time before entering directly into the territory of veterinarians and physicians.

Until 1876, fully five years after his new laboratory was completed, Pasteur continued to devote its resources and his own energies to studies on beer and to persistent controversies over his work on fermentation and spontaneous generation. By the time he did make the infectious diseases the focus of his research, beginning with anthrax, the germ theory of disease had already made substantial headway through Joseph Lister's dramatic campaign for "antiseptic" surgery and through Robert Koch's just published work on the etiology of anthrax in sheep, which won immediate acclaim and went a long way toward raising anthrax to its special status as the first major lethal disease of large animals widely admitted to be microbial in origin.

Once Pasteur did enter the veterinary and medical arena, he enjoyed rapid and ultimately spectacular success. His work on the etiology of anthrax, though much less significant than that of Koch and the German school of bacteriologists, did extend and fortify the latter. More important, Pasteur and his French disciples quickly revealed the practical benefits to be gained from research on immunity and prophylaxis against microbial diseases—in a word, from vaccination.

The French government materially encouraged such efforts. In May 1880, shortly after Pasteur announced the discovery of a vaccine against chicken cholera, the city of Paris gave him access to some unoccupied land near his laboratory. On this site, which belonged to the old Collège Rollin, Pasteur made extensive provisions for the care and shelter of the many animals used in his experiments. Later that year, his annual budget for research expenses, a mere 2,000 francs in the early 1860s and fixed at 6,000 francs in 1871, was increased nearly tenfold through a supplementary annual credit of 50,000 francs from the Ministry of Agriculture.[29] In granting the full increase for which Pasteur had appealed, the Ministry was recognizing and abetting the success of his new research on vaccines against animal diseases.

On Christmas Day 1881, by which time Pasteur had also announced the discovery of a vaccine against anthrax, he asked the French government to create a state laboratory for the manufacture of this vaccine. He further proposed that he be named director of this laboratory, with assistance from his two senior collaborators at the time, Charles Chamberland and Emile Roux. By its support for this project, wrote Pasteur, the French state would gain prestige and gratitude as anthrax disappeared. In return, he asked only that he and his family "be freed of material preoccupations."[30] In the event, the government rejected Pasteur's proposal, and his own laboratory became the center for the manufacture of anthrax vaccines. A new annex of the laboratory, located on the rue Vauquelin two blocks away from the Ecole Normale, was turned over to Chamberland and devoted entirely to the production of this and other vaccines.

In 1884, when Pasteur was in hot pursuit of a rabies vaccine, a governmental commission (convened at his request) recommended the establishment of a large kennel for the housing and surveillance of his experimental dogs. The site initially chosen, in Meudon Park, was abandoned in the face

Table 2.2 List of Pasteur's Major Prizes and Honors

1853	Chevalier of the Imperial Order of the Legion of Honor
1853	Prize on racemic acid, Société de pharmacie de Paris
1856	Rumford Medal, Royal Society (for work in crystallography)
1859	Montyon Prize for Experimental Physiology, Académie des sciences
1861	Zecker Prize, Académie des sciences (chemistry section)
1862	Alhumbert Prize, Académie des sciences
1862	Elected member of the Académie des sciences (mineralogy section)
1866	Gold Medal, Comité central agricole de Sologne (for work on diseases of wine)
1867	Grand Prize Medal of the Exposition universelle (Paris), for method of preserving wine by heating
1868	Honorary M.D., University of Bonn (returned during Franco-Prussian War, 1870–1871)
1868	Promoted to commander of the Legion of Honor
1869	Elected fellow of the Royal Society of London
1871	Prize for silkworm remedies, Austrian government
1873	Commander of the Imperial Order of the Rose, Brazil
1873	Elected member of the Académie de médecine
1874	Copley Medal, Royal Society of London (for work on fermentation and silkworm diseases)
1874	Voted national recompense of 12,000 francs
1878	Promoted to grand officer of the Legion of Honor
1881	Awarded Grand Cross of the Legion of Honor
1882	Grand Cordon of the Order of Isabella the Catholic
1882	National recompense augmented to 25,000 francs
1882	Elected to Académie française
1886	Jean Reynaud Prize, Académie des sciences
1887	Elected perpetual secretary, Académie des sciences (resigned because of illness in January 1888)
1892	Jubilee celebration at the Sorbonne

Source: Geison 1974, p. 351.

of vigorous protests from local inhabitants who wished to avoid the nuisance and danger of having noisy rabid dogs in their neighborhood. Similar local protests erupted upon the selection of a second site—in the park of Villenevue l'Etang, near St.-Cloud, an enormous state domain a dozen kilometers west of central Paris that had once belonged to Emperor Louis Napoleon. Although these protests helped delay a legislative appropriation of 100,000 francs, they were ultimately ineffectual. By May 1885 the old stables of the Château de St.-Cloud had been converted into an enormous paved kennel with accommodations for sixty dogs. A laboratory was also soon established, and modest living quarters nearby were renovated for Pasteur's private use.[31]

With the triumphant success of his rabies vaccine, first applied to a human case in July 1885, Pasteur and his laboratory were deluged with an outpouring of grateful donations from private individuals and organizations throughout the world. A formal subscription was soon organized, and the contributions easily surpassed two million francs by November 1888, when the magnificent new Institut Pasteur was officially inaugurated.[32] Pasteur, by then sixty-five years old and gradually failing in strength, had achieved world renown and a string of major national honors in the decade since turning his attention to disease, including the Grand Cross of the Legion of Honor, awarded to him in 1881, and election to the Académie française in 1882. Pasteur died on 28 September 1895; a week later, on 5 October, the French state honored his passing with a grand public funeral worthy of its latest fallen hero.

SCIENTIFIC RESEARCH

With this chronological sketch of Pasteur's career now in place, we can better survey some of the more general features of his life and work. His scientific research deserves first claim on our attention. What is perhaps most striking is the apparent simplicity and accessibility of most of Pasteur's work. His genius lay not in ethereal subtlety of mind. Although often bold and imaginative, his work was characterized mainly by clearheadedness, extraordinary experimental skills, and tenacity—almost obstinacy—of purpose. His contributions to basic science were extensive and very significant, but less revolutionary than his reputation suggests.

Pasteur's most profound and most original contributions to science are also the least famous, and they came at the very outset of his career. Beginning about 1847, he carried out an impressive series of investigations into the relation between optical activity, crystalline structure, and chemical composition in organic compounds, particularly tartaric and paratartaric acid. Early on in this work came his dramatic discovery of a new form of paired compounds—optical isomers, substances identical in every respect except in their opposite effect on polarized light (their "optical activity") and in tiny details of crystalline form which made them mirror images of each other. In pursuing this topic, Pasteur became convinced that optical activity and microstructural asymmetry were somehow peculiarly associated with life, a position that remains broadly valid despite significant alterations in the details of Pasteur's conceptions.

From crystallography and structural chemistry, Pasteur moved on to the controversial and interrelated topics of fermentation and spontaneous generation. He was drawn to a biological or "germ" theory of fermentation from the outset. Because the products of fermentation are often optically active, and since he had already linked optical activity with life, he was predisposed to link fermentation with life in the form of microbes or "germs." He did more than any single figure to promote the microbial theory of fermentation and to discredit the doctrine of spontaneous generation. But the profound influence of his work on these problems owed less to conceptual originality than to experimental ingenuity and polemical virtuosity, which served him well throughout his career. He did broach and contribute importantly to fundamental questions in microbial physiology, including the relationship between microorganisms and their environment, but he was readily distracted from such basic issues by more practical concerns—the manufacture of wine, vinegar, and beer; the diseases of silkworms; and the etiology and prophylaxis of diseases in general.

To some extent, Pasteur's interest in practical problems evolved naturally from his basic research, especially that on fermentation. The germ theory of fermentation carried quite obvious implications for industry and medical doctrine. By insisting that each fermentative process could be traced to a specific living microorganism, Pasteur drew attention to the purity and special nutritional and oxygen needs of the microbes employed in industrial processes. He also suggested that the primary industrial product could be preserved by appropriate sterilizing procedures, labeled "pasteurization" almost from the outset. Furthermore, the old and widely accepted analogy between fermentation and disease made any theory of the former immediately relevant to the latter. The germ theory of fermentation virtually implied a germ theory of disease as well. This implication was more rapidly exploited by others, particularly Joseph Lister and Robert Koch, but Pasteur also perceived it from the first and devoted his last twenty years almost

Table 2.3 **Chronological Outline of Pasteur's Major Research Interests**

1847–1857	Crystallography: optical activity and crystalline asymmetry
1857–1865	Fermentation and spontaneous generation; studies on vinegar and wine
1865–1870	Silkworm diseases: *pébrine* and *flacherie*
1871–1876	Studies on beer; further debates over fermentation and spontaneous generation
1877–1895	Etiology and prophylaxis of infectious diseases: anthrax, fowl cholera, swine erysipelas, rabies

Source: Geison 1974, p. 351.

exclusively to working out some of the practical consequences of the germ theory of disease.

No one insisted more strongly than Pasteur himself on the degree to which his pragmatic concerns grew out of his prior basic research. He saw the progression from crystallography through fermentation to disease as not only natural but almost inevitable; he had been "enchained," he wrote, by the "almost inflexible logic of my studies."[33] Yet it is clear that his work could have taken many other directions with equal fidelity to the internal logic of his research.[34] To some extent, Pasteur chose to pursue the practical consequences of his work at the expense of his potential contributions to basic science. Without disputing the immense value and fertility of the basic research he did accomplish, it is fascinating to speculate on what might have been. Late in life, Pasteur indulged in similar speculation and expressed regret that he had abandoned his youthful researches before fully resolving the relationship between asymmetry and life. Had he succeeded in his once hopeful quest for a "cosmic asymmetric force," he would surely have fulfilled his ambition of becoming the Galileo or Newton of biology.[35]

By taking another direction, however, Pasteur revealed the enormous economic and medical potential of experimental biology. He developed only one treatment directly applicable to a human disease—his treatment for rabies—but his widely publicized and highly successful efforts on behalf of the germ theory were quickly credited with saving much money and many lives. No one, at least no experimental biologist, had done so much to show that scientific research could pay off so handsomely in practical results. It is for this reason above all that Pasteur was recognized and honored during his lifetime as few scientists indeed ever have been and that his name remains a household word.

As his correspondence makes clear, Pasteur chose his path under the impulse of complex and mixed motives. Apart from the internal logic of his research, these motives included ambition for fame and imperial favor, his desire to serve humanity or at least his country, and his concern for financial support and security. In the highly competitive academic world of mid-nineteenth-century France, he was unabashedly ambitious and opportunistic. While rejecting his father's admonition to set more modest goals for himself, he did accept his advice to cultivate important friends as well as knowledge. His letters are filled with references to academic politics and with appeals for support from his influential friends—notably, at first, the famous physicist Jean Baptiste Biot and the well-placed chemist Jean Baptiste Dumas, and later a number of important ministers and government officials, including Emperor Louis Napoleon and Empress Eugénie.

Pasteur's ambition was joined with enormous self-confidence, which emerged early on and only increased with the years. When not yet thirty, he consoled his rather neglected wife by telling her that he would "lead her to posterity."[36] In controversy, his combative self-assurance could be devastating to the point of cruelty. He so offended one opponent, an eighty-year-old surgeon, that the latter actually challenged him to a duel—which, happily for both, never took place.[37] He claimed to prefer thoughtful criticism to sterile praise, but in fact he almost always exploded whenever criticism was directed his way, whether that criticism was responsible or not. Pasteur shared with many of his peers a rather simpleminded and absolutist notion of scientific truth, rarely conceding the possibility of its being multifaceted and relative. He generally gave credit to others only grudgingly and mistrusted those who claimed to have reached similar views independently. He insisted that he was willing to await the verdict of posterity on his work, but spent considerable time and effort seeking to establish the priority of his concepts and discoveries, particularly his process for preserving wines.

In these respects, it should be stressed, Pasteur was hardly alone. Ambition for fame or priority, and sometimes for a measure of fortune as well, has ever been a feature of modern intellectual life, and Pasteur was not vastly more susceptible to the claims of self-interest than many other scientists then or since. What set him apart from his rivals was the consummate success with which he deployed his polemical talents, rhetorical skills, and institutional advantages. In a highly competitive and contentious environment, he was particularly bold and successful in the art of self-advertisement. By appeal to public demonstrations—most spectacularly in the sheep vaccination experiments at Pouilly-le-Fort—and by frequent recourse to "judiciary" commissions of the Académie des sciences, Pasteur nearly always won public and quasi-official sanction for his views.[38] Whatever else one may think of Pasteur's polemical inclinations and talents, they were a major factor in his success.

So, it should never be forgotten, was his awesome capacity for work. The mature Pasteur himself always ascribed his success to hard work, perseverance, and tenacity. He arose at dawn, arrived at his laboratory early in the morning almost every day, including Sundays, and he usually stayed there into the evening hours. His assistants and collaborators stressed his ability to concentrate intensely on one problem for long stretches of time—an ability so pronounced that he seemed almost to drift into a trance at such times.[39] Of the other factors invoked to account for Pasteur's success, perhaps the most surprising is his nearsightedness. There is, it seems, no known biological basis for the alleged virtues of myopia. Yet it is curious to note that Pasteur shared this visual defect with his great German rival Robert Koch and their distant predecessor, the pioneering seventeenth-century Dutch microscopist Antonie van Leeuwenhoek, among other great microscopists. And Pasteur's collaborators insisted that his myopia so enhanced his close vision that, in an object under the microscope or between his hands, he really could see things that were hidden to normally-sighted people around him.[40]

Blind luck has also sometimes been used to explain, or rather to dismiss, Pasteur's success. Two examples have been repeatedly invoked to illustrate the role of serendipity in Pasteur's research—his early discovery of optical isomers in the tartrates, and his discovery of a chicken cholera vaccine in 1880. In the first case, it is claimed, his discovery of optical isomers might never have come had he begun his research on any compounds other than the tartrates and paratartrates. In these compounds, the relation between chemical composition, crystalline structure, and optical activity is atypically—perhaps even uniquely—clear and straightforward. It is also said that the discovery depended on the weather in Paris at the time of the research, for the asymmetric forms of the paratartrate in which he discovered optical isomers do not precipitate out except under quite special conditions, especially with regard to temperature.[41] But this conception of the story severely minimizes Pasteur's chemical artistry, his ability to produce crystals of different sizes and shapes by delicate manipulations of the crystallizing conditions. In some cases, as we shall see in Chapter Three, Pasteur displayed a magus-like capacity for almost literally "creating" crystals of a sort that would confirm his alleged correlation between optical activity and crystalline asymmetry.

The second example of a "lucky" discovery by Pasteur is much more familiar. Indeed, this example—the discovery of the chicken cholera vaccine—has become a stock item in discussions of the role of serendipity in scientific discovery. According to standard accounts, which can be traced to Pasteur's collaborator Emile Duclaux, an attenuated strain of the chicken cholera microbe—in a word, a "vaccine" against the disease—emerged only because Pasteur's collaborators forgot or neglected his instructions to recultivate the microbe at short intervals during a summer vacation that he spent, as usual, at the familial home in Arbois. As the cultures sat on the shelf unattended, they underwent attenuation and proved to induce immunity against chicken cholera when injected into experimental animals.[42]

By this account, a lack of diligence during a summer vacation was thus a major factor in the discovery of the first laboratory-produced vaccine, the only other vaccine at the time being the naturally occurring cowpox virus that Jenner had deployed against smallpox. Unfortunately for advocates of serendipity, Antonio Cadeddu has recently destroyed this appealing legend by analyzing Pasteur's notebooks from the time. Cadeddu shows that the chicken cholera vaccine did not emerge "by accident" at all, but rather was the product of a prolonged, complex, and quite deliberate program of research undertaken by Emile Roux without Pasteur's knowledge.[43] Perhaps that is why Duclaux's version of the story does not appear in Pasteur's quasi-autobiography of 1883, which elsewhere reveals his willingness to indulge such popular stories of the path to his discoveries.[44]

But even in the absence of this new revelation—even if the chicken cholera vaccine had been an example of "accidental" discovery—it would still be a mistake to dismiss this or other examples of Pasteur's achievements as the result of sheer luck. Such a conclusion would ignore the fact that Pasteur creatively seized the opportunities that seemed to come to him "accidentally," and that he did so repeatedly. Repeated strokes of "luck" render the word meaningless. There is real wisdom in Pasteur's own famous maxim that "chance favors only the prepared mind."[45]

At a more prosaic level, Pasteur's success certainly did depend crucially on financial support from the government. He sometimes complained bitterly of the neglect of science by the French state, and he resented the need to make constant appeals to the bureaucracy for research expenses, describing the process as "antipathetic to a scientist worthy of the name."[46] Yet appeal he did, and rarely did he fail. Especially once his concern with practical problems became manifest, he enjoyed truly remarkable success at getting whatever he sought—a new or expanded laboratory, additional personnel, a larger research budget, even national railroad passes for himself and his assistants. Among the governmental sources he tapped were the Ministries of State, Agriculture, Public Instruction, Public Works, and even the Imperial House itself, where the more pragmatic aspects of his work on wine and disease received personal support and encouragement from

Emperor Louis Napoleon and Empress Eugénie.[47] Nor did the emperor's abdication and the coming of the Third Republic do anything to interrupt the flow of government funds. The work on vaccines was especially well funded. Some German scientists, including Robert Koch, may have fared just as well, but the support given Pasteur was spectacularly generous by French standards. By the early 1880s, when the vaccines against chicken cholera and anthrax emerged from his laboratory, Pasteur may have been the recipient of 10 percent or more of the annual governmental outlay for *all* scientific research in France.[48]

Yet Pasteur had never been content to rely solely on the generosity of the French state. From the beginning of his career he competed actively for monetary awards from scientific societies, one modestly lucrative example of early success being the prize of 1,500 francs he won in 1853 from the Société de pharmacie de Paris for his work on racemic acid. He also paid close attention to announcements of monetary prizes by foreign governments, winning 5,000 florins (roughly 8,500 francs) from the Austrian government in 1871 for his efforts against the silkworm blight. By far the most spectacular such award for which he competed—in this case unsuccessfully—was a prize of 625,000 francs that the government of New South Wales announced in 1887 for practical measures to reduce its excessive rabbit population.[49] Still other financial support came from industrialists and other wealthy or not-so-wealthy individuals. His studies on beer were supported in part by brewers, and he coaxed a check for 100,000 francs out of Madame Boucicaut, whose late husband had founded the enormously successful Parisian department store, the Bon Marché.[50] Other private donors, including even some poor ones, were certainly the main source of funds for Pasteur's most enduring monument, the Institut Pasteur in Paris, which cost well over 2.5 million francs to build and equip.

Finally, Pasteur derived major support from revenues on patents and licenses, despite his occasional qualms that it was not quite proper for a scientist to benefit from the commercial exploitation of his discoveries. As early as 1857, he took out a patent for a process of alcoholic fermentation, and he later received patents for a bacterial filter (the Chamberland-Pasteur filter) and for his methods of manufacturing and preserving wine, vinegar, and beer.[51] No adequate account exists of the fate of these patents. Some, perhaps most, were deliberately allowed to fall into the public domain or went unexploited for other reasons. But the patent on the bacterial filter apparently was exploited and probably yielded significant revenues. Still larger returns were realized through commercialization of the anthrax vaccine, thanks especially to foreign licenses and sales. If one estimate from the mid-1880s can be believed, the *annual* net return on the anthrax vaccine amounted to 130,000 francs,[52] more than twice as much as even Pasteur had ever managed to wrest from government sources for his annual research expenses. The French state, by then, may have regretted its negative response to Pasteur's proposal of Christmas Day 1881 to create a state vaccine factory in return for assurances that he and his family would be freed of "material preoccupations."[53]

We may never know exactly how much of this income found its way into the private hands of Pasteur, his family, and his co-workers. Most of the revenue seems to have gone to the French state or to the budget of the Institut Pasteur. Reportedly at the urging of his wife and family, Pasteur eventually did take some of the income from his patents and licenses, but the amount is unknown. Thus far, we know only that his last will and testament provided that his wife receive "all that the law allows."[54]

But there can be no doubt that Pasteur and his family enjoyed a very handsome annual income, especially after 1882, when the French state awarded him an annuity of 25,000 francs. From that source alone, Pasteur received nearly twice as much income as the average university professor in Paris and perhaps ten times as much as the typical "white-collar" employee of a Parisian department store.[55] By then, his *annual* salary was worth at least half of the estimated *total* assets of his parents at the time he married. Pasteur had come a long way from his petit bourgeois roots, and money (or, at least, the security it offered) was very definitely important to him. Yet it is said that he paid surprisingly little attention to the details of his own financial circumstances, and he surely could have made even more money had he been less scrupulous. He did not, it seems, accumulate a vast personal fortune. His claim that he worked solely for the love of science, country, and humanity and his enduring image as a *savant désintéressé* are decidely exaggerated. But they carry rather more conviction than attempts to depict him as a scientific prostitute. Compared, for example, to the German chemist Justus von Liebig, he was a model of commercial restraint.[56]

Obsessed with science and its applications, Pasteur devoted little thought to religious, philosophical, or political questions. His beliefs in these areas were basically visceral or instinctive. At the center of his public views on religion and philosophy lay his insistence on an absolute separation between matters of science and matters of faith. Although he was reared and died a Catholic, he was by no means so "devout" as he is sometimes portrayed. Even as a schoolboy, he confessed to the sacrilege of reading moral philosophy during Mass, and he later abandoned most religious practices entirely. Neither religious ritual nor the details of theological doctrine held much attraction for him.[57] He cared as little for formal philosophy. By his early forties, he had read only a few "absurd passages" in Comte, and he described his own philosophy as one "entirely of the heart."[58] Throughout his life he disdained materialists, atheists, freethinkers, and positivists. In 1882, in his inaugural address at the Académie française, Pasteur found wanting the positivistic philosophy of Emile Littré, whom he was replacing. For Pasteur, the failures of positivism included its lack of real intellectual novelty, its confusion of the true experimental method with the "restricted method" of observation, and above all its disregard for "the most important of positive notions, that of the Infinite," one form of which is the idea of God. Pasteur never expressed doubt about the existence of the spiritual realm or the immortal soul. In that sense, and in his opposition to philosophical materialism, he was a "spiritualist." Indeed, in his inaugural address at the Académie française, he spoke of the service his research had rendered to "the spiritualist doctrine, much neglected elsewhere, but certain at least to find a glorious refuge in your ranks."[59]

Pasteur's chief contribution to the "spiritualist doctrine" was his campaign against spontaneous generation. He stressed the religious and philosophical implications of this campaign from time to time, all the while denying that any such concerns influenced his own work. In any truly scientific question, he insisted, neither spiritualism nor any other philosophical school had a place. Only the "experimental method" could arbitrate scientific disputes.[60] Yet we shall see that Pasteur did not hesitate to bolster his experiments against spontaneous generation with thinly veiled appeals to reigning religious and philosophical orthodoxies. Throughout the 1860s and 1870s, when many French thinkers regarded Darwinism, spontaneous generation, and philosophical materialism as threats to church and state, Pasteur's published scientific work lent support to a "vitalistic" position that enjoyed philosophical and theological respectability in France. In public, he dismissed speculation on the ultimate origin and end of things as beyond the realm of science. In private, however, he did not refrain from speculation on the origin of life. As we shall see more fully below, Pasteur even tried to create or modify life himself as part of his "mad" quest for a "cosmic asymmetric force" in the early 1850s. We shall find reason to believe that Pasteur's whole approach to the question of the origin of life was strongly conditioned by an intertwined set of philosophical, religious, and political interests.

Pasteur's public positions on religious and philosophical questions dovetailed neatly with his basic political instincts. Despite his youthful flirtation with republicanism during the Revolution of 1848, Pasteur was essentially conservative, not to say reactionary. His political instincts found their most faithful reflection in his admiration for the least liberal phases of the Second Empire. He considered strong leadership, firm law enforcement, and the maintenance of domestic order more important than civil liberty or even democracy, which he distrusted lest it lead to national mediocrity or vulgar tyranny. Yearning above all for the past glory of France, which he (like his father) traced to Napoleon Bonaparte, he hoped that the hero's nephew, Louis Napoleon, might somehow restore it.[61]

From the coup d'état of 2 December 1851, by which Louis Napoleon dissolved the Constituent Assembly, Pasteur declared himself a "partisan" of the new leader.[62] Partly through the chemist Jean-Baptiste Dumas—his former mentor and patron, whom Louis Napoleon had named a senator—Pasteur developed personal relations with the imperial household, to which he sent copies of his works on fermentation and spontaneous generation. Especially after 1863, when Dumas presented him to Louis Napoleon, Pasteur openly sought to encourage imperial interest in his research. In 1865 the emperor invited him to Compiègne, the most elegant of the imperial residences. In breathless letters to his wife during the week he spent there, Pasteur betrayed his fascination with imperial power, pomp, and wealth.[63] The next year he dedicated his book on wines to the emperor, who returned the favor by promoting Pasteur to commander of the Legion of Honor in 1868. Louis Napoleon's abdication in 1870 filled Pasteur with sorrow. It also nullified an imperial decree of 27 July 1870 by which he would have been awarded a national pension and made a senator for life.

But Pasteur was no mere political opportunist. He continued to acknowledge his association with and indebtedness to the imperial household even after the abdication—even in the face of advice that it could be politically imprudent to do so.[64] In 1875 Pasteur was asked by friends in his hometown of Arbois to run for the Senate. Saying that he had no right to a political opinion because he had never studied politics, he nonetheless agreed to run as a conservative. Presenting himself as the candidate of science and patriotism, he made it his central political pledge "never [to] enter into any combinations the goal of which is to upset the established order of things."[65] Lest that be taken as a commitment to the new republican government, Pasteur emphasized that the Third Republic was by law a temporary experiment that should be continued only if it succeeded at achieving internal order and external prestige. Pasteur's electoral rivals exploited his links with the Second Empire and his suspected Bonapartist loyalties. In response, Pasteur merely noted that Louis Napoleon had died owing him 4,000 francs and disclaimed any link with organized Bonapartist groups. Pasteur was crushed in the election, with the Arbois electorate giving him only sixty-two votes, nearly four hundred fewer than each of the two successful republican candidates. Asked at least twice during the 1880s to run again for the Senate, Pasteur declined while his strength for scientific work remained. By then he referred to politics as ephemeral and sterile compared with science, a view that can only have been reinforced by his hostile reception on a visit to Arbois in 1888.[66] In 1892, no longer strong enough for research, he began soliciting support for a place in the Senate but eventually withdrew.[67]

However firm Pasteur's loyalty to the Second Empire and to political conservatism, his general patriotism was even stronger. Sometimes, indeed, it took the form of chauvinism. In 1871, despite tempting offers from Milan and Pisa, Pasteur remained in France, partly because of his wife's unwillingness to expatriate but especially because he felt it would be an act of desertion to leave his country in the wake of its crushing defeat in the Franco-Prussian War.[68] That defeat and the excesses of the Prussian army so outraged Pasteur that he vowed to inscribe all of his remaining works with the words, "Hatred toward Prussia. Revenge! Revenge!"[69] Also in 1871 he returned in protest an honorary M.D. degree awarded in 1868 by the University of Bonn. In an exchange of letters with the dean of the medical faculty there, which he published as a brochure, Pasteur screamed out against the "barbarity" being visited upon his country by Prussia and its king. In another brochure of 1871, "Some Reflections on Science in France," he emphasized the disparity between the state support of science in France and in Germany, tracing the defeat of France in the war to its excessive tolerance toward the "Prussian chancre" and to its neglect of science during the preceding half-century. In 1873, when he patented a process for manufacturing beer that he mistakenly hoped would pose a challenge to the superior German breweries, he stipulated that the beer made by his method should bear in France the name "Bières de la révanche nationale" and abroad the name "Bières françaises."[70] Chauvinism doubtless played some part in his refusal to grant permission for a German translation of his studies on beer and in his bitter and protracted dispute with Robert Koch in the 1880s. It probably also helps to explain his insistence on using the term "microbiology" instead of "bacteriology," which he considered a constricting "Teutonic" label.[71] Even on the eve of his death, Pasteur's memories of the Franco-Prussian War remained so strong that he declined the Prussian Ordre Pour le Mérit.

THE PRIVATE PASTEUR

There is grist for the psychobiographer's mill in Pasteur's life, career, and personality. His precarious health was a constant source of concern to his family. Nearsightedness was the least serious of his physical infirmities. As his father's letters to him make clear, young Louis had never been robust.

Excessive physical and mental exhaustion further undermined his constitution. On 19 October 1868, when he was forty-five and in the midst of the silkworm studies that took so much of his time between 1865 and 1870, Pasteur suffered a cerebral hemorrhage or stroke. By the next day, the left side of his body was totally paralyzed. Treated initially by bleeding with leeches, later by electricity and mineral waters, Pasteur regained most of his powers, though he did spend the remaining three decades of his life with a hemiplegia severe enough to impair his speech, gait, and manual dexterity. He continued to design and direct experiments with his typical care and ingenuity, but he could perform by himself only the simplest procedures and thus required assistants and collaborators to execute most of the experiments he designed. For nearly twenty years after his first major stroke, Pasteur's health remained fairly stable. By the autumn of 1886, however, he began to display unmistakable signs of cardiac deficiency, and in October 1887 he suffered another stroke. Although less serious than the attack of 1868, it further impaired his speech and mobility. From then on his strength faded steadily, and he was visibly feeble by the time he moved into the new Institut Pasteur in 1888 at the age of sixty-five. He later expressed brief enthusiasm for Charles Brown-Séquard's controversial injections of testicular extract, but in 1894 he suffered another setback, probably a third stroke. At his death a year later, in his seventy-second year, Pasteur was almost completely paralyzed.[72]

Of Pasteur's intimate life, there is little to say and none of it titillates. He was a paragon of bourgeois respectability. He ate and drank moderately, having surprisingly little interest in wine for a Frenchman and no taste for beer whatever. His wife did sometimes feel neglected. On their thirty-fifth wedding anniversary, she wrote to their daughter: "Your father, very busy as always, says little to me, sleeps little, and gets up at dawn—in a word, continues the life that I began with him thirty-five years ago today."[73] But as she well knew, Louis went scarcely anywhere except to his laboratory. Their nephew, Adrien Loir, who worked in the master's laboratory for six years in the mid-1880s, tells us that Pasteur almost never ventured beyond the Latin Quarter on the Left Bank in Paris, where he moved between the family's apartment, his laboratory, the Sorbonne, the Académie des sciences, and the other scientific and educational institutions that abound in the Latin Quarter and do so much to determine its special character. Even in Madame Pasteur's company, Louis rarely went out on the town. His nephew could not recall their going even once to the theater, and a visit to the Right Bank for any purpose was a real excursion. Evenings were almost always spent at home with Madame Pasteur reading the daily newspaper aloud to her husband.[74] So when she did feel ignored, it was only because of her husband's passion for science. She did also sometimes wonder, or so it is said, why his great and useful discoveries did not bring more money into their home. Perhaps she felt this more acutely than he, since it was she who managed the family income, disbursing a regular "allowance" to her preoccupied husband.[75]

In any case, throughout the forty-five years between their marriage in 1849 and his death in 1895, Madame Pasteur served as her husband's devoted helpmate, supportive partner, and dedicated stenographer or secretary—so effectively, in fact, that Pasteur's most famous co-worker, Emile Roux, once called Madame Pasteur her husband's most important collaborator.[76] She also bore five children, including three daughters who died before reaching maturity. Their only son, Jean-Baptiste (1850–1908), became a member of the French diplomatic delegations in Rome and Copenhagen. Much to Pasteur's dismay, Jean-Baptiste's marriage produced no children and the family name thus ended with him.[77] In 1879 Pasteur's one surviving daughter, Marie-Louise (1858–1938), married René Vallery-Radot, a popular writer of conservative cast who soon became his father-in-law's enthusiastic and most famous biographer. Their distinguished son, Louis Pasteur Vallery-Radot, became in his turn the guardian of his grandfather's reputation and private papers until his own death in 1971.

Both at home and in his laboratory, Pasteur was the very model of the patriarch. He was sometimes severe with his son and son-in-law, and his students, assistants, and collaborators must often have felt as if they too were his children. The rigid authoritarianism that marked, and ultimately ended, his reign as "disciplinarian" at the Ecole Normale remained undimmed at home and in his laboratory. Among a host of surviving portraits and photographs, exceedingly few show even a hint of a smile. What strikes one instead is the firm-set jaw of the youthful Pasteur and the somber mien and penetrating eyes of his later years. Even his most loyal disciples conceded that he lacked the gift of repartée, or anything like a sense of humor. He was instead profoundly serious, almost dour, and more than a little cool and aloof toward those outside his select circle. Obsessed with his work, he brooked no interference with it. His celebrated affection for children was sincere, but he could be insensitive and exploitative toward others, including his closest disciples. He was so secretive about the direction of his research that even his most trusted collaborator, Emile Duclaux, complained of his "Olympian silence."[78] He was reportedly reluctant to let anyone else record the experimental notes or even to label the animal cages in his laboratory. Duclaux compared him to "a chief of industry who watches everything, lets no detail escape him, wishes to know everything, to have a hand in everything, and who, at the same time, puts himself in personal relation with all his clientele."[79]

A compulsive administrator and fastidious organizer, Pasteur's passion for tidiness and cleanliness approached the eccentric. It is said that fear of infection made him wary of what was then seen as a peculiarly English ritual of *le handshake*.[80] Before eating, he routinely recleansed his utensils and examined his food minutely. He also observed a highly regular, even regimented, schedule of daily life. At least some of his co-workers welcomed his conscientious participation in meetings of the learned academies and societies to which he belonged, for it left them free to relax and to indulge such vices as smoking cigarettes for an hour or so most afternoons. For one of them, the bon vivant Charles Chamberland, Pasteur's annual late-summer holiday at the familial home in Arbois was an occasion for rejoicing or at least for avoiding the constant diligence that the master's close supervision ordinarily entailed.[81]

Not everyone was eager to work under such conditions or for such a man, however great his fame and however lavish his facilities and budget. No ambitious young scientist who thought of joining the Pastorian team could ignore the likelihood that he would be cast in the master's shadow. And in any case his career prospects would perforce be bound up with the future success of the Pastorian program. That was especially so because there was no immediately obvious link between the somewhat idiosyncratic and unorthodox Pastorian program and the rest of French academic science. Pasteur himself once ascribed his delay in entering directly into medical research partly to the difficulty of securing a "courageous and devoted collaborator,"[82] and there is other evidence to suggest that he did not always or easily attract the talent he sought. Pierre Duhem, the future physicist and historian of science, was one young graduate of the Ecole Normale who briefly pondered the possibility of joining Pasteur's laboratory staff as an *agrégé-préparateur*, only to decide that this would be too risky a path to success in the world of French science.[83]

By that point, in the mid-1880s, Pasteur had spent a decade away from the classroom, where his beautifully organized if not quite spellbinding lectures had brought him an excellent reputation as a teacher. Now, however, he taught only by precept and example in the laboratory, and those who joined him there were required simultaneously to contribute to his work and to meet his exacting standards and demands. It is perhaps not surprising that very few of his assistants and collaborators became distinguished scientists in their own right. Even the two most distinguished among them, Emile Duclaux and Emile Roux, tend to be treated as toilers in the Pastorian vineyard, whose work depended heavily on the insights of the master and whose most lasting contribution was their transmission of the Pastorian legacy to others who then fanned out to establish and to man the more than one hundred research institutes and centers that now bear Pasteur's name. There are more than a few hints of strain in Pasteur's relationship with his assistants and collaborators, including the worldly Charles Chamberland and especially the ascetic yet mercurial Emile Roux. In the end, though, not even Roux ever gave public expression to any sense that his own contributions were being unduly appropriated to Pasteur's name, and the strains and tensions within the Pastorian camp remained almost entirely a "family affair."

That phrase has been chosen with some care, for the Pastorian circle exhibited some of the features and values of the "mom-and-pop stores," those state-protected family enterprises that are so familiar to observers of French and Italian life.[84] For well-behaved members of the Pastorian enterprise, job security seems to have been commonplace, and there is a striking pattern of employment of several members of the same family (sometimes into succeeding generations), especially but by no means exclusively at the level of low-level technicians or custodial staff.[85] Among the earliest custodial employees at the Institut Pasteur were the two peasant boys who first submitted to his treatment for rabies. The striking paternalism of nineteenth-century firms, which was as evident in the huge Bon Marché department store as in the small, single-family shops,[86] also found abundant expression in Pasteur's laboratory. His wife served as his personal secretary, and he hired his own nephew as his personal research assistant, at least partly to ensure that his private views were kept strictly and literally within the family.[87] When the widow of the founding father of the Bon Marché contributed 100,000 francs to Pasteur's research, it cannot have hurt that both enterprises so fully embodied the paternalism and other family-centered values of the nineteenth-century French bourgeoisie.

But perhaps the most distinctly "familial" feature of the Pastorian circle was its fierce and unbending public loyalty to the head of the family—and indeed to the legend that he helped to shape, as we shall see more fully in the concluding chapter. In death as in life, Pasteur's reputation was jealously protected by his family and associates, including even his sometimes disenchanted collaborators. And the Pastorian legend also quickly became an entrenched part of the folklore of French patriotism. To be sure, Pasteur did not entirely escape the withering eye of criticism. His early chemical and crystallographic research aroused little opposition, but when he turned his attention to fermentation, and then to spontaneous generation and disease, controversy followed him everywhere. He aroused fervent antagonism in some quarters, and his adversaries included several distinguished scientists, including some in France and Justus von Liebig and Robert Koch in Germany. As he accumulated ever greater power within French governmental and scientific circles, as he attracted wholly unprecedented levels of state support for his research, and as he focused his attention on practical and especially medical problems, Pasteur's critics grew ever more shrill. A certain portion of the medical profession and of what he liked to denigrate as the "so-called scientific press" vilified him as an egomaniacal and intolerant representative of "official" science and as an unscrupulous, secretive, and greedy opportunist. Some heatedly denied that his work had brought all of the immense industrial, agricultural, and medical benefits claimed for it. There is exaggeration in all of this, but some of the claims advanced by Pasteur's critics do deserve vastly more serious and more detailed examination than they have yet received. Happily their validity can now be tested against the wealth of newly available manuscript materials by and about Pasteur.

PART II

From Crystals to Life

THREE

The Emergence of a Scientist: The Discovery of Optical Isomers in the Tartrates

[WITH JAMES A. SECORD]

IN APRIL 1848 the streets of Paris still echoed with the shock waves set off by the revolutionary "February days," during which King Louis Philippe had abdicated and a provisional republican government had been formed. Among those who played a minor role in defense of the new provisional government was a twenty-five-year-old chemist named Louis Pasteur. During his brief service in the 200,000-man National Guard—a city militia charged with the maintenance of civil order and the protection of municipal liberties—Pasteur apparently experienced no hostile action, nor even any serious disruption in his chemical research. At the Ecole Normale Supérieure, where he had received a doctorate the previous August and now served as a sort of teaching assistant to Professor Antoine Jérôme Balard, Pasteur continued to bend over his laboratory bench, examining crystals of the tartrates, a group of well-studied organic compounds long associated with wine-making and tanning.

It was on one of these April days in 1848—or so legend has it—that the young chemist suddenly "rushed out of the laboratory, not unlike Archimedes [when he yelled 'Eureka'] . . . met a curator in the passage, embraced him as he would have embraced [his best friend], and dragged him out with him into the Luxembourg Gardens to explain his discovery." And "never," continues René Vallery-Radot in his heroic *Life of Pasteur*, "was there greater or more exuberant joy on a young man's lips."[1] Later biographers sometimes display more restraint in their descriptions of Pasteur's response, but none disputes Vallery-Radot's account of Pasteur's path to his first major discovery. In the standard sources on Pasteur, this discovery is presented as an elegant solution to a puzzle posed in 1844 by the German crystallographer, Eilhard Mitscherlich (1794–1863).[2]

Mitscherlich was among the most distinguished of Pasteur's predecessors in the close study of the tartrates and their chemical relatives. This group of compounds offered one of the few sets of examples then known of the phenomenon of isomerism—the existence of two (or more) different substances with the same chemical formula. By 1830 it had been established that racemic acid (or, as it was henceforth also called, paratartaric acid) had the same chemical formula as tartaric acid, though the substances were otherwise easily distinguishable through their crystalline forms and other physical properties. The known salts of these two acids, roughly a score in number, displayed corresponding differences in their form and other properties despite their identity in chemical composition. As in other cases of isomerism, these differences were ascribed to differences in the spatial arrangement of otherwise identical atoms.[3]

What Mitscherlich announced in 1844 was his discovery of an exception to this pattern. He had found that sodium-ammonium tartrate and sodium-ammonium paratartrate not only had the same chemical formula, but were also identical in every other respect save one—solutions of the tartrate rotated a plane of polarized light to the right, while solutions of the paratartrate exerted no effect on polarized light. In chemical shorthand, the sodium-ammonium tartrate was "optically active" to the right, while the sodium-ammonium paratartrate was "optically inactive." Here then, insisted Mitscherlich, are two substances that differ in their effect on polarized light despite complete identity in every other respect—despite even their identity in "the nature, number, arrangement and distances of the atoms."[4] This claim posed a challenge to the received definition of chemical "species." For, as Michel Eugène Chevreul had proposed in 1823, species of compound bodies are identical when the nature, proportion, and arrangement of the atoms are the same.[5] By Chevreul's criteria, Mitscherlich's sodium-ammonium tartrate and sodium-ammonium paratartrate should have been totally indistinguishable.

Pasteur resolved the difficulty—and thus rescued the notion of stable chemical species—by showing that the optical difference between the two sodium-ammonium salts could be correlated with a subtle structural difference that Mitscherlich had missed. Despite his well-deserved reputation as a skillful observer, Mitscherlich was mistaken in his claim that the two salts were identical in crystalline form. In fact, the sodium-ammonium paratartrate could be separated into two distinct crystalline forms—identical in every respect except that they were mirror images of each other (see fig. 3.1). One form displayed microscopic hemihedral facets on its right

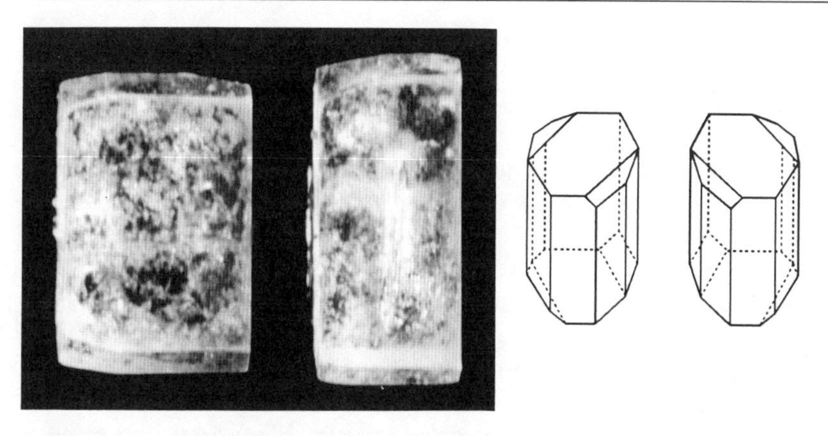

Figure 3.1. Hemihedral crystals of sodium ammonium tartrate. *Left:* Photograph of the actual crystals produced by George B. Kauffman and Robin D. Myers in a fascinating attempt to replicate Pasteur's experiment. See Kauffman and Myers, "The Resolution of Racemic Acid: A Classic Stereochemical Experiment for the Undergraduate Laboratory," *Journal of Chemical Education* 12(1975):777–781. As this article and its illustrations make clear, the production and detection of asymmetric forms in the tartrates is by no means a simple or straightforward reading of "nature." (Photograph and drawing courtesy of George B. Kauffman) *Right:* Idealized drawings of the right-handed (a) and left-handed (b) forms.

edge, the other on its left: the two forms were related to each other as our right hand is related to our left. The right-handed form was identical to ordinary sodium-ammonium tartrate; the other form was a hitherto unknown left-handed version of the same compound. Furthermore, solutions of these two crystalline forms rotated the plane of polarized light in equal but opposite directions—one being optically active to the right, the other to the left. When equal weights of these left- and right-handed crystals were dissolved separately and then combined, the result was sodium-ammonium paratartrate, which exerted no effect on polarized light. The equal but opposite optical activities of the two crystalline forms had canceled each other.

With the announcement of this discovery, on 15 May 1848, Pasteur became a force to be reckoned with in the scientific world. No one before him had observed left- and right-handed hemihedral forms in a soluble substance. No one had probed more deeply into the relation between the crystalline form and the internal structure of a chemical compound. Much greater fame would come his way, but it was through his meticulous examination of what he once called "the arid details of crystal form"[6] that Pasteur took the first major step on his journey to scientific glory.

THE STANDARD STORY: PASTEUR'S LECTURES OF 1860

This has been an abbreviated account of the standard story of Pasteur's first major discovery. A much richer and significantly different version emerges from a closer examination of Pasteur's path to that discovery. Here, as elsewhere in this book, the most striking revisions emerge from a detailed analysis of Pasteur's unpublished laboratory notes. They give ample reason to doubt René Vallery-Radot's "Eureka" story of Pasteur racing from his laboratory at the Ecole Normale and dragging a startled technician with him into the nearby Luxembourg Gardens to share his excitement. This legend is surely nothing but one of many examples of the literary license so evident throughout Vallery-Radot's *Life of Pasteur*. It is, however, typical of other popular attempts to telescope the usually extended and sometimes tedious process of scientific discovery into a single dramatic moment of illumination. Interestingly, Pasteur himself allowed this simplistic version of the story to stand when he corrected the galley proofs of his son-in-law's earlier biography, "The Story of a Scientist by a Layman."[7] In this and many other ways throughout his career, Pasteur displayed his appreciation for popular stories about the genesis of scientific discoveries, especially when these narratives combined conceptual lucidity with human interest and drama.

In fact, Pasteur had already provided a dress rehearsal for René Vallery-Radot's Eureka version of his first major discovery. In a famous pair of lectures to the Parisian Société de chimie, delivered in late January and early February 1860, Pasteur constructed a "history" of his discovery that has been repeated with only minor variations ever since. Pasteur's lectures even anticipate Vallery-Radot's formulaic use of human drama—albeit with a halfhearted apology for violating "the custom of our times" by including "personal reminiscences . . . in a scientific discussion."[8]

Pasteur was referring here to his account of how his discovery brought him "naturally into communication with Monsieur Biot," the distinguished French scientist Jean-Baptiste Biot (1774–1862), whose earlier investigations of optical activity were familiar to all contemporary scientists. According to Pasteur, Biot asked him to repeat his experiments in his presence. While Biot watched, Pasteur prepared the sodium-ammonium paratartrate, separated it into its left- and right-handed crystals, and indicated which of the two piles of crystals would rotate the plane of polarization to the right and which to the left. Biot then declared that he would complete the experiments. After preparing two solutions of equal weight from these crystals, Biot put the more interesting solution—that of the hitherto unknown left-handed crystals—into his polarizing apparatus or "polarimeter" to test for

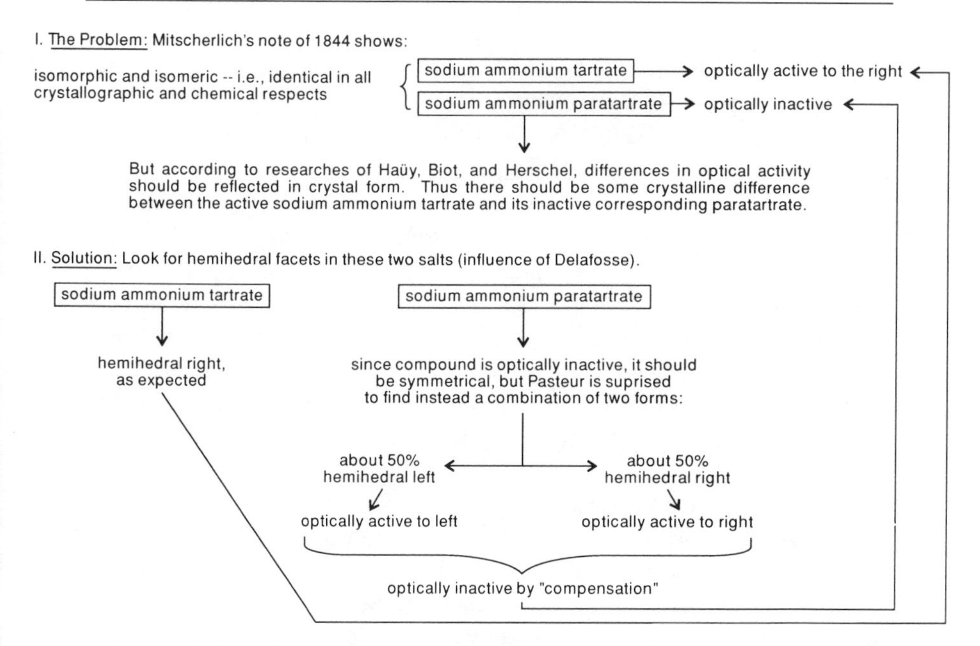

I. The Problem: Mitscherlich's note of 1844 shows:
isomorphic and isomeric -- i.e., identical in all crystallographic and chemical respects
sodium ammonium tartrate
optically active to the right
sodium ammonium paratartrate
optically inactive
But according to researches of Haüy, Biot, and Herschel, differences in optical activity should be reflected in crystal form. Thus there should be some crystalline difference between the active sodium ammonium tartrate and its inactive corresponding paratartrate.
II. Solution: Look for hemihedral facets in these two salts (influence of Delafosse).
sodium ammonium tartrate
sodium ammonium paratartrate
hemihedral right, as expected
since compound is optically inactive, it should be symmetrical, but Pasteur is suprised to find instead a combination of two forms:
about 50% hemihedral left
about 50% hemihedral right
optically active to left
optically active to right
optically inactive by "compensation"

Figure 3.2. The path to Pasteur's discovery of optical isomers—the standard story.

the predicted optical activity. Pasteur's prediction was instantly confirmed. "Then," said Pasteur from his podium in an oft-quoted passage, "the excited old man seized my hand and said: 'My dear child, I have all my life so loved this science that I can hear my heart beat for joy.' "[9]

It would be a mistake to dismiss this scene as a mere literary flourish on Pasteur's part. It was, instead, a dramatically personal way of encapsulating a message conveyed more prosaically elsewhere in Pasteur's lectures of 1860. Those lectures, delivered a dozen years after Pasteur's first major discovery, projected that achievement onto a stage where the leading characters were Mitscherlich and Biot, and where the major conceptual issue was the relationship between optical activity and crystalline form.

A RECONSTRUCTION OF THE DISCOVERY: THE LABORATORY NOTES

In the rest of this chapter, the above retrospective version of the story is disputed mainly on the basis of evidence from Pasteur's first laboratory notebook—the only notebook missing from the unpublished Pasteur collection now deposited at the Bibliothèque Nationale in Paris; its current whereabouts is a mystery. Happily, J. D. Bernal reproduced several crucial pages of it photographically for an essay of 1953. More important, a microfiche copy of the entire notebook is in the possession of Seymour Mauskopf, who has deposited a duplicate copy in Firestone Library at Princeton University and whose generosity in sharing that copy has made the writing of this chapter possible. In fact, Mauskopf's own insightful analysis of Pasteur's first notebook is the point of departure for the interpretation that follows.[10] In both, the now obscure French chemist Auguste Laurent (1807–1853) is restored to a central role that was denied him in Pasteur's retrospective accounts of his first major discovery. In both, Laurent's decisive influence on Pasteur is traced more specifically to their shared concern with the phenomena of isomorphism and dimorphism, rather than the more familiar issue of the relationship between crystal form and optical activity.

This interpretation disputes no part of Mauskopf's impressive study so far as it goes. But it does try to provide a fuller and more tightly connected account of Pasteur's concerns at each step in the program of research that led up to his first great discovery, giving especially close scrutiny to those pages of the notebook that immediately precede the recording of the discovery. The resulting story emphasizes more strongly than Mauskopf does that Pasteur's discovery emerged from a complicated sequence of investigations in which the issue of optical activity did not enter until late in the game. It thus draws full and explicit attention to the sense in which Pasteur's retrospective lectures of 1860 misrepresent the actual route to his discovery, and it suggests that Pasteur's supposedly "strange and oblique disavowal of Laurent"[11] was motivated in part by his quest for a secure place in the French scientific establishment. More generally, the following interpretation seeks (1) to remind us that we should always be skeptical of participants' "historical" (i.e., retrospective) accounts; (2) to illustrate the point that the vocabulary and assumptions we ordinarily use in historical accounts of scientific discovery tend to collapse a complex process into a single event; (3) to suggest how perceptual experience of the material world enters into the discovery process; and (4) to argue that the allegedly mysterious process of scientific discovery can be analyzed in a coherent way, even if it cannot be reduced to a set of universal epistemic rules.[12]

THE LAURENTIAN BACKGROUND TO PASTEUR'S FIRST MAJOR DISCOVERY

Auguste Laurent's reputation as a brilliant chemist, political radical, and difficult personality preceded his arrival at the Ecole Normale, where he worked for several months from late 1846 to April 1847. He was just then on the verge of publishing a textbook on crystallography firmly rooted in a great French tradition stemming from the work of the Abbé Rene-Just Haüy (1743–1822). By the time Laurent joined this tradition, its representatives had to contend with certain phenomena that seemed to contradict Haüy's view that each chemical substance (or "species") possessed a unique (or "fixed") form. In particular, Mitscherlich had emphasized examples of both isomorphism (identical crystalline forms between substances of different chemical composition) and of its inverse, dimorphism (differences in crystalline form between substances of the same chemical composition). Defenders of the Haüyian tradition responded to these anomalies by developing more flexible definitions and conceptions of chemical species and fixed crystalline forms.[13]

Of special interest here are Laurent's efforts to blur the distinction between isomorphism and dimorphism. By the time he came to the Ecole Normale, Laurent had developed what might be called isomorphism *à peu près*, a looser conception of the phenomenon that embraced substances with slightly different crystalline forms. In keeping with the tradition of Haüy, Laurent looked upon crystalline form as an outward expression of internal structure.[14] When substances had closely similar crystalline forms, Laurent supposed that they must have in common a "fundamental radical," an inner hydrocarbon that determined the basic characteristics of the outward crystalline form. This radical, and the associated crystal form, could then be modified within certain limits by the addition of water molecules or acids. In a sense, such substances were at once isomorphic and dimorphic: despite slight differences in their crystalline form, they could still be considered broadly isomorphic. These slight differences would most likely emerge only in the details of the outer edges and extremities of a crystal.

A scientist influenced by this Laurentian scheme would thus pay very special attention to isomorphism, the number of water molecules in a compound, and the modifying faces and edges of each crystal. These are precisely the concerns that underlay Pasteur's initial efforts at original research. He owed very substantial intellectual debts also to Biot and especially to Gabriel Delafosse, another representative of the Haüyian tradition who had emphasized crystalline hemihedrism (or asymmetry) in the course Pasteur took from him at the Ecole Normale. Still other chemists and mineralogists had already drawn attention to the effect of waters of crystallization upon crystal form; indeed, Mitscherlich himself had displayed such a concern in his work on dimorphism.[15] But Laurent, as we shall see, played the central role in pointing the way toward the path Pasteur took.

When Laurent came to work in Professor Balard's laboratory at the Ecole Normale, Pasteur had scarcely begun research on the two theses (one each in chemistry and physics) that he was to complete for his degree of *docteur-ès-sciences*. For his chemistry thesis, Pasteur originally proposed to work under Balard's supervision on the effects of pressure on chemical reactions and crystalline form. For his physics thesis, he projected a study of "a multitude of densities of which I will have need in order to later undertake a work in chemistry on atomic volumes."[16] In the event, neither thesis turned out as planned.

From the text of Pasteur's completed chemistry thesis, presented in August 1847, we know that he had switched topics at Laurent's suggestion. His new subject was the saturation capacities of arsenious acid and several of its salts. The first section of the thesis announced the discovery of a hitherto unknown type of arsenious acid (a "dibasic" type, as distinct from the known "monobasic" type). In the second part of the thesis, Pasteur sought crystallographic confirmation of the more strictly chemical conclusions of the first part. He had managed to find—as expected on Laurentian principles—two slightly different crystalline forms of arsenious acid corresponding to the monobasic and dibasic types. He also claimed that these two forms of arsenious acid were isomorphic with corresponding types of antimonious acid. Saying that he would establish that "arsenious and antimonious acid are at once dimorphous and isomorphous," Pasteur described this conclusion as "a very reasonable induction" from still unpublished research by Laurent on tungstic acid. More generally, he acknowledged Laurent's influence by expressing his gratitude for "the kindly advice of a man so distinguished both by his talent and by his character."[17]

Pasteur's completed thesis in physics, also presented in August 1847, explores issues that seem more obviously central to his famous discovery of April 1848. The original topic of atomic volumes was abandoned in favor of "phenomena relating to the rotatory polarization of liquids," as the completed thesis was entitled. Pasteur here stressed the contributions that the techniques of crystallography and physics could make to the most interesting problems in chemistry, "those relating to the molecular constitution of bodies." He directed particular attention to the value of optical activity as a guide to chemical structure, saying that Biot's important papers on the chemical activity of liquids had been "too much neglected by chemists." Guided by these papers, and using Biot's own polarimeter, Pasteur had investigated several problems, including notably the relationship between optical activity and crystalline form. His conclusion, based on two pairs of isomorphic substances, was that substances of the same crystalline form had the same optical activity.[18]

In retrospect, Pasteur's physics thesis seems to display what Mauskopf has called a "remarkable prevision of the reasoning which was to lead him to his [first major discovery] the following year."[19] Quite apart from its concern with optical activity, the thesis referred specifically to the tartrates. One passage addressed the structural implications of the difference between tartaric and paratartaric acid, and one of the two isomorphic pairs that Pasteur had examined most closely with Biot's polarimeter belonged to the tartrates (potassium-ammonium tartrate and simple potassium tartrate).[20] It is easy to assume that Pasteur was now clearly on the path to his discovery of left- and right-handed crystals in sodium-ammonium paratartrate. Moreover, the prominence given here to Biot's work on optical activity seems in keeping with Pasteur's retrospective account in his famous lectures of 1860.

In fact, however, the story is much more complicated. For one thing, Pasteur's physics thesis had been put together in a hurry, and he did not attach great significance to it at the time. A letter written in July 1847—scarcely a month before he presented completed versions of both of his theses—suggests that he had just begun the physics thesis. "I will do a little something in physics," he wrote, describing the projected thesis as "only a program for some very useful researches which I will undertake next year and which I have only begun in the thesis."[21] Pasteur's training in the use of the polarimeter, and especially his concern with the relationship between optical activity and crystalline form in the tartrates, were obviously crucial to the major discovery he announced less than a year later. But more immediately important was his continuing concern with isomorphism and dimorphism. That concern is common to his two theses, and it remained a central theme in his first published papers. Biot and Delafosse obviously had a powerful impact on Pasteur, but Laurent's role was both more immediate and more pervasive. Even in Pasteur's physics thesis, for all of its references to Biot, the basic conclusion—that substances of the same crystalline form have the same optical activity—had already been advanced by Laurent for other compounds.[22]

Following the completion of his two theses for his doctorate degree, Pasteur undertook a systematic investigation of dimorphism. Substances such as calcite and aragonite, which had identical chemical formulas and yet crystallized in different ways, had long been a thorn in the side of those who sought to use crystal form as an index of chemical composition. As in the second half of his chemistry thesis, Pasteur hoped to remove the thorn by showing that all dimorphic substances were really isomorphic. The short abstract of his first paper on dimorphism, delivered before the Académie des sciences on 20 April 1848, shows Pasteur on the eve of his first great discovery, conducting his research according to the theoretical precepts of Laurent.[23]

THE PROBLEM OF THE TARTRATES

Pasteur ended the preliminary abstract of his paper on dimorphism with a note indicating the direction of his latest research. He claimed to have experimental proof that a group of eight tartrates was broadly isomorphic and could be crystallized together in any proportion.[24] The announcement was a striking one, as Pasteur noted, for these eight tartrates belonged to two theoretically incompatible crystallographic systems. Five of the compounds were right rectangular prisms, while the other three were slightly oblique. Pasteur's claim that they had in fact been crystallized together promised powerful new support for Laurent's flexible concept of isomorphism *à peu prés*.

But Pasteur soon doubted the accuracy of his assertion, and he omitted the passage on the isomorphism of the eight tartrates from the full-length version of his "Researches on Dimorphism," published in 1848 in the *Annales de chimie*.[25] These uncertainties emerge much more explicitly in his private laboratory notebooks. In fact, these notebooks reveal that Pasteur's concern about the isomorphism of the eight tartrates became the principal theme in the research leading up to his first great discovery. The desire to validate a claim made in the public forum of the Académie des sciences presumably concentrated the young scientist's mind wonderfully, and he immediately attacked the problem of the tartrates with great energy and success.

In Pasteur's surviving notebooks, the earliest reference to the isomorphism of the tartrates comes immediately after an outline draft of the full version of his dimorphism paper. This undated note, probably written early in 1848, began by emphasizing the discrepancies between the chemical formulas and outward crystalline forms of the eight tartrates (see fig. 3.3). According to Laurent's theory, compounds possessing not only analogous chemical formulas, but also equal waters of crystallization, should have almost identical crystal forms. But as Pasteur remarked, the five tartrates possessing one molecule of water were actually divided among the two possible crystal systems: all had closely analogous formulas, yet some crystallized as right rectangular prisms, while others formed slightly oblique rectangles.[26]

Pasteur also mentioned a related anomaly involving a separate criterion for isomorphism in this early notebook entry. He referred to the German chemist Hermann Kopp, who had defined "atomic volume" as the ratio between the molecular and specific weights of a compound. Similarity in atomic volumes, Kopp maintained, was a necessary (if not sufficient) indicator of true isomorphism. Using this idea as his point of departure, Pasteur

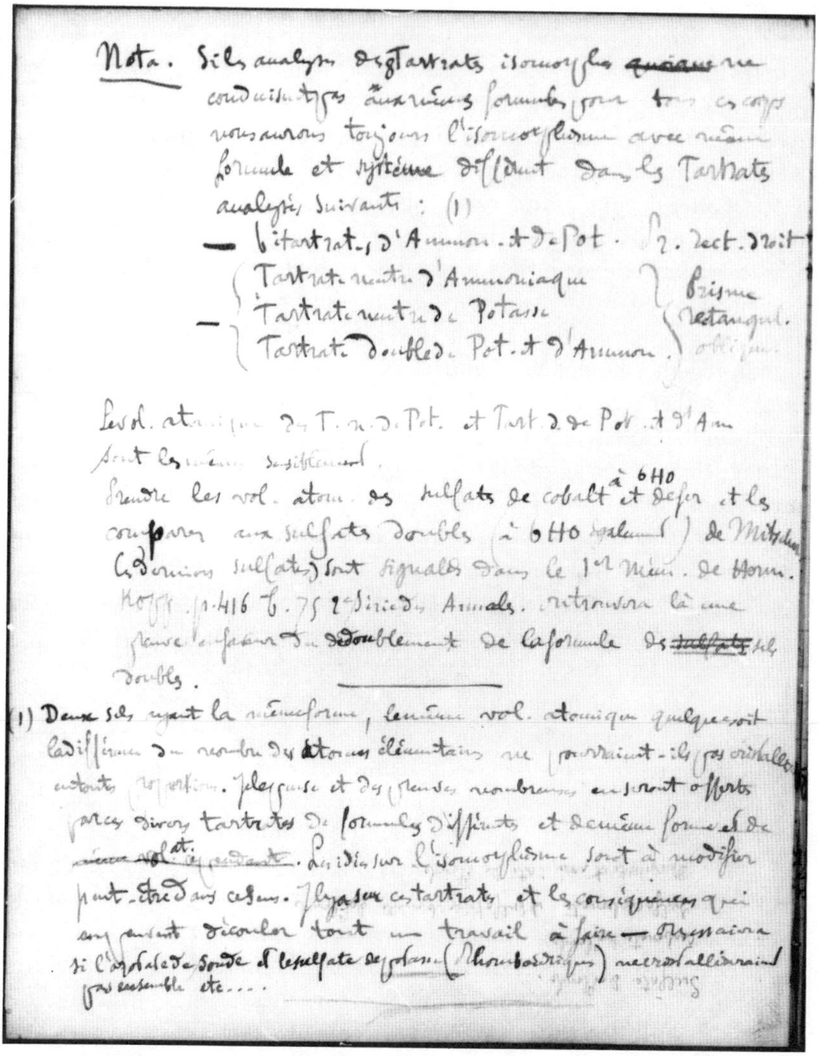

Figure 3.3. From Pasteur's first laboratory notebook, *Notes divers*, 2/5/2 on the microfiche owned by Seymour Mauskopf.

noted that the atomic volumes and crystal forms of two of his tartrates—potassium tartrate and potassium-ammonium tartrate—were virtually identical. But these apparently isomorphic salts "cannot crystallize [together] in all proportions," Pasteur wrote—thus contradicting his earlier claim before the Académie des sciences. He then began to wonder if this dilemma could be resolved through a detailed examination of the other tartrates, with their differing chemical formulas and similar crystal forms. "There is in these tartrates and the consequences which derive from them," he wrote prophetically, "the wellspring of an entire work to be accomplished"—research that could offer fundamental insights into the nature of isomorphism.[27]

Pasteur returned to the detailed analysis of the tartrates in the second half of April, after dealing with a group of sulfates connected with the work of Kopp.[28] His plan, if indeed he had one, seems to have been to go through the tartrates one by one, using any method he could to clarify the isomorphism problem. He began with sodium tartrate, determining the relation of its crystal axes to one another. He planned to crystallize a pair of tartrate compounds together as a way of illuminating their composition and crystal forms. Another page has only this heading: "Determination of the specific weights of the isomorphic tartrates." Although nothing is written below that heading, the next page is filled with calculations on the subject.[29] These calculations indicate that Pasteur hoped to use the method of specific weights to determine whether ammonium tartrate crystallized with one water molecule, or with none at all. And it was his interest in the presence or absence of this water molecule—the reasons for which will soon become clear—that led him into a more systematic investigation of the tartrates.

Because Pasteur was groping for a way to handle the isomorphism question, it is hardly surprising that these early pages of his notebook seem uncertain and tentative, a jumbled mixture of calculations, measurements, and title pages. He responded to this confused situation by making a list of what he knew, with a possible program of resolution. Accordingly, the next notebook page is headed "Tartrates (Questions to resolve)."[30]

Pasteur began by listing the accepted chemical formulas of the eight tartrates that he had once claimed could be crystallized together, and then laid out a detailed agenda for future research (see fig. 3.4 and table 3.1). On this and the next two notebook pages he characterized four anomalies concerning the relationship between the chemical formulas, crystal forms, and the alleged isomorphism of these eight tartrates. In every case, his research plans were guided by a concern—almost surely derived from Laurent—with the effect of the number of water molecules on the crystal forms of his allegedly isomorphous tartrates.[31] All four of his "anomalies," in fact, shared one feature: in these tartrates, the number of waters of crystallization did not correlate with their crystal form.

One consequence of Laurentian theory was that substances that could be crystallized together (thus displaying a basic isomorphism) should possess equal numbers of waters of crystallization. On this theory, otherwise analo-

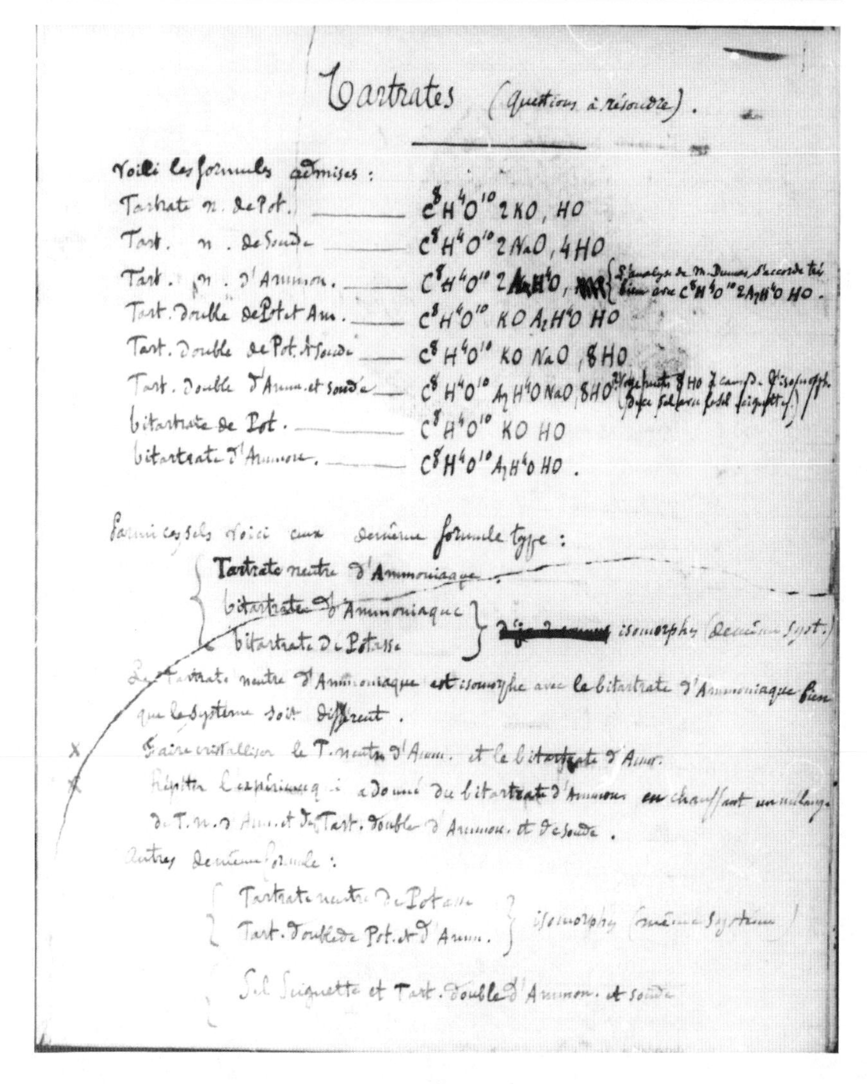

Tartrates (questions à résoudre).

Voici les formules admises :

Tartrate n. de Pot. ———— $C^8H^4O^{10}\,2KO, HO$

Tart. n. de Soude ———— $C^8H^4O^{10}\,2NaO, 4HO$

Tart. n. d'Ammon. ———— $C^8H^4O^{10}\,2AzH^4O$, {L'analyse de M. Dumas s'accorde très bien avec $C^8H^4O^{10}\,2AzH^4O\,HO$.

Tart. double de Pot. et Am. ———— $C^8H^4O^{10}\,KO\,AzH^4O\,HO$

Tart. double de Pot. et Soude ———— $C^8H^4O^{10}\,KO\,NaO, 8HO$

Tart. double d'Amm. et Soude ———— $C^8H^4O^{10}\,AzH^4O\,NaO, 8HO$

bitartrate de Pot. ———— $C^8H^4O^{10}\,KO\,HO$

bitartrate d'Ammon. ———— $C^8H^4O^{10}\,AzH^4O\,HO$.

Parmi ces sels voici ceux derniers formule type :

{Tartrate neutre d'Ammoniaque
bitartrate d'Ammoniaque
bitartrate de Potasse} isomorphes (derniers syst.)

Le Tartrate neutre d'Ammoniaque est isomorphe avec le bitartrate d'Ammoniaque bien que le système soit différent.

Faire cristalliser le T. neutre d'Amm. et le bitartrate d'Amm.

Répéter l'expérience qui a donné du bitartrate d'Ammon. en chauffant un mélange de T. n. d'Amm. et de Tart. double d'Ammon. et de Soude.

Autres derniers formule :

{Tartrate neutre de Potasse
Tart. double de Pot. et d'Amm.} isomorphes (même système)

{Sel Seignette et Tart. double d'Ammon. et soude

Figure 3.4. Pasteur, "Notes divers," 1/3/1.

gous compounds that differed in number of water molecules should be incapable of crystallizing together. When substances violated this set of expectations—in other words, when anomalies arose—there were two ways of resolving the difficulty: (1) the chemical formulas could be challenged in such a way as to preserve the correlation between numbers of water molecules and crystalline form; or (2) the alleged isomorphism of such compounds

Table 3.1 **Pasteur's List of Eight Tartrates**

Tartrate	*Chemical Formula*	*No. of Water Molecules*[a]	*Crystal System*
Neutral tartrate of potassium	$C^8H^4O^{10}2KO, HO$	1	oblique
Neutral tartrate of sodium	$C^8H^4O^{10}2NaO, 4HO$	4	right
Neutral tartrate of ammonium	$C^8H^4O^{10}2AzH^4O$, ~~HO~~[b]	0 or 1?	oblique
Double potassium-ammonium tartrate	$C^8H^4O^{10}KOAzH^4O, HO$	1	oblique
Double sodium-potassium tartrate (Seignette salt)	$C^8H^4O^{10}KONaH^4O, 8HO$	8	right
Double sodium-ammonium tartrate	$C^8H^4O^{10}AzH^4ONaO, 8HO$[c]	8	right
Bitartrate of potassium	$C^8H^4O^{10}KO, HO$	1	right
Bitartrate of ammonium	$C^8H^4O^{10}AzH^4O, HO$	1	right

Source: Adapted from Pasteur, "Notes divers," 1/3/1.

[a] Note again that the accepted formula for water was HO instead of our H_2O.

[b] The HO in the original formula has been crossed out, and written next to it is "L'analyse de M. Dumas s'accorde très bien avec $C^8H^4O^{10}2AzH^4O$ HO."

[c] Pasteur commented on this formula: "Je mets 8HO a cause de l'isomorph. de ce sel avec le sel seignette." This was later crossed out.

could be denied by showing that their capacity to crystallize together was in fact illusory.

Pasteur's first anomaly had to do with ammonium tartrate and the bitartrates of ammonium and potassium. These three substances had closely analogous chemical formulas (in Laurentian terms, they belonged to the same "formula type"), and the accepted formulas showed one water of crystallization for all three. In his announcement of 20 April 1848 to the Académie des sciences, Pasteur had included these three compounds among the eight tartrates that "crystallized together in all proportions." Yet now, in his notebook agenda, Pasteur focused on a crystallographic difference between these tartrates that seemed hard to reconcile with their alleged ability to crystallize together (see figs. 3.5 and 3.6 and table 3.2). When crystallized separately, the ammonium tartrate belonged to a different crystallographic system (oblique rectangular prism) from that of the two bitartrates (right rectangular prisms). Despite his earlier public announcement that these three compounds could crystallize together, Pasteur seemed unable to dismiss this difference in crystallographic system between the tartrates and the bitartrates. It was at this point, it seems, that he began to wonder whether this crystallographic difference could be correlated with a difference in

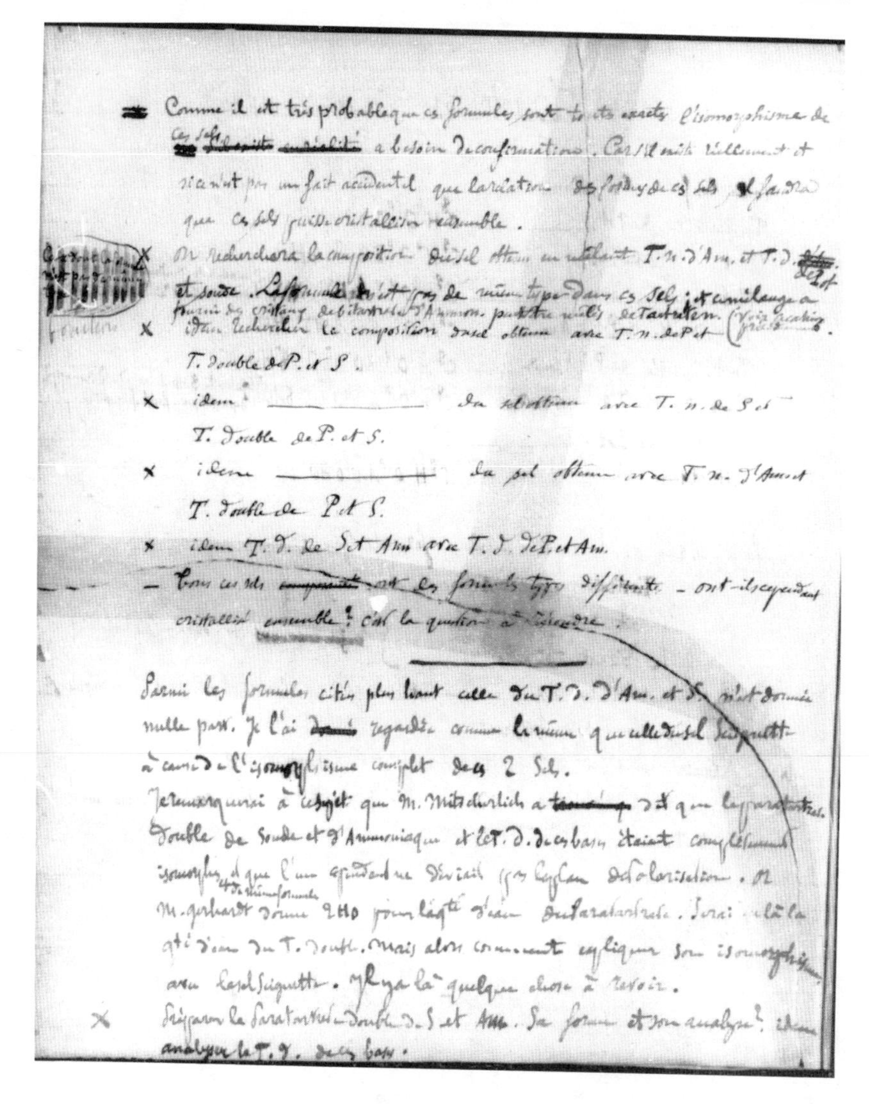

Figure 3.5. Pasteur, "Notes divers," 1/3/2.

waters of crystallization. We can thus make sense of his use of the method of specific weights to determine the presence or absence of a water molecule in the ammonium tartrate. If that tartrate had proved to lack the water of crystallization possessed by the bitartrates, Pasteur would have found a chemical difference to match the difference in their crystallographic systems. But, apparently unable to resolve the anomaly in this way, Pasteur

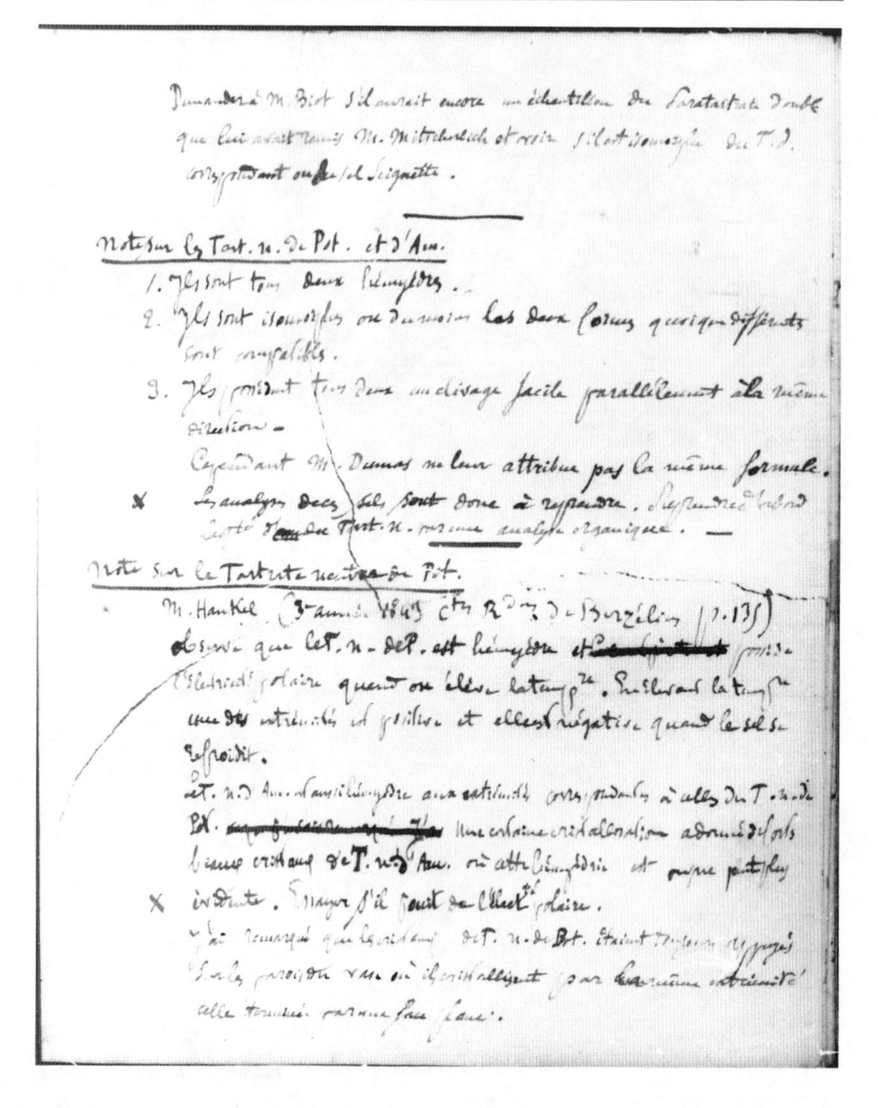

Figure 3.6. Pasteur, "Notes divers," 1/3/3.

now projected new joint crystallization experiments that would reexamine the alleged isomorphism of these three tartrates.[32]

Pasteur's second anomaly widened the inquiry to include four more of the eight listed tartrates. He divided these four compounds into two pairs of "formula types," with one pair (potassium tartrate and potassium-ammonium tartrate) containing one water molecule each, while the other pair (Seignette salt, or sodium-potassium tartrate, and sodium-ammonium tar-

Table 3.2. "Tartrates (Questions to resolve)." Pasteur's four anomalies in the relation between crystalline isomorphism and waters of crystallization.

	Water	Salt	Relation	
A	One HO	neutral tartrate of ammonium	isomorphic: diff. crystal forms (neutral tartrate of ammonium, bitartrate of ammonium)	
		bitartrate of ammonium	isomorphic: same crystal forms (bitartrate of ammonium, bitartrate of potassium)	
		bitartrate of potassium		
B	One HO	neutral tartrate of ammonium	oblique rectangular prisms	should be isomorphic, crystallizable in all proportions
		neutral tartrate of potassium		
		double potassium-ammonium tartrate		
		bitartrate of ammonium	right rectangular prisms	
		bitartrate of potassium		
	8 HO	double sodium-potassium tartrate		
		double sodium-ammonium tartrate		
C	2 HO or 8 HO	double sodium-ammonium tartrate	isomorphic (double sodium-ammonium tartrate, double sodium-potassium tartrate)	
		double sodium-potassium tartrate	isomorphic (Mitscherlich's note) (double sodium-potassium tartrate, double sodium-ammonium paratartrate)	
		double sodium-ammonium paratartrate		
D	No HO?	neutral tartrate of potassium	virtual identity of crystal forms, even to hemihedry	
	One HO	neutral tartrate of ammonium		

Source: Adapted from Pasteur, "Notes divers," 1/3/1–3.

trate) contained eight water molecules each. The two tartrates in each pair were considered completely isomorphic with each other, belonging to the same crystallographic system, but there was a difference in the crystallographic system between the two pairs (oblique rectangular prisms vs. right rectangular prisms). Immediately after listing these four tartrates, Pasteur wrote in his notebook: "Since it is very probable that these formulas are perfectly correct [*touts exacts*], the isomorphism of these salts ["if it really exists" is crossed out] is in need of confirmation. For if [the isomorphism] really exists, and if the relation between the form of these salts is not accidental, these salts must be capable of being crystallized together."[33]

In other words, as with the three tartrates with which he had begun his agenda, Pasteur was now reassessing his claim that these four were broadly isomorphic and could crystallize together. His special concern with waters of crystallization is clear from the set of projected experiments listed below the passage just quoted. He proposed to begin by reexamining "the composition of the salt obtained by mixing the neutral tartrate of ammonium with the double tartrate of potassium and sodium." The formulas of these two salts, Pasteur noted, were "not of the same type." What he did not make explicit, but clearly appreciated full well, was that these two salts were believed to differ in their number of waters of crystallization—the ammonium tartrate had one while the sodium-potassium tartrate had eight. A difference in waters of crystallization also characterized each of the other mixtures that Pasteur now proposed to reexamine: potassium tartrate (one water molecule) with, once again, sodium-potassium tartrate (eight); sodium tartrate (four) with, yet again, sodium-potassium tartrate (eight); and sodium-ammonium tartrate (eight) with potassium-ammonium tartrate (one). In effect, Pasteur was here proposing to reexamine the alleged isomorphism of six of his eight listed tartrates by seeing if each of them could be crystallized together with a tartrate that differed from it in waters of crystallization. As Pasteur wrote at the end of this list of projected experiments, "All these salts are of different formula types. Did they nonetheless crystallize together? That is the question to resolve."[34]

Pasteur's third anomaly concerned the tartrates in which he would eventually find left- and right-handed crystals. Referring first to sodium-ammonium paratartrate, Pasteur noted that its formula was not listed anywhere in the existing literature. But he presumed that it must be closely analogous to that of the sodium-potassium tartrate (the Seignette salt), its isomorphic counterpart. At this point in his notebook Pasteur recalled Mitscherlich's famous note of 1844—the note that was to become so prominent in his retrospective accounts of his discovery: "M. Mitscherlich has said that the sodium-ammonium paratartrate and the sodium-ammonium tartrate are completely isomorphic, and of the same formula—and yet, the former did not deviate the plane of polarization [i.e., the plane of polarized light; in other words, it was optically inactive]."[35]

In view of what preceded and followed this statement in the notebook, this reference to optical activity was clearly incidental. The more central question in Pasteur's mind at this point—the anomaly he wished to resolve—concerned the extent of chemical and crystallographic identity between sodium-ammonium tartrate and the corresponding paratartrate. These two salts supposedly had identical crystalline forms and identical chemical formulas. If so, the Laurentian correlation between waters of crystallization and crystalline form meant that the two salts should also have equal molecules of water at crystallization. It was generally agreed that the sodium-ammoniuim tartrate contained eight waters of crystallization. "Yet," Pasteur wrote in his notebook, "M. Gerhardt gives 2HO as the quantity of water in the paratartrate." For a moment, Pasteur considered the possibility that the sodium-ammonium tartrate might also contain two waters of crystallization instead of eight. "But," he continued, "how then to explain its isomorphism with the Seignette salt [i.e., sodium-potassium tartrate, with eight waters of crystallization]? There is something here to be reexamined."[36]

In other words, Pasteur was puzzled by Gerhardt's claim that the sodium-ammonium *paratartrate* contained two waters of crystallization despite its crystalline identity and close chemical analogies with sodium-ammonium *tartrate* and sodium-potassium tartrate (the Seignette salt), both of which were believed to contain eight waters of crystallization. "Prepare the double paratartrate of sodium and ammonium," he wrote. "Its form and its [chemical] analysis? Ditto analyze the tartrate of these bases [i.e., the sodium-ammonium tartrate]." Pasteur apparently had none of the paratartrate on hand at the time, and he now reminded himself to "ask M. Biot if he still has a sample of the double [sodium-ammonium] paratartrate that M. Mitscherlich gave him and see if it is isomorphic with the corresponding double tartrate or with the Seignette salt."[37]

Nothing in Pasteur's notebook suggests that he had as yet attached any special significance to this third anomaly. In projecting a close study of sodium-ammonium tartrate and its corresponding paratartrate, Pasteur was merely extending the list of experiments that grew out of his expectation that truly isomorphic substances should have equal waters of crystallization. At this point in his research, he was not especially concerned about the relation between optical activity and crystalline form, but was focusing instead on the relation between waters of crystallization and crystalline form.

Before he undertook any experiments to resolve this third anomaly, Pasteur recorded yet a fourth anomaly in his lab notebook. The crystalline forms of potassium tartrate and ammonium tartrate were remarkably similar, extending even to the modification of their extremities by microscopic hemihedral facets. What was odd, given the virtual identity of their outward forms, was that Dumas had assigned them to different formula types. Pasteur's first reaction was to doubt the accuracy of the accepted chemical formulas. "The analyses of these salts [i.e., the potassium tartrate and the ammonium tartrate] need to be redone," he wrote. But he also considered a second method for exploring the precise degree of similarity between these two tartrates. He remembered that a German chemist named Hankel had drawn a connection between the presence of hemihedral facets in potassium tartrate and its "pyroelectricity," a measure of electrical polarity in a crystal under certain conditions of changing temperature. Pasteur now wondered whether ammonium tartrate also exhibited pyroelectricity (thus confirming its hemidedry), especially since he had once obtained some "beautiful crystals" of this tartrate in which its alleged hemihedrism (thus far reported only in an important memoir by de la Provostaye) was "no longer evident at all."[38]

Pasteur, it seems, was now planning to search for a crystallographic difference between the ammonium and potassium tartrates if further analyses confirmed Dumas's claim that they belonged to different formula types.

Pasteur's laboratory notes do not explicitly reveal why these two tartrates seemed "anomalous" to him; by some accounts they were at once completely isomorphic with each other and equal in waters of crystallization (with one each). But he was obviously struck by their alleged difference in formula type, and remarks elsewhere in his notebook suggest that there was some doubt whether the ammonium and potassium tartrates crystallized with one molecule of water or none at all.[39] There is thus good reason to suppose that Pasteur's concern with the relation between waters of crystallization and crystal form also lay behind his fourth anomaly, as it more obviously did in the first three cases.

Pasteur had now completed his programmatic list of four anomalies, but he appended to it a note concerning "the hemihedry of tartrates in general." The hemihedrism of the ammonium and potassium tartrates had presumably alerted him to the value of this property as a way of getting at other anomalies in the tartrates. In addition to the ammonium and potassium tartrates, he now listed two hitherto unmentioned compounds in which de la Provostaye had reported the existence of hemihedrism—"the emetic of ammonium and potassium" and "the emetic of ammonium with several equivalents of water."[40] From this notebook entry, it seems likely that Pasteur was beginning to suspect that hemihedrism was a property common to all of the tartrates and their derivatives.

Taken as a whole, Pasteur's notebook agenda of late April 1848 suggests that he was now paying very close attention to those subtle differences in crystal form and behavior that he had minimized in his earlier efforts to show that dimorphic substances were really isomorphic in the broad sense. That effort to blur the distinction between isomorphism and dimorphism had itself been inspired by Laurent, as Pasteur emphasized when he told the Académie des sciences that he had achieved the joint crystallization of eight tartrates belonging to two theoretically incompatible crystallographic systems.[41] But this claim, which Pasteur very quickly abandoned, could not easily be reconciled with another Laurentian precept: that differences in waters of crystallization should be correlated with specific differences in crystal form. It was the latter precept that dominated Pasteur's notebook agenda.

In a sense, then, Pasteur had exchanged one set of Laurentian spectacles for another: where once he had seen his eight "isomorphic" tartrates crystallize together, he now focused on small but suddenly crucial differences in their crystal forms, extending his gaze beyond the angles between crystal faces to the edges of the crystals, where (according to Laurent and others in the Haüyian tradition) even tiny hemihedral facets could express differences in the chemical or physical properties of otherwise analogous compounds. It was only by doubting what he had once seen with his first set of Laurentian spectacles—those focused on waters of crystallization—that Pasteur found his way to the path that would lead to his discovery of optical isomerism in the tartrates. In the third and fourth of his anomalies he was already directing special attention to the microscopic hemihedral facets that would give him the key to unlocking the "hidden" structural difference between the sodium-ammonium tartrate and paratartrate.

Yet it would be a mistake to suppose that even this first step had already been taken. Up to this point, in fact, Pasteur had only outlined a rather elaborate research program in his notebook. There is no indication that he had thus far carried out any of the projected experiments. In fact, he could well have been in the library, surrounded by the published works of Gerhardt, Dumas, de la Provostaye, Delafosse, and especially Laurent. True, Laurent's name does not appear in the pertinent pages of this first laboratory notebook,[42] but his influence is clear in the expectations that Pasteur brought to the study of the tartrates. Those expectations emerge more sharply here, in this notebook of 1848, than they do in any of Pasteur's published works. In the public record, and especially in his famous lectures of 1860, the issue of optical activity—and thus the role of Biot—ascends as the issue of waters of crystallization and the influence of Laurent fade from view. The 1848 notebook conveys a different picture. There Pasteur projected an all-out experimental assault on four clearly formulated anomalies, each involving a specific set of tartrate compounds. But these were problems only within the framework he had inherited from Laurent: for most chemists, Pasteur's four "anomalies" did not even exist.

PASTEUR IN THE LABORATORY: THE DISCOVERY OF OPTICAL ISOMERS IN THE TARTRATES

Pasteur began his actual laboratory research with the last of his four anomalies, working through an organic analysis of potassium tartrate, breaking and heating the crystals to determine whether or not the compound had a molecule of water at crystallization. "It is very probable from this," he concluded in his notebook, "that M. Dumas's analysis is correct."[43] Pasteur then subjected the potassium tartrate to a systematic crystallographic analysis, confirming the accepted view that it formed slightly oblique prisms on crystallization. There is no direct evidence that Pasteur thought this program of research resolved his fourth anomaly, but his final position on the chemical formulas and crystal forms of the ammonium and potassium tartrates were consistent with Laurentian expectations: these two

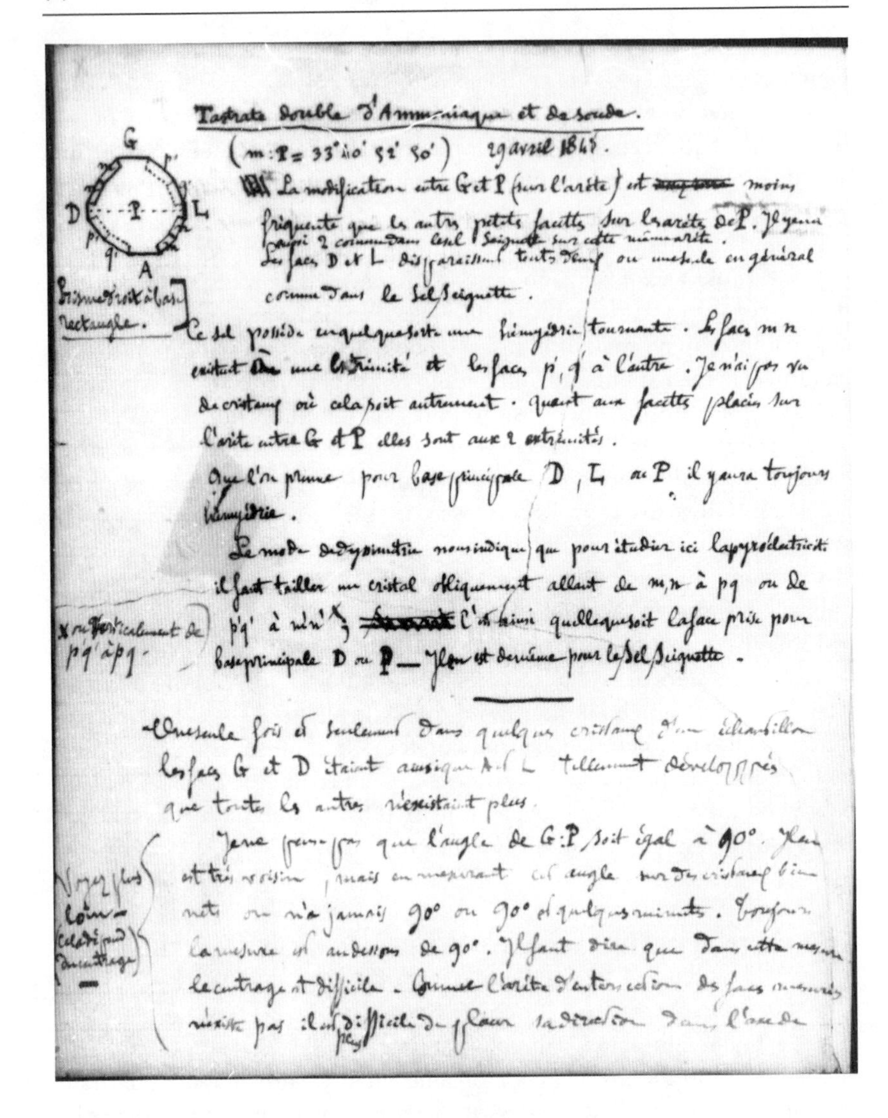
Tartrate double d'Ammoniaque et de soude.

Figure 3.7. Pasteur, "Notes divers," 1/3/7.

completely isomorphic salts were ultimately assigned one water of crystallization each.[44]

Pasteur then continued to work backwards through his agenda (see figs. 3.7 and 3.8). A specific date (29 April 1848) now appears in the notebook for the first time—though it may well have been added later, after Pasteur had discovered the special significance of his third anomaly. Here he began

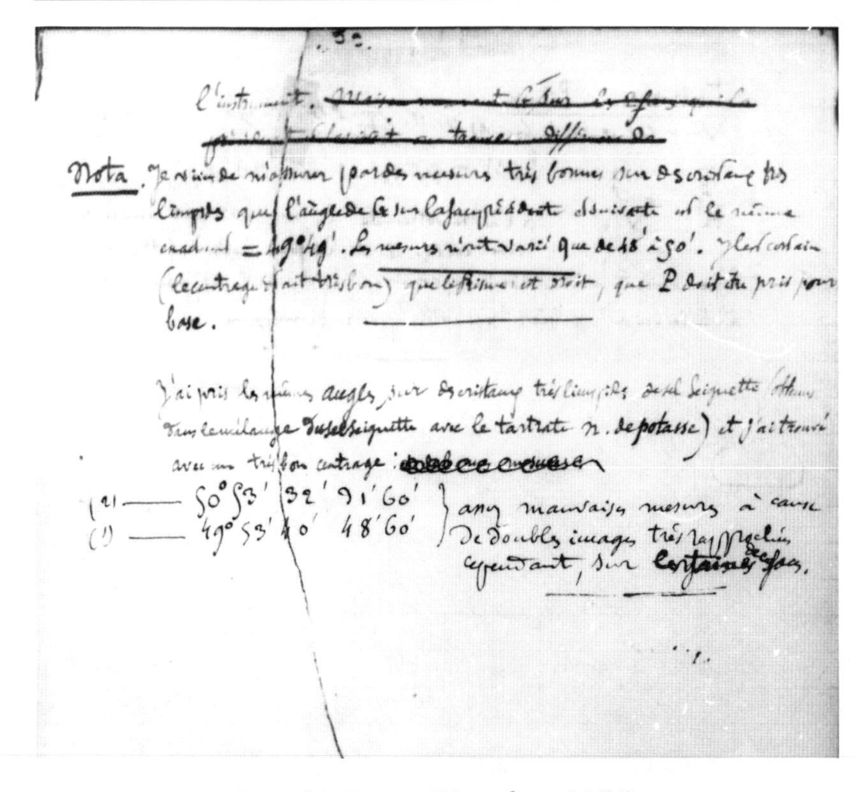

Figure 3.8. Pasteur, "Notes divers," 1/3/8.

with a crystallographic examination of the sodium-ammonium tartrate, systematically comparing it with the sodium-potassium tartrate (the Seignette salt). He was, it seems, trying to determine if these two salts really were completely isomorphic with each other, as one would expect from their accepted chemical formulas, which assigned eight waters of crystallization to each. Pasteur focused particular attention on the small modifying facets of the sodium-ammonium tartrate crystal and almost immediately discovered that it was hemihedral. De la Provostaye had not included this salt among his four hemihedral tartrates, so Pasteur had now made a truly novel observation. As a means of checking it, he indicated plans to cut some crystals of the sodium-ammonium tartrate so that they could be tested for pyroelectricity, the property previously used by Hankel to identify hemihedrism in potassium tartrate. Pasteur's increasingly general concern with hemihedrism in the tartrates is also apparent in the separate notebook page he now devoted to this claim by Hankel that the potassium tartrate was hemihedral. Due to the difficulty of producing good crystals of this

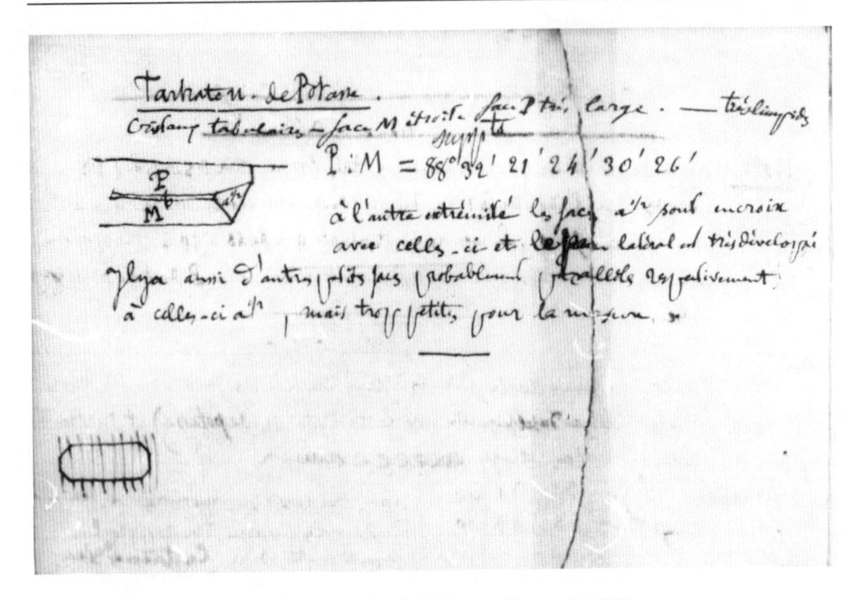
Tartrate de Potasse.
Cristaux tabulaires — faces M étroites faces P très larges. — très limpides
P : M = 88°32' 21' 24' 30' 26'
à l'autre extrémité les faces ... sont ...
avec celles-ci et les pans latéral est très développé
Il y a aussi d'autres petites faces probablement parallèles respectivement à celles-ci ..., mais trop petites pour la mesure.

Figure 3.9. Pasteur, "Notes divers," 1/3/9.

compound, de la Provostaye had been unable to detect any hemihedrism in it, but Pasteur had managed to produce some well-formed crystals that exhibited the tiny facets.[45]

Having thus confirmed the existence of hemihedrism in the potassium tartrate, Pasteur returned to the sodium-ammonium tartrate, in which he had shortly before discovered hemihedral facets. In the interim, however, his view of the subject had changed in a crucial way. Two pages earlier in the notebook, he did not seem concerned with the specific orientation of the crystals of the sodium-ammonium tartrate. He had displayed no interest in the direction, right or left, in which the hemihedry of these crystals turned. But now, by orienting the crystal faces according to a consistent convention, he found that the hemihedrism was always in one direction—to the left, according to the convention he then adopted. After examining a large number of these crystals, he became convinced of the constancy of this left-handedness, going so far as to claim that "it would suffice that there be one case which offers a hemihedral face to the right . . . for the hemihedry not to exist, or at least to be doubtful" (see figs. 3.9 and 3.10).[46]

Pasteur now approached what was to become the crucial case of the *sodium-ammonium paratartrate*. He clearly hoped that its crystals would somehow differ from the left-handed hemihedral crystals he had just detected in the sodium-ammonium *tartrate*. According to his Laurentian model, the reported difference in the waters of crystallization of the two compounds

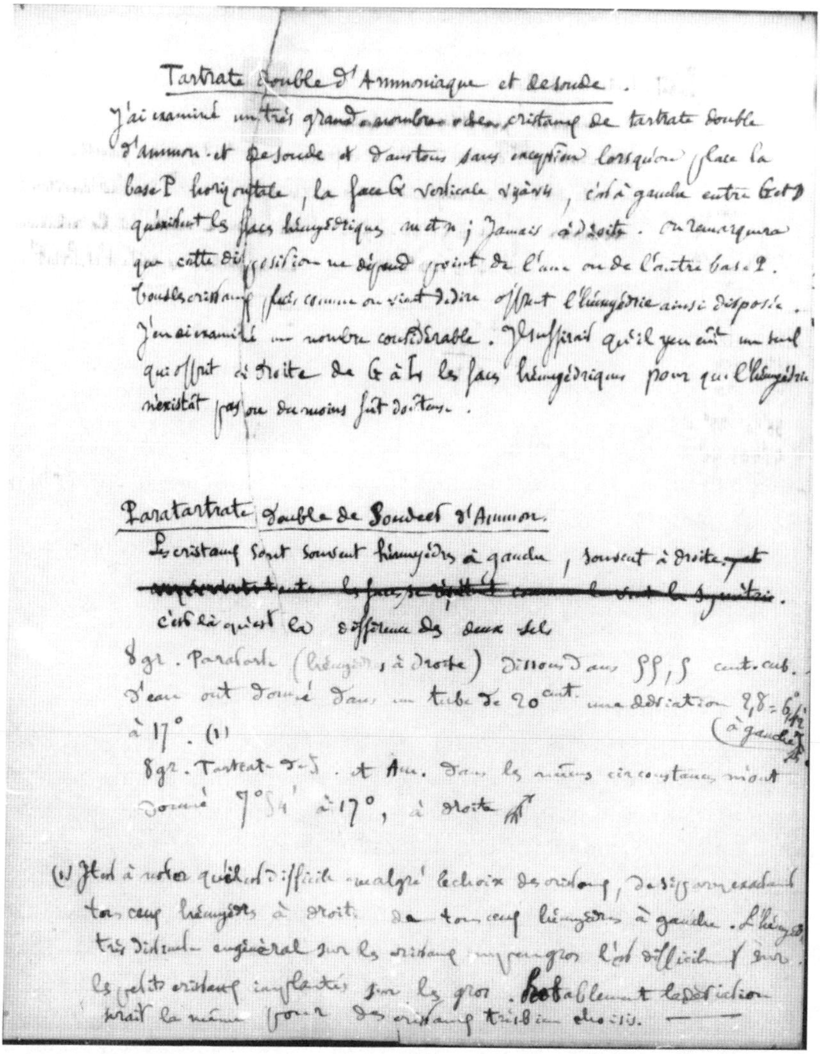

Tartrate double d'Ammoniaque et de soude.

J'ai examiné un très grand nombre de cristaux de tartrate double d'ammon. et de soude et toujours sans exception lorsqu'on place la base P horizontale, la face G verticale vis-à-vis, c'est à gauche entre G et P qu'existent les faces hémiédriques m et n ; jamais à droite. On remarquera que cette disposition ne dépend point de l'une ou de l'autre base P. Tous les cristaux placés comme on vient de dire offrent l'hémiédrie ainsi disposée. J'en ai examiné un nombre considérable. Il suffirait qu'il y en eût un seul qui offrît à droite de G à P les faces hémiédriques pour que l'hémiédrie n'existât pas ou du moins fût douteuse.

Paratartrate double de Soude et d'Ammon.

Les cristaux sont souvent hémièdres à gauche, souvent à droite ~~[illegible]~~.

C'est là qu'est la différence des deux sels

8 gr. Paratart. (hémièdr., à droite) dissous dans 99,5 cent. cub. d'eau ont donné dans un tube de 20 cent. une déviation 9,8 = 6 1/2 à 17°. (1) (à gauche)

8 gr. Tartrat. d. S. et Am. dans les mêmes circonstances m'ont donné 7°54' à 17°, à droite

(1) Il est à noter qu'il est difficile malgré le choix des cristaux, de s'en [illegible] tous ceux hémièdres à droite ou tous ceux hémièdres à gauche. L'hémiédrie très distincte en général sur les cristaux un peu gros l'est difficilement sur les petits cristaux implantés sur les gros. Probablement la déviation serait la même pour des cristaux triés [illegible].

Figure 3.10. Pasteur, "Notes divers," 1/3/10.

(eight for the tartrate, two for the paratartrate) should show up as a microscopic difference in the extremities of their crystal forms. He therefore began looking for hemihedral facets in the sodium-ammonium paratartrate, orienting its crystals according to the same convention he had used with the corresponding tartrate. The results were confusing. "The crystals," Pasteur recorded in his notebook, "are frequently hemihedral to the left, frequently to the right." "And sometimes," he added in a crucial passage that he later crossed out, "all the faces repeat themselves according to the laws of symmetry." (See fig. 3.10.) For a while—until he crossed out that passage—Pasteur apparently decided to give preference to his observations of the latter crystals, those in which "all the faces repeat themselves according to the laws of symmetry." During that brief period, he thought he could resolve his third anomaly in this way: the paratartrate was symmetrical and the tartrate was hemihedral. "Therein lies the difference between the two salts," he concluded.[47]

THE ROLE OF EMPIRICAL NOVELTY

At this point, we should examine more closely the expectations that Pasteur brought to his study of the sodium-ammonium paratartrate. The question is an important one, for this was the compound in which he would shortly find left- and right-handedness at the level both of crystal form and optical activity. In his 1860 lectures on these early experiments, Pasteur claimed that he had expected to find crystalline symmetry in the (optically inactive) sodium-ammonium paratartrate—precisely because it was optically inactive.[48] Such an expectation could only have been based on an a priori belief that optically active substances, and only optically active substances, were asymmetric (hemihedral) in their crystal forms. Pasteur's retrospective accounts of his first major discovery developed the story in exactly that way.

But there are reasons to doubt that Pasteur was thinking along these lines when he first examined the crystal form of the sodium-ammonium paratartrate. For one thing, he had not mentioned optical activity in connection with the corresponding tartrate in the immediately preceding pages of his notebook. Instead, his only concern had been the precise nature and orientation of its hemihedrism. Second, the convention of orientation that he had then adopted for this tartrate made it hemihedral to the left, whereas Mitscherlich and Biot had described it as optically active to the right. If Pasteur had already correlated optical activity and hemihedrism, it seems odd that he did not orient the crystals in such a way that their hemihedry turned to the right, matching their optical activity.

Up to this point, then, Pasteur did not display any special concern with the optical activity of his crystals. Rather, he had sought and found a crystallographic expression of the difference in waters of crystallization between the (hemihedral) sodium-ammonium tartrate and the (apparently symmetrical) paratartrate. He could thus feel satisfied that he had resolved the third of the four anomalies set out in his agenda. For he had found a stable crystallographic difference between the sodium-ammonium tartrate, with its eight waters of crystallization, and the paratartrate, with its two molecules of water of crystallization: in their crystal forms, the tartrate was asymmetric (hemihedral), while the paratartrate was symmetric.

Why, then, did Pasteur eventually reject this conclusion that the crystal form of sodium-ammonium paratartrate was symmetric in the usual sense? Why and when, in his first laboratory notebook, did he cross out the phrase "and sometimes all the faces repeat themselves according to the laws of symmetry"? Why did he decide to give new attention and credence instead to the immediately preceding phrase: "The crystals are frequently hemihedral to the left, frequently to the right"? Why, in short, did Pasteur decide that the sodium-ammonium paratartrate was *not* symmetric in the usual sense—that it was composed instead of equal portions of left- and right-handed crystals? And why and when did he turn to measurements of optical activity in support of this new conclusion?

These questions bring us face-to-face with one of the deepest problems in science studies. How can we understand the process of discovery? How can we account for the emergence of novelty in the production of scientific knowledge? This chapter has emphasized Pasteur's indebtedness to his apprenticeship under Auguste Laurent. Even here, on the threshold of Pasteur's first major discovery, it is tempting to rehearse and pursue that line of argument. Recall, yet again, that Laurent belonged to the Haüyian French tradition that taught Pasteur to pay close attention to the correlation between chemical composition and crystal form, even at the level of microscopic hemihedral facets. Recall, more specifically, that Laurent insisted on this precept even in cases where substances differed only in their waters of crystallization—and thus directed Pasteur's attention to the four "anomalies" that he listed under the heading "Problem of the Tartrates" in his first laboratory notebook. And recall, finally, that Laurent had also insisted on a correlation between crystal form and optical activity: for Laurent, substances with the same crystalline form should have the same optical activity.[49]

But if Laurent thus remains central to our story, we should also note that his speculative molecular models did not predict any particular crystallographic feature at the level of detail represented by Pasteur's discovery of left- and right-handed hemihedral facets in the sodium-ammonium paratartrate. Laurent's models were flexible enough to accommodate *any* consistent crystallographic difference between the sodium-ammonium tartrate and its corresponding paratartrate. In the Laurentian scheme, there was no obvious reason to choose between the crystalline symmetry that Pasteur initially thought he had found in the paratartrate and the mixed left- and right-handed hemihedrism on which he ultimately settled. Simply put, Laurentian models did not necessarily predict left- and right-handedness in the sodium-ammonium paratartrate. No such preconceived idea required the change of view that led to Pasteur's first major discovery. We must therefore look elsewhere to explain the discovery, insofar as we can do so at all.

Historians, philosophers, and sociologists of science have become increasingly reluctant to explain such discoveries as Pasteur was about to make simply by pointing to the empirical evidence at hand—to the "nature of things" or to "the real world out there." Indeed, this chapter and much of the rest of this book are meant to suggest just how pliable the supposedly hard evidence of the natural world can be. And yet, for all of that, we do sometimes bump up against situations that ask us to give credence to our historical actors' perception of the empirical world.[50] For no obvious reason to be found in his a priori theoretical commitments or other interests, Pasteur became convinced that he could detect left- and right-handed crystals in the sodium-ammonium paratartrate.

In conceding this point, however, we should not ignore the extent to which Pasteur constructed the empirical world in which he made his first major discovery—nor the extent to which that discovery depended on the "privileged material" represented by the tartrates.[51] Even for one so observant as he, the separation of left- and right-handed crystals in the paratartrate posed a difficult and delicate task. For the smaller specimens, it required considerable training and skill to detect the hemihedral facets at all. But Pasteur managed to produce larger crystals in which the existence of the two forms of hemihedrism was more apparent and sometimes even striking. The presence of so many left- and right-handed crystals contrasted sharply with the situation found in most symmetric compounds. In fact, as we have been told by those in a position to know, it is only in the tartrates that the relationship between chemical composition, optical activity, and crystalline asymmetry is so consistent and so visible.[52]

We cannot know for sure, but it may simply have been Pasteur's capacity to construct unusually visible and persistent hemihedral forms of the sodium-ammonium paratartrate that led him to doubt its crystalline symmetry. In any case, Pasteur now began to consider a different and much more interesting possibility: one where the paratartrate was composed of two forms—one hemihedral to the right, the other to the left. Such a result, though unexpected, would nonetheless also serve to resolve the third anomaly in Pasteur's Laurentian agenda: the existence of left- and right-handed hemihedrism in the paratartrate would still provide a crystallographic distinction (albeit of an unexpected sort) between the sodium-ammonium tartrate (with its eight waters of crystallization) and the corresponding paratartrate (with its two molecules of water at crystallization).

It was surely now, while pondering this fascinating crystallographic difference between the sodium-ammonium tartrate and its corresponding paratartrate, that Pasteur began to focus on the possibility of a correlation between optical activity and crystalline hemihedrism. The first step was a small one: optical activity provided an obvious means of confirming his resolution of the anomaly that he had just been investigating—that is, the relation between crystalline isomorphism and waters of crystallization. This anomaly arose partly from Mitscherlich's claim that the crystal forms of the tartrate and paratartrate were identical.

Once Pasteur had detected a crystallographic difference (in the form of tiny hemihedral facets) between the two salts that his distinguished German predecessor had missed, he surely also expected that Mitscherlich's characterization of the optical activities of the two substances would also require revision. Mitscherlich had claimed that the dissolved tartrate rotated the plane of polarized light to the right, while the paratartrate was optically inactive. But Pasteur, who had now resolved the paratartrate crystals into left- and right-handed hemihedral forms, was ready to challenge Mitscherlich at a deeper level—at the level of the relation between optical activity and microscopic hemihedral facets. In the case of the sodium-ammonium tartrate, there was no need to revise Mitscherlich's conclusion: it should indeed deviate the plane of polarized light to the right. But in the case of the corresponding paratartrate, now recognized as a combination of left- and right-handed microscopic hemihedral forms, the two halves of the paratartrate should deviate the plane of polarized light in equal amounts and opposite directions—one to the left, the other to the right. Its optical inactivity would thus be revealed as only apparent, the result of two separate and compensating optical activities.

As Pasteur's first laboratory notebook shows, it was only now that he used the polarimeter to compare the optical activity of a dissolved sample of the sodium-ammonium tartrate with a similar reading for an equal amount of its mirror image form in the paratartrate. Such a measurement would provide a quick means of checking Pasteur's surmise. For if his resolution of the third anomaly was correct, the two samples should deviate the plane of polarized light in opposite directions, but by the same amount. In the event, the result was not quite so decisive as Pasteur doubtless had hoped: his notebook records that the tartrate deviated the plane 7° 54′ to the right, while its mirror image in the paratartrate gave a result of 6° 42′ to the left.

I. The Problem: According to Laurent, chemical constituents of compounds should be expressed externally in crystalline form. Further, substances that crystallize together should be isomorphic and belong to the same chemical types. But Pasteur has announced that a group of eight tartrates crystallize together, even though they belong to different chemical types. In particular, several differ in their number of waters of crystallization. Thus, wrote Pasteur in his laboratory notes: "The isomorphism of these salts is in need of confirmation." Makes a list of four anomalies in the relation between crystalline isomorphism and waters of crystallization. (See Table 3.2, with list of anomalies A-D. Hereafter number of waters of crystallization indicated by HO, which was the conventional formula for water at that time, rather than our H_20.)

II. Pasteur in the Lab: Begins working backwards through list of anomalies.

Anomaly D.	Potassium tartrate	No HO?	Identity of crystal forms, even to hemihedrism
	Ammonium tartrate	One HO	

Apparently resolved by establishing that potassium tartrate also has one HO

Anomaly C.	Sodium ammonium tartrate	8 HO	Isomorphic (with Seignette salt)
	Sodium potassium tartrate (Seignette salt)	8 HO	Isomorphic (with sodium ammonium paratartrate)
	Sodium ammonium paratartrate	2 HO	

Sequence of experiments and significant results:

1. Close crystallographic study of sodium ammonium tartrate shows that it is hemihedral:
   A novel finding that Pasteur checks with pyroelectricity (Hankel's technique).

2. Searches for hemihedrism in sodium ammonium paratartrate, trying to see if it is really isomorphic with the sodium ammonium tartrate. Results are confusing; sometimes crystals seem hemihedral to the left, at other times hemihedral to the right, and sometimes symmetrical. Initial conclusion: paratartrate is symmetrical, which thus gives a crystallographic difference between the tartrate, with its eight HO's, and the paratartrate with its two HO's.

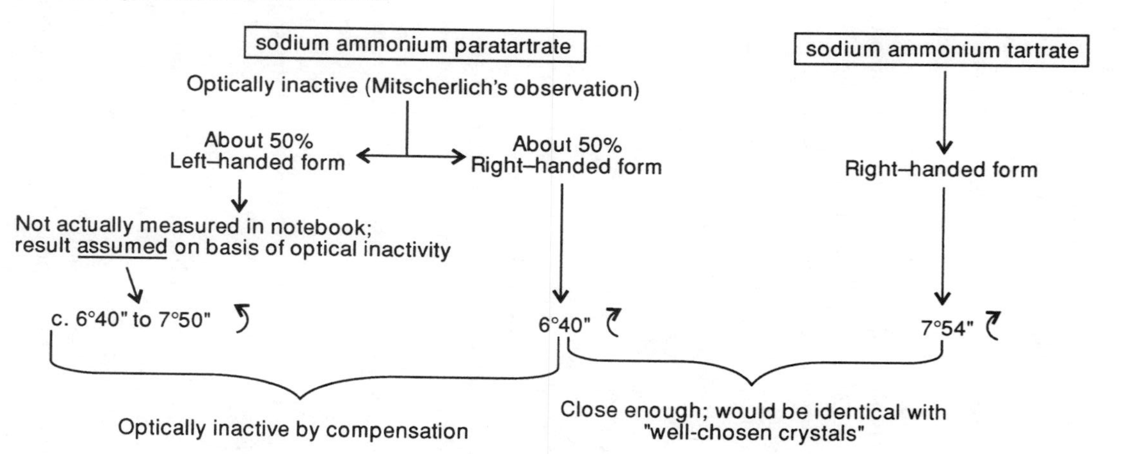
3. Yet Pasteur still uncertain. Produces larger crystals of the sodium ammonium paratartrate and is persuaded that his initial conclusion is mistaken. Crosses out passage that says that the crystals are (sometimes) symmetric.
(Role of new emprical evidence, unpredicted at this level of detail by Laurentian precepts.)
Discovery of left– and right–handed forms in sodium ammonium paratartrate.
Only now is measurement of optical activity by polarimeter used to confirm discovery. (Mitscherlich's note of 1844).
Results of polarimetric measurments:
sodium ammonium paratartrate
Optically inactive (Mitscherlich's observation)
About 50%
Left–handed form
About 50%
Right–handed form
sodium ammonium tartrate
Right–handed form
Not actually measured in notebook;
result assumed on basis of optical inactivity
c. 6°40" to 7°50"
6°40"
7°54"
Optically inactive by compensation
Close enough; would be identical with
"well-chosen crystals"

Figure 3.11. The path to Pasteur's discovery of optical isomers—as reconstructed from his laboratory notes.

Pasteur minimized this difference—in effect, he explained it away—by pointing to the difficulty of completely separating the two hemihedral forms of the paratartrate. The deviation would "probably be the same for very well-chosen crystals," he now claimed.[53]

In other words, Pasteur first drew a link between crystalline form and optical activity in the context of his ongoing program of research on isomorphism, dimorphism, and their relation to waters of crystallization. He was indeed inspired by Mitscherlich's famous note, as all the standard sources report, but not quite in the way that he retrospectively claimed. At the outset, the significance of Mitscherlich's note for Pasteur was that it pointed toward an anomaly—the third on his research agenda—in the presumed (Laurentian) relationship between waters of crystallization and crystal form. It also pointed toward the measurement of optical activity with the polarimeter as a ready means of checking his attempt to resolve that anomaly. It was only then, and in that context, that Pasteur began to conceive of a broader link between optical activity and crystalline asymmetry.

But the discovery of two hemihedral forms in the paratartrate did indeed rapidly lead Pasteur beyond his initial list of anomalies and his interest in waters of crystallization. There was one other case, he knew, of a crystal having both left and right hemihedral forms. Quartz had been shown by Haüy, Herschel, and Biot to have right-handed hemihedral crystals that were correlated with optical activity to the right, and left-handed hemihedral crystals correlated with optical activity to the left. Delafosse thought that the hemihedry of the aggregated quartz crystal indicated a corresponding asymmetry in its constituent molecules, while Biot argued that it resulted from an asymmetrical aggregation of individually symmetric quartz molecules. In his physics thesis, Pasteur had agreed with Biot that aggregate asymmetry in a solid crystal like quartz could mask an underlying symmetry at the molecular level, whereas substances that were optically active in solution must be composed of individually asymmetric molecules.[54]

Once Pasteur came to believe that the sodium-ammonium paratartrate consisted of left- and right-handed crystals, he recognized that he had a situation like that of quartz, with one crucial difference. Unlike quartz, the paratartrate was highly soluble. In this case, then, crystalline hemihedrism could be correlated with an asymmetry in the soluble constituent molecules. Pasteur's polarimeter measurement for the right-handed half of the paratartrate thus not only aided him in resolving his experimental anomaly; it also pointed to unprecedented insights into the relation between crystal form and molecular structure. To establish fully the existence of a correlation between hemihedrism and optical activity, Pasteur would need a polarimeter reading for the left-handed half of his paratartrate sample as well, though this was in many respects a formality. The expected outcome of this measurement was already clear from the polarimeter reading for the first hemihedral form, since Pasteur already knew, from Mitscherlich's research, that the resultant of these two polarimetric measurements of the "inactive" form of the compound should be zero. As a matter of fact, Pasteur's laboratory notebook does not record his measurement of the left-handed part of the paratartrate, even though he clearly appreciated its importance and presumably made it before disclosing his discovery in a public forum.

Pasteur summarized his findings before the Académie des sciences on 15 May 1848—and, in doing so, took the crucial step of orienting his crystals so that the direction of hemihedry would be the same as the direction of optical activity:

> Tartaric acid and the tartrates deviate the plane of polarization; they are all hemihedral. They deviate always to the right, and are hemihedral in the same direction. The paratartrates do not deviate [the plane of polarized light]; they are not hemihedral. One of them [the sodium-ammonium paratartrate] does deviate and is hemihedral. It deviates sometimes to the right, sometimes to the left: this is because it is hemihedral—sometimes in one direction, sometimes in the other.[55]

With this superficially simple announcement, Pasteur was well on the way to fulfilling his goal of joining the Parisian scientific elite. The rest of his brief paper did nothing to reveal the complex process by which he had reached this striking conclusion. In this case, Pasteur was glad to conform to the already formulaic style of the published scientific paper.

NEW PATHS AND NEW DIRECTIONS: PASTEUR IN THE WAKE OF HIS FIRST MAJOR DISCOVERY

From Pasteur's notebook of 1848, it is clear that his research plans shifted radically in the very course of making his first major discovery. Rather than moving methodically to the next item on his list of anomalies by checking the ability of his eight allegedly isomorphic tartrates to crystallize together, he abandoned his elaborate series of projected experiments. Instead, he searched through a series of other tartrates and paratartrates, eager to show that the former were always hemihedral and the latter always symmetrical except when composed of both left- and right-handed crystalline forms. In this quest, Pasteur frequently used pyroelectricity as a test for hemihedry, and his interest in this phenomenon is further illustrated by nine pages of notes taken from works by Hankel and other authors. But the case of the sodium-ammonium paratartrate proved to be unique. Typical instead was potassium paratartrate: "[It] does not possess the facets which make hemihedrism possible in this substance. Upon cooling it does not, or at least does not appear to be, pyroelectric." Pasteur was unable to find another case of left- and right-handed crystals in the paratartrates: they were all symmetrical.[56]

After jotting down a few plans for future experiments,[57] Pasteur began composing an account of his discovery. The first draft, inscribed in his notebook during the first two weeks of May 1848, amply revealed the Laurentian context in which the discovery had been made. Reasserting his claim that all of the tartrates were broadly isomorphic, Pasteur insisted that this similarity of form made it "impossible to doubt that a certain molecular group remains constant in all these salts; that the water of crystallization, the bases . . . modify it at the extremities only, hardly touching the central molecular arrangement and there only in the difference of angles observed between the facets."

In keeping with this Laurentian molecular model, with its notion of a stable "radical" core, Pasteur maintained that the only crystallographic difference between the tartrates and paratartrates was to be found at the extremities of the crystals, where "all of the tartrates are hemihedral," whereas the edges of the paratartrate were symmetrical. "There is a difference in the molecular arrangement of the tartrates and paratartrates," Pasteur wrote in this first draft, "but this difference seems only to stem from a regular distribution of the oxide or water of crystallization molecules in the one case [the paratartrates] and a disymmetrical distribution in the other [the tartrates]." The sole exception to symmetry in the paratartrates—the sodium-ammonium salt—could be accommodated to this Laurentian scheme by recognizing that it could be separated into two "veritable" tartrates, one hemihedral to the left, the other to the right.

In this first (unpublished) draft of Pasteur's discovery, he made it clear how much he owed to Laurent. Within weeks, in the brief note of 15 May 1848 that he published in the *Comptes rendus* of the Académie des sciences, Pasteur already began to tone down his allusions to Laurentian molecular models. And soon thereafter, in the first long paper describing his discovery, Pasteur completely eliminated Laurent's name from his text.[58] From that point on, Laurent's name, so prominently cited in Pasteur's chemistry thesis and in his earliest published papers, virtually disappeared from his publications and was not mentioned even once in Pasteur's famous 1860 lectures on the discovery.

CONCLUSION

In his famous lectures of 1860, Pasteur was fully swept up in the excitement of trying to establish a new correlation—no longer just a correlation between optical activity and crystalline form, but more grandly between optical activity and life itself. By comparison, his initial Laurentian concern with the correlation between waters of crystallization and the details of crystalline form must have seemed distant in time and minor in significance. Especially given Pasteur's preoccupations by 1860, it is not surprising that his lectures of that year simplified and telescoped the process by which he had reached his first major discovery twelve years before—and, in doing so, erased Laurent from the scene.

In fact, Pasteur's laboratory notes of 1848 agree with his 1860 account in ways that suggest how hard it is to attach a single unequivocal meaning to his "discovery" and how misleading it would be to ascribe his achievement solely to the influence of Laurent. The notebook of April 1848, like the lectures of 1860, attests to the significance of Mitscherlich's note of 1844, to an initial belief in the symmetry of the sodium-ammonium paratartrate crystals, and to the influence of de la Provostaye, Delafosse, and Biot. More than that, Pasteur's notebook of 1848 allows us to recognize the sense in which he went beyond his initial Laurentian program in the very process of making his discovery. For, in fact, that achievement—described in abbreviated and somewhat misleading language as the discovery of left- and right-handed forms in the tartrates—consisted of two major elements, of which only the first emerged directly out of Laurentian concerns.

Pasteur's Laurentian concerns, including specifically the relation between waters of crystallization and the details of crystalline form, led him to pay very special attention to the edges of the sodium-ammonium tartrate and paratartrate crystals and thus to detect hitherto unknown hemihedral facets in both salts. But the second element in his discovery—the recognition of both left- and right-handed forms in the paratartrate—was not predicted by Laurentian theory and won Pasteur's full confidence only after he had used optical activity to explore the precise character of the first element in his discovery. When both elements of the discovery were fully in place, Pasteur immediately abandoned his Laurentian research program to pursue an entirely new correlation between optical activity and crystalline hemihedry. It is as if one "preconceived idea"—the Laurentian correlation between waters of crystallization and crystalline form—now gave birth to another, Pasteur's "law of hemihedral correlation," which immediately became the guiding theme in his future research. Even in 1848, Pasteur had thus already emerged from the Laurentian program with which he had begun his research and had staked out important new territory of his own.

The manuscript record of this story reminds us to be ever skeptical of participants' retrospective accounts of scientific discovery. In this case it can also be argued that political considerations of a sort lay behind Pasteur's move to erase any trace of Laurent's early influence. During the few months he spent working side-by-side with Laurent at the Ecole Normale, Pasteur's references to him in his private correspondence were fully in keeping with the respectful, admiring tone of his acknowledgments in his doctoral thesis in chemistry. In letters to his schoolboy friend, Charles Chappius, Pasteur expressed his disappointment upon learning the news that Laurent might leave the Ecole Normale and described him as one who was "destined to occupy the first rank among chemists."[59]

But Pasteur's discovery of April 1848 had a profound and immediate impact on his career and on his relation to Laurent. The discovery brought him recognition by the French scientific community, altered his position within its elaborate social system, and made such revered figures as Biot and Dumas seem henceforth more appropriate patrons than the troubled, controversial, and fading Laurent. In February 1852, in a letter to Dumas, Pasteur wrote disparagingly of Laurent's early influence on him: "I had worked under the guidance of the good M. Laurent . . . at an age when the mind is fashioned by the model which is presented to it. I was enveloped by hypotheses without basis, by a redaction that was completely devoid of precision, and I ruined the exposition of new and interesting facts. I was quickly enlightened by your advice."[60]

It is telling that Pasteur should have made his defection from Laurent explicit in a letter to Dumas, who had long been in conflict with Laurent and his molecular models.[61] This theoretical shift can be made "isomorphic" with a contrast in the respective political positions and fortunes of Dumas and Laurent. In the centralized and fiercely competitive environment of nineteenth-century French academic life, Dumas was the quintessential establishment scientist: minister of agriculture from 1850 to 1851 and a distinguished professor at several institutions in Paris, including the Sorbonne, the Ecole Polytechnique, and the Ecole Centrale des Arts et Manufactures, which he had co-founded in 1829. Dumas's political conservatism also sat well with the Second Empire of Louis Napoleon, so much so that the new emperor named him a senator.[62]

Meanwhile Laurent, with his radical political sympathies and difficult personality, never attained the position or influence that his early promise had once seemed to foretell. Consigned for most of his career to the provinces, mainly at the University of Bordeaux, Laurent's periodic attempts to obtain a position in Paris brought no permanent success. He died in April 1853 at the age of forty-six, in many ways a pitiful and broken figure. By then, Pasteur's youthful flirtation with republicanism had given way to his declaration of partisanship toward the new emperor and to the political conservatism that marked the rest of his days.[63] Small wonder that he was no longer eager to link Laurent's name with his own. Emulating Laurent was no way to get ahead in the Parisian scientific scene. In fact, Pasteur's accounts of his discovery of optical isomerism offer but one example of a correlation between the scientific and political dimensions of his career. The most blatant example, perhaps, is to be found in his work on spontaneous generation. But there are several others—enough to begin to suggest a pattern to those alert to it.

In the end, however, it would be fruitless to insist that academic politics or Pasteur's self-interest determined the direction of his research or the remarkable success with which he pursued it. The correlations between Pasteur's scientific work and his political inclinations or unabashed careerism cannot be established with anything like the precision or power of his "natural" correlations between optical activity and crystalline hemihedry, between optical activity and life, or between life, fermentation, and disease. In the case at hand—the discovery of optical isomers in the tartrates—Pasteur's success depended not only on theoretical precepts derived from his politically well-placed patrons, but also on his remarkable visual acuity, his manual dexterity, his creativity in manipulating privileged materials to yield the results he sought, and his formidable rhetorical talents. In this case and others, Pasteur won patrons and allies because he was a superb performer across the full range of roles that make up the scientific form of life.

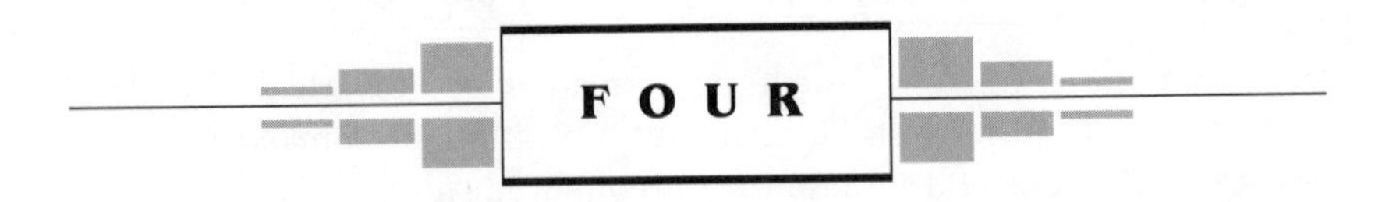

FOUR

From Crystals to Life: Optical Activity, Fermentation, and Life

ON 3 AUGUST 1857, three years after he had been named professor of chemistry and dean of the newly established Faculty of Sciences at Lille, Pasteur delivered a now famous paper on the lactic fermentation to the Société des sciences, d'agriculture, et des arts de Lille.[1] This paper announced a major shift in Pasteur's research interests—a shift, briefly put, from crystallography to fermentation. More than that, it laid out the central theoretical and technical precepts that marked all of his subsequent work on fermentation. It was, in effect, the opening salvo in Pasteur's campaign on behalf of the biological or "germ" theory of fermentation—and, by extension, his even more celebrated efforts to establish the germ theory of disease.

Pasteur's paper of August 1857 was astonishing in its audacity and scope. Ostensibly, it concerned lactic fermentation—that is, the transformation of sugar into lactic acid, a process whose most familiar consequence is sour milk. Pasteur's central claim was that he had discovered a specific new ferment—a "lactic yeast"—that was invariably present when sugar was transformed into lactic acid. Pointing to a wide range of suggestive, if less than conclusive evidence, Pasteur insisted that this new lactic yeast was a living microorganism and that its vital activities were essential to lactic fermentation.

In fact, Pasteur's brief discussion of lactic fermentation introduced the basic convictions and techniques that would thereafter guide his study of fermentation in general: (1) his biological conception of fermentation as the result of the activity of living microorganisms; (2) his notion of specificity, according to which each fermentation could be traced to a specific microbe; (3) his belief that the fermenting medium provided nutrients for the impliciated microbe and must therefore be adapted to its nutritional requirements; (4) his claim that particular chemical features of the fermenting medium could promote or impede the development of any given microbe in it; (5) his insight that different microbes competed for the aliments contained in the medium; (6) his insistence that ordinary atmospheric air was the source of the microbes that appeared in fermentations of whatever sort; and (7) his technique of "sowing" the microbe that was presumed responsible for a given fermentation in order to isolate and purify it.[2]

Actually, this paper of August 1857 lacked only two important features that were to become part of Pasteur's mature conception of fermentation. The first missing feature would become a crucially important technique—namely, his method of cultivating fermentative microbes in a purely mineral medium. The second missing feature was more overtly theoretical—the notion of fermentation as "life without air" (*vie sans air*). By 1863 all of these elements of Pasteur's mature theory of fermentation were in place. From that point on, his work in this domain consisted mainly of repetitive defenses of his theoretical precepts and sometimes crudely empirical efforts to preserve and improve the quality of wine and beer—along with time-consuming battles to defend his priority for those precepts and practices.

Pasteur's sterilizing techniques quickly acquired the label "pasteurization," and they have had a powerful impact on everyday life ever since. Only a fervent French patriot would dare to suggest that Pasteur made any headway in his efforts to make French beers superior to German (or English) rivals, but no one doubts that his research was crucial to the rise of the germ theory of fermentation and disease. There is, however, an interesting divergence of opinion about the origins of and motivations for Pasteur's interest in fermentation. Basically, this disagreement has to do with the relative importance for Pasteur of his a priori theoretical commitments and his pragmatic or industrial concerns. One goal of this chapter is to resolve this issue, relying in part on Pasteur's correspondence, manuscripts, and laboratory notebooks. In this case, however, these private sources will be used mainly to develop rather than to challenge Pasteur's own published accounts of the origins of his work on fermentation and the motivations behind it.

THEORETICAL VERSUS INDUSTRIAL CONCERNS IN THE ORIGINS OF PASTEUR'S WORK ON FERMENTATION

Observers of nineteenth-century French academic life, ever alert to the magnetic power of Paris and its Académie des sciences, have sometimes expressed surprise that Pasteur chose the modest provincial society at Lille as the initial audience for his paper of 3 August 1857 on lactic fermentation.

Not until more than three months later, on 30 November 1857, did the Académie des sciences in Paris hear a much abbreviated version of the paper. In the meantime, in October 1857, Pasteur had moved from Lille to Paris to become director of scientific studies at his alma mater, the Ecole Normale. For his son-in-law and hagiographer René Vallery-Radot, Pasteur's decision to deliver the first version of this paper at Lille was an expression of "much delicacy of feeling"—presumably feelings of gratitude toward the friends and colleagues at Lille that he was about to leave behind.[3] Yet the published version of Pasteur's August 1857 paper at Lille displayed no special "delicacy of feeling" toward his provincial colleagues and patrons.

What that paper emphatically does reveal is Pasteur's felt need to explain the shift in the focus of his research from crystallography to fermentation. He was clearly aware that this move would come as a surprise to his growing audience among the scientific elite. He emphasized the extent to which his new research on fermentation grew out of his earlier research on the relations between crystal structure and optical activity, and he expressed the hope that he would soon be able to find a way to link these two domains. Even then, in 1857, and more emphatically in retrospect, Pasteur was convinced that this first major shift in his scientific interests had its origins in the "inflexible" internal logic of his research.[4]

In this case, however, Pasteur's insistence on the internal logic of his research met a posthumous challenge from an unexpected source—his son-in-law René Vallery-Radot, who offered a more dramatic version of the story in his standard biography of 1900. Vallery-Radot's account is surely the original source of the enduring legend that Pasteur turned to the study of fermentation in response to the needs of the brewing industry in Lille. Like most legends, this one is plausible enough on the surface, and it is not hard to see why it has endured so long.

The newly established Faculty of Sciences at Lille, located at the heart of the most flourishing industrial region in France, was indeed expected to bring scientific knowledge to bear on local chemical and brewing industries. In his inaugural address as dean of the new Faculty, Pasteur strongly supported this goal.[5] True, he did resist pressure from the Ministry of Public Instruction to emphasize applied subjects at the expense of basic science in the curriculum at Lille. But even in his own courses, Pasteur taught the principles and techniques of bleaching, of extracting and refining sugar, and especially of the manufacture of beetroot alcohol, a major industry in Lille. Among his students was the son of a local beetroot alcohol manufacturer, one M. Bigo, whose distillery had been afflicted of late with inexplicably disappointing results. In 1900, in his canonical biography of Pasteur, René

Vallery-Radot linked Bigo's industrial concerns with Pasteur's work on fermentation in the following way:

> In the summer of 1856 a Lille manufacturer, M. Bigo, had, like many others that same year, met with great disappointments in the manufacture of beetroot alcohol. He came to the young dean for advice. The prospect of doing a kindness, of communicating the results of his observations, . . . caused Pasteur to consent to make some experiments. He spent some time almost daily at the factory. . . .
>
> M. Bigo's son, who studied in Pasteur's laboratory, has summed up in a letter how these accidents of manufacture became a starting point for Pasteur's investigations on fermentation, particularly alcoholic fermentation. "Pasteur had noticed through the microscope that the globules were round when fermentation was healthy, that they lengthened when alteration began, and were quite long when fermentation became lactic. This very simple method allowed us to watch the process and to avoid the failures in fermentation which we so often used to meet with. . . ." Young Bigo indeed remembered the series of experiments, the numerous observations noted, and how Pasteur, whilst studying the causes of those failures in the distillery, had wondered whether he was not confronted with a general fact, common to all fermentations. Pasteur was on the road to a discovery the consequences of which were to revolutionize chemistry.[6]

Twenty-five years later, in his spectacularly popular book *The Microbe Hunters*, Paul De Kruif seized on this presumed association between M. Bigo, the industrial interests in Lille, and Pasteur's turn to fermentation. In typically overheated prose, De Kruif wrote that "it was in this good solid town of distillers and sugar-beet raisers and farm implement dealers [i.e., Lille] that [Pasteur] began his great campaign, part science, part drama and romance, part religion and politics, to put microbes on the map. . . . He came to Lille and fairly stumbled on the road to fame—by offering help to a beet-sugar distiller [i.e., Bigo]."[7]

Another generation later, in 1953, the well-known crystallographer and committed Marxist J. D. Bernal told much the same story in his book, *Science and Industry in the Nineteenth Century*. According to Bernal, Pasteur in 1856 "threw himself with enthusiasm into the service of the chemical industries of the district—largely distilling and vinegar-making." It was, Bernal continued, "a failure of a Mr. Bigo in making beetroot alcohol that led Pasteur to make the decisive step that drew him away from physics and chemistry into the unknown field of microbiology. He used the microscope to distinguish between the round yeast globules of alcoholic fermentation and the long vibrios of the unwanted lactic fermentation. He had at once found a practical test that any brewer could use and had started the study of the minute organisms responsible, he firmly believed, for the chemical processes of fermentation."[8]

A few years later, René Dubos, in his abbreviated scientific biography of 1960, *Pasteur and Modern Science*, also traced Pasteur's interest in fermentation to Bigo's request for aid with his brewing problems—and, in doing so, somehow forgot the countervailing evidence that he had presented ten years before in his own full-bodied biography, *Pasteur: Free Lance of Science*.[9] As recently as 1985, the legend was repeated yet again by a leading American scholar of nineteenth-century French science, Harry Paul, who used it in support of his claim that "the teaching and the research of the professors were directly conditioned by the local demands at Lille." In terms very reminiscent of those used almost a century before by René Vallery-Radot, Paul wrote:

> In Lille the faculty of science proclaimed an overtly utilitarian ideology. Its practical orientation was demonstrated from the start in the work of Pasteur. . . . In the first semester of 1856–7 Pasteur gave a weekly lecture on chemistry applied to the industries of the Nord, especially the manufacture of alcohol and sugar from beets. It was Bigo-Daniel, a student in Pasteur's chemistry class, who got Pasteur interested in the study of fermentation. Bigo *père*, who owned one of the leading factories (at Esquermes) for producing alcohol from sugar beets, wanted to find out why abnormal fermentation in some of his vats resulted in lower alcohol production.[10]

Nothing more need be said to illustrate the persistence of the legend that it was problems in M. Bigo's beetroot alcohol factory—and thus, more generally, the needs of the distilling industry in Lille—that inspired the shift in Pasteur's research interests from crystallography to fermentation. The legend is obviously appealing to those who want to draw a tight link between Pasteur's theoretical and practical concerns, and it has a solid evidential basis in this indisputable fact: Pasteur did indeed regularly visit M. Bigo's factory in the spring of 1856. The most crucial piece of evidence here is Pasteur's brief account of his visits to Bigo's factory in one of his laboratory notebooks on fermentation.[11]

And yet, for all of that, the legend dissolves under close scrutiny. Indeed, Pasteur himself never endorsed the version of the story as handed down by his son-in-law in the standard biography of 1900. The Bigo story found no place in the first "biography" of Pasteur that René Vallery-Radot had published almost two decades before, in 1883, under Pasteur's direct and close supervision. That first "biography," which was in effect Pasteur's ghostwritten autobiography, is rather vague about the exact origin of his interest in fermentation. Elsewhere, there is ample evidence to suggest that Pasteur was concerned with fermentation almost from the outset of his career—not surprisingly, since his first great discovery, of optical isomers in soluble substances, had its origins in his study of the tartrates, which are themselves by-products of the manufacture of wine.[12]

Some of the most compelling evidence against the Bigo legend is to be found in Pasteur's own published papers—where, not so incidentally, the name Bigo never appears. In fact, there is reason to believe that the concerns that lay behind Pasteur's shift from crystallography to fermentation were almost the opposite of pragmatic or industrial. Here, in fact, we are following Pasteur during the most boldly theoretical phase of his career, where it was not utilitarian goals but rather "preconceived ideas" that were most at play.

AMYL ALCOHOL, THE "LAW OF HEMIHEDRAL CORRELATION," AND THE ORIGINS OF PASTEUR'S RESEARCH ON FERMENTATION

In the public record, the best place to begin the search for the roots of Pasteur's interest in fermentation is the famous Lille memoir of 3 August 1857. In the introduction to that memoir, which René Vallery-Radot and later epigones of the Bigo legend chose to ignore, Pasteur provided an admirably clear and concise account of how he came to focus his attention on fermentation:

> I think I should indicate in a few words how I have been led to occupy myself with researches on the fermentations. Having hitherto applied all my efforts to try to discover the links that exist between the chemical, optical, and crystallographic properties of certain compounds [*corps*] with the goal of clarifying their molecular constitution, it may perhaps be astonishing to see me approach a subject of physiological chemistry seemingly so distant from my earlier works. It is, however, very directly connected with them.
>
> In one of my recent communications to the Académie [des sciences], I established that amyl alcohol, contrary to what had been believed hitherto, was a complex substance formed of two distinct alcohols, one deviating the plane of polarization of light to the left, the other devoid of all [optical] activity. But what gave [these alcohols] a special value in the direction of the studies that I have adopted is that they offered the first known exception to the law of correlation between hemihedrism and the molecular rotatory phenomena [i.e., optical activity]. I therefore resolved to make a close study of these two amyl

> alcohols in order to determine, if possible, the causes of their simultaneous production and their true origin, about which certain preconceived ideas led me to doubt the common opinion.[13]

This "common opinion" traced the optical properties of amyl alcohol to the optical properties of the sugar from which it presumably derived. But Pasteur had come to believe that the optical properties of his two amyl alcohols could only be explained on the assumption that asymmetry, and thus life, somehow intervened in their production during the process of fermentation. As he put it at the end of his introductory remarks in the Lille memoir of 1857:

> I repeat, these are preconceived ideas [*idées préconçues*]. But they sufficed to determine me to study what influence the ferment might have in the production of these two amyl alcohols. For one always sees these alcohols take birth in the operation of fermentation, and that was yet a further invitation to persevere in the solution of these questions. In fact, I ought to admit [*avouer*] that my researches have been dominated for a long time by this thought that the constitution of substances [*corps*]—considered from the point of view of their molecular asymmetry or non-asymmetry, all else being equal—plays a considerable role in the most intimate laws of the organization of living organisms [*êtres*] and intervenes in the most hidden of their physiological properties.
>
> Such has been for me the occasion and the motive of [my] new experiments on the fermentations. But, as often happens in such circumstances, my work has grown little by little and has deviated from its initial direction—so much so that the results that I publish today seem distant from my prior studies. The link will become more evident in that which follows. Later I hope to be able to draw a connection between the phenomena of fermentation and the character of molecular asymmetry peculiar to organic substances.[14]

This autobiographical account of August 1857 as to the origins of Pasteur's interest in fermentation gains in credence when we examine the relatively obscure paper on amyl alcohol to which he refers in the first of the two passages just quoted. In that brief paper, published in August 1855, Pasteur traced his interest in amyl alcohol to 1849, when his mentor and patron Jean-Baptiste Biot informed him that amyl alcohol was optically active. Upon receipt of this information, Pasteur reported, he had immediately undertaken a study of amyl alcohol but soon abandoned the topic because of the obstacles he encountered in his efforts to crystallize pure amyl alcohol. This published account is confirmed by Pasteur's laboratory notebooks, where his initially disappointing efforts to crystallize and investigate amyl alcohol are recorded at least as early as April 1850.[15]

In the meantime, however, Pasteur had returned to the problem and now, in August 1855, claimed that he could "easily" produce large quantities of pure amyl alcohol by repeated crystallizations with barium sulfate. More than that, these renewed efforts had revealed the existence of two isomeric forms of amyl alcohol, one optically active and one optically inactive, which could be separated from one another by virtue of slight differences in their solubility.[16] Here, too, Pasteur's notebooks lend general support to the published account, although it is clear from his laboratory notes and from manuscript drafts of his published work that the task of isolating the two optically distinct forms of amyl alcohol had been far from "easy" and that efforts to replicate the feat did not routinely succeed.[17]

In any case, by June 1856, Pasteur had decided that the most interesting feature of amyl alcohol was that it represented the first "legitimate" exception to his "law of hemihedral correlation," which was, in effect, an extension to the microscopic and molecular level of the long-standing French (Haüyian) tradition that insisted upon a correlation between internal chemical structure and external crystalline form. According to Pasteur's law, which had its origins in his discovery of left- and right-handed hemihedral forms in the tartrates, any substance that was asymmetric at the molecular level should be asymmetrically hemihedral in its crystalline form and optically active in solution, since optical activity was the external, visible sign of an invisible and otherwise undetectable internal molecular asymmetry. In the case of amyl alcohol, despite a careful crystallographic study of its two isomers and their derivatives, Pasteur had been unable to find any evidence of hemihedral facets in the optically active form of amyl alcohol, as his law required.[18]

For Pasteur, it should be emphasized, this was no trivial concession. Quite the opposite. For the law of hemihedral correlation had been the leading theme in his research ever since the discovery of optical isomers in the tartrates in 1848. From that point through 1856, his laboratory notebooks are filled with painstaking efforts to establish and extend the law, beginning with a long series of experiments designed to prove that the right-handed tartrate form (or "dextro-racemate") that he had manually separated out from the sodium-ammonium paratartrate was, in fact, identical to naturally occurring right-handed tartaric acid. In this and similar cases, Pasteur set an exceptionally high standard for himself and his law. As he put it in the first of the laboratory notebooks now deposited at the Bibliothèque Nationale, the optical activity (*pouvoir rotatoire*) of truly isomorphic substances "must be *precisely* (*rigoureusement*) the same if the law is true."[19]

By that demanding standard, Pasteur encountered numerous "exceptions" to his law of hemihedral correlation from the outset. Indeed, it is hard, if not impossible, to find a single confirmation of the law, strictly speaking, in his laboratory notebooks. But it is fascinating and instructive to follow his efforts. Among other things, these early notebooks are a powerful reminder of Pasteur's awesome capacity for work, his meticulous attention to detail, and his ingenuity of mind and hand. They also testify to his Olympian self-confidence and to the fertility of the cluster of "preconceived ideas" that he carried with him as he moved from one material or problem to another.

In fact, until 1856, when he conceded defeat in the case of amyl alcohol, Pasteur had managed to show, to his own satisfaction, that the ever-expanding list of "exceptions" to his law of hemihedral correlation were merely apparent and could be explained away. He displayed a remarkable equanimity in the face of evidence that seemed to contradict his preconceived ideas. When, for example, his measurements of optical activity failed to yield the precise amount—or sometimes even the direction—that his law required, Pasteur pointed to a wide range of circumstances that could affect measurements of optical activity. Throughout his early laboratory notebooks, one can watch him struggle to control the following "perturbing" factors among others: (1) the density of the solution being measured;[20] (2) the ambient temperature;[21] (3) the period of time that elapsed from the beginning to the end of the polarimetric observations;[22] (4) the effects of the purity of the samples being measured;[23] (5) the effects of various acid solutes in place of water;[24] (6) the position of the illuminating lamp;[25] and, (7) perhaps most interestingly, the size and calibration of the instruments used to measure optical activity.[26]

To an "impartial" observer, one who had no commitment to Pasteur's alleged law of hemihedral correlation, the number and range of such perturbing influences would have been a source of confusion and uncertainty. And Pasteur's efforts to adjust and control them in keeping with his law would have looked very strained and ad hoc. For Pasteur, however, the same wide range of perturbing factors provided a measure of comfort and interpretive flexibility in the face of inconvenient empirical results.

Some of the most "inconvenient" results emerged as early as 1850–1851, when Pasteur turned his attention from the tartrates to asparagine and its derivatives (aspartic acid, malic acid, the aspartates and malates). These compounds, which were analogous in many ways to the tartrates, were also among the very few optically active substances from which crystals could be obtained in sizes and amounts adequate for Pasteur's investigations into the relations between chemical composition, crystal structure, and optical activity. Most of the aspartates and malates displayed hemihedral facets as well as optical activity—and, in this respect, fulfilled Pasteur's expectations. But some of these compounds unexpectedly rotated the plane of polarized light in a direction opposite to the direction of their hemihedrism. Worse yet for Pasteur's law of hemihedral correlation, other aspartates and malates displayed hemihedrism in the absence of optical activity, while still others displayed optical activity in the absence of hemihedral crystals. Even in cases where the relation between optical activity and crystal form in the aspartates and malates did seem to conform to Pasteur's law, the evidence was much more ambiguous than in the paradigmatic case of the tartrates.

Not surprisingly, Pasteur overlooked some of the "exceptions" to his law of hemihedral correlation. His response to those exceptions that he did acknowledge was sometimes brilliant, sometimes evasive, but always ingenious. For cases of hemihedrism in the absence of optical activity, he had a ready explanation derived from the case of quartz. Like quartz, he argued, such substances must possess not true molecular asymmetry but merely a fortuitous or "accidental" asymmetry in the form of their crystal as a whole. More generally and more importantly, Pasteur insisted that specifiable features of the conditions of crystallization could mask the existence of correlations between molecular asymmetry, crystalline asymmetry, and optical activity. When, for example, optically active substances seemed to display no hemihedrism, Pasteur managed to uncover crystalline asymmetry by changing the conditions of crystallization in such a way that he was able to produce—almost literally to create—crystals in which previously "hidden" hemihedral facets became manifest.

In 1856, however, Pasteur ran hard up against the recalcitrant case of amyl alcohol and its derivatives, the amylates. Here the optically active forms resisted even Pasteur's virtuoso efforts to tease out the hemihedral facets that his law required. Here, as Pasteur conceded in June 1856, the optically active form of amyl alcohol crystallized under such special conditions that he was unable to find, or even to imagine, any way to uncover any "hidden" asymmetries of the sort he had found in the case of earlier "exceptions" to his law of hemihedral correlation.[27]

If Pasteur was disappointed by his failure to cajole amyl alcohol into conformity with his supposed law of hemihedral correlation, he did not dwell on it for long. Instead, beginning with his Lille memoir of 3 August 1857, he turned all of his energy to the study of fermentation in general, with consequences that reverberate still. Once Pasteur was attracted to the general problem of fermentation, as his laboratory notebooks clearly show, he paid little attention to the details of crystal form.[28] He now focused instead on the question of why fermenting liquids were optically active. As noted above, the "common opinion" (actually, it was Justus von Liebig's idea) traced the optical activity of fermenting saccharine liquids to the sugar (also optically active) that served as the starting material. In his Lille paper of August 1857, Pasteur rejected this view in the specific case of amyl alcohol for reasons that could easily be extended to other cases.

No one has conveyed this chain of reasoning more clearly than Pasteur's own favorite disciple, Emile Duclaux, who in 1896, within a year of the master's death, published a brilliant scientific biography that included an insightful account of his turn to fermentation. Like Pasteur himself, Duclaux said nothing of M. Bigo or industrial interests in Lille. Instead, he took as his point of departure Pasteur's memoir of 3 August 1857, specifically its introduction, which he paraphrased and embellished as follows:

> In many of the industrial fermentations, we meet, as a secondary product, amyl alcohol, a substance endowed with rotary power [optical activity] and capable, furthermore, of forming several crystalline combinations which do not show any hemihedrism. It was the first exception which Pasteur had encountered in this law of correlation between hemihedrism and the rotary power. Now, according to the [prevailing] ideas of the epoch, fermentation was a disintergration: it was the breaking up of a molecule by decay, the debris of which, still voluminous, formed new molecular edifices which were the products of the fermentation. Consequently, [according to] the theory of Liebig, the edifice of amyl alcohol must form some part of the framework in the molecule of the sugar in order to resist dismemberment, and [since] it preserves the rotary power in its optical action, must be derived from that of sugar.
>
> This idea was repugnant to Pasteur. He had seen, for example in malic and maleic acids, that the least injury to the structure of the molecule made its [optical activity] disappear. "Every time," he says, "that we try to follow the rotary power of a body into its derivatives we see it promptly disappear. The primitive molecular group must be preserved intact, as it were, in the derivative, in order that the latter may continue to be active, a result which my researches permit me to predict, since the optical property is entirely dependent on a dissymmetrical arrangement of the elementary atoms. But I find that the molecular group of amyl alcohol is [too much unlike] that of sugar, if derived from it, for it to retain therefrom a dissymmetrical arrangement of its atoms."
>
> The origin of this alcohol must, therefore, be more profound, and, recalling the before-mentioned fact that life alone is capable of creating full-fledged new dissymmetries, and thinking that his objection would no longer have a *raison d'être*, if between sugar and the amyl alcohol a living organism were interposed, Pasteur found himself led quite naturally to think of fermentation as a vital act.[29]

As Duclaux thus suggests, Pasteur drew a link between crystalline asymmetry, optical activity, and a "biological" conception of fermentation at some fairly early point in his career. The question of exactly when he drew these connections has given rise to a sometimes passionate debate—surely because the answer carries deep epistemic implications about the research of one of the indisputably great scientists of all time. At stake, crudely put, is this issue: was Pasteur committed a priori to a link between optical activity and life, or did he reach that conclusion a posteriori, as a result of his own empirical research. My own position, based on Pasteur's testimony, published and unpublished, is in keeping with the a priorists. In other words, I believe that Pasteur began his empirical research already expecting to find a correlation between crystalline asymmetry, optical activity, and life.

That is not to suggest that Pasteur's "a priori" commitments came out of the blue. They were based on solid, if less than definitive, empirical evidence that had already been gathered by other scientists, including notably two of Pasteur's own mentors, Jean-Baptiste Biot and Auguste Laurent. By the time Pasteur began the research for his doctorate at the Ecole Normale, Biot and Laurent had already drawn attention to evidence that many natural substances—for example, camphor, sugars, oil of turpentine, nicotine, and above all tartaric acid itself—displayed optical activity in solution, while no inorganic substance had been found to possess this property when dissolved. More than that, Laurent had pointed out in the early 1840s that certain organic alkalis—for example, nicotine—were optically active in their natural state, but optically inactive in their synthetic or artificially prepared forms.[30] In his typically bold fashion, Pasteur very swiftly extended the more limited claims of Biot and Laurent into a fundamental division of the natural world into optically active and optically inactive substances. And for Pasteur, as we shall see, this distinction could be linked with a still more fundamental division of the world into the realms of the living and dead. For Pasteur, optical activity was associated with life itself, while optical inactivity was associated ultimately with death and decay.

One index of Pasteur's early commitment to the correlation between optical activity and life was his immediate and dramatic response to the work of Victor Dessaignes, who announced in 1850 that he had prepared aspartic acid by heating optically inactive starting materials (maleic and fumaric acids). Since the only known aspartic acid was optically active, Dessaignes's discovery seemed to imply the "artificial" creation of an optically active compound. From Pasteur's reaction to Dessaignes's announcement, it is clear that he was already, by mid-1850, firmly committed to another preconceived idea, namely, that optically active (i.e., "live") substances could not be created by ordinary chemical processes beginning from optically inactive (i.e.,"dead") starting materials. This story offers powerful—indeed, in my view, decisive—evidence against the claim that Pasteur was interested in Dessaignes's work not because of anything to do with the correlation between optical asymmetry and life, but rather because Dessaignes's result raised stoichiometric questions about the number of atoms in the final product as compared to its precursors. That claim, which need not detain us here, has been a central part of the argument of those who wish to see Pasteur as operating in an "objective," a posteriori mode.

In the late summer of 1850, as soon as he learned of Dessaignes's announcement, Pasteur went to Dessaignes's laboratory in Vendôme, where he was generously allowed to collect samples of the new acid in the form of one of the salts from which it could be derived. Pasteur returned to Paris with this precious gift and quickly took it to the laboratory at the Ecole Normale. In his laboratory notebook from this period, in an entry dated 7 8bre 1850, Pasteur reported that he had dissolved five grams of very white crystals of the aspartic acid salt ("given me by M. Dessaignes") in eighty-five grams of dilute nitric acid. The salt dissolved easily, and Pasteur then put the solution into a 20-centimeter tube of his "apparatus Soleil," which was a sort of "perfected" polarimeter designed to measure optical activity on cloudy days. Although the notebook entry was laconic as usual, I suspect that Pasteur was delighted to record that the solution of Dessaignes's salt displayed "no detectable optical activity [*pas de pouvoir rotatoire sensible*]."[31] On 11 November 1850 he "repeated" the experiment, with such "minor" differences as using a different instrument to measure the optical activity and with even more decisive and satisfying results: no trace of optical activity ("*pas trace de deviation*").[32]

But of course the possibility remained that this optically inactive aspartic acid might be "racemic" in character—that is, that its optical inactivity resulted from a compensation between left-handed and right-handed forms. In a memoir of 1852, Pasteur rejected this possibility on the ground that such "racemic" acids could be synthesized only from "racemic" starting materials, while the available evidence suggested that neither the malic nor the fumaric acids with which Dessaignes had begun could possess such a constitution.[33]

Having rejected this explanation for the inactivity of Dessaignes's aspartic and malic acids, Pasteur boldly suggested that they belonged to an entirely new class of compounds (which he designated by the prefix *méso*)—namely, compounds whose original asymmetry had been "untwisted" so that they had become inactive by total absence of any asymmetry, "inactive by nature" rather than "inactive by compensation."[34] We need not pursue this part of the story here, though it does deserve mention that Pasteur's new, *méso* class of compounds allowed him to maintain his faith in the maxim that optical activity could not be produced by ordinary chemical means from inactive starting materials. For if, as Pasteur put it in a famous lecture of 1860, Dessaignes's malic acid had been inactive by compensation between left-handed and right-handed forms, he would have performed the remarkable feat of producing not just one but two optically active substances from inactive starting materials.[35]

Pasteur's commitment to the special link between optical activity and asymmetry emerged in other contexts over the next decade. In 1852, for example, he reported that optically active bases could react with racemic acid in such a way as to favor the crystallization of one only of the left-handed and right-handed forms which together compose the racemic acid.[36] In other words, he pointed to the capacity of optically active substances to make a choice, as it were, between two asymmetric forms in a racemic substance. In December 1857, by which time he was deeply engaged in studies of fermentation, Pasteur argued that living microorganisms could also possess this "discriminatory" capacity. He had found that a microbe (identified in 1860 as *Penicillium glaucum*) that was responsible for the fermentation of ammonium paratartrate selectively metabolized the right-handed component of the paratartrate while leaving the left-handed component intact. In his initial, very brief announcement of this discovery, Pasteur referred only to its practical value as a means of separating left-handed from right-handed components in racemic substances.[37] But in March 1858, he described how he had made his discovery by following the fermentation of ammonium paratartrate with a polarimeter, which showed that the fermenting fluid displayed increasing optical activity to the left until, eventually, the fluid yielded only left-handed ammonium tartrate. Pasteur now emphasized that this discovery, by its connection with the biological process of fermentation, demonstrated for the first time that molecular asymmetry (represented by the microorganism) could intervene to modify "chemical reactions of a physiological sort."[38] By this point, Pasteur's investigations of asymmetry, optical activity, and fermentation were thoroughly intertwined. In 1860, Pasteur produced celebrated works in each domain.

PASTEUR'S LECTURES OF 1860 ON "THE ASYMMETRY OF NATURALLY OCCURRING ORGANIC COMPOUNDS"

For roughly a decade, Pasteur confined his more cosmic speculations about the link between asymmetry and life to his private notebooks and correspondence. His published papers were more circumspect. Not until 1860 did he give anything like full public voice to his conviction that the world could be divided into "live" and "dead" substances. His initial public "confession" came in the form of two famous and frequently reprinted lectures of late January and early February 1860 to the Chemical Society of Paris under the general title, "On the Asymmetry of Naturally Occurring Organic Compounds."[39] That title should alert us to a very special feature of Pasteur's conception of the world. For his division of the world into live and dead substances was by no means the same as our standard chemical distinction between inorganic and organic compounds, according to which only the latter contain carbon.

Like other scientists, Pasteur did believe that the inorganic realm was made up of dead substances. But in the case of organic substances, he drew a sharp and decidedly unusual distinction between "secondary products" and "naturally occurring" organic compounds. In assigning particular organic compounds to one or the other of these two categories, Pasteur paid no attention to their chemical composition. He claimed to rely instead on their role in vital processes. In practice, his criterion was optical activity. Optically active substances were assigned to the category of "naturally occurring" organic compounds, while optically inactive carbonaceous substances were considered "secondary products." Simultaneously, Pasteur insisted that optically active substances were "essential to life."[40]

The fact that racemic substances could be separated into their left- and right-handed components under asymmetric conditions (whether chemical, as in the case of optically active bases, or physiological, as in the case of living microorganisms) led Pasteur to speculate on "the mysterious cause which presides over the asymmetric arrangement of the atoms in natural organic substances":

> Why this asymmetry? Why any particular asymmetry rather than its inverse?. . . Indeed, why right or left [substances]? Why not only nonasymmetric [substances], like those in dead nature. There are evidently causes of this curious behavior of the molecular forces. To indicate them precisely would certainly be very difficult. But I do not believe I am wrong in saying that we know one of their essential characteristics. Is it not necessary and sufficient to admit that at the moment of the elaboration of the immediate principles in the vegetable organism, an asymmetric force is present? For we have just seen that there is only one case in which right-handed molecules differ from left-handed ones, the case in which they are subjected to actions of an asymmetric order. Can these asymmetric actions be connected with cosmic influences? Do they reside in light, electricity, magnetism, heat? Could they be related to the movement of the earth, the electrical currents by which physicists explain the terrestrial poles.[41]

In other words, Pasteur here boldly compared his "asymmetric forces" to physical forces at work in the universe at large. At one point in these lectures, he even suggested that "it seems logical to me to suppose that [artificial or mineral substances] can be made to present an asymmetric arrangement in their atoms, as natural products do."[42] From these two passages, considered in isolation, it might be supposed that Pasteur was here suggesting that asymmetric molecules (and thus life) could be produced artificially under the influence of physical asymmetric forces—that abiogenesis could occur under purely "mechanistic" conditions.

Nonetheless, the dominant thrust of the 1860 lectures was to insist upon the distinction between (asymmetric) "living nature" and (symmetric) "dead nature," and to insist that asymmetric molecules could not be produced from symmetric starting materials by ordinary chemical procedures. Indeed, Pasteur so closely linked asymmetry and "living nature" as to deny that *symmetric organic* substances (including oxalic acid and urea) could be considered "natural" in the same sense as asymmetric substances. For him, symmetric organic substances should be considered "excretions rather than secretions, if I may so express it."[43] In effect, Pasteur *defined* substances as "naturally occurring organic compounds" and thus as "the immediate principles of life" only when those substances were optically active in solution or the fluid state. Optically *inactive* substances were consigned, *by definition*, to the realm of dead nature.

Nor did Pasteur express much conviction at that point that the barrier between living and nonliving would soon (if ever) fall. If it seemed "logical" to suppose that symmetric molecules could become asymmetric, it remained to be discovered how this could be accomplished. If it seemed plausible to ask whether asymmetric forces might be related to physical forces in the universe, it was "not even possible at present to offer the slightest suggestions" as to the answer. If it was "essential" to conclude that "asymmetric forces exist at the moment of the elaboration of natural organic products," it was equally clear that these forces "would be absent or without effect in our laboratory reactions, whether because of the violent action of these phenomena or because of some other unknown circumstance." In the end, the molecular asymmetry of natural organic products remained "perhaps the only well-marked line of demarcation that we can at present draw between the chemistry of dead and living nature."[44] Because Pasteur used his concept of asymmetric force in such a context, and because he stated it so tentatively and elusively, others regarded it as little more than another "vitalistic" attempt to erect a barrier between the living (or asymmetric) and the nonliving (or symmetric). And the perception of Pasteur as a "vitalist" was only reinforced by his biological or "germ" theory of fermentation, which was summed up in another famous memoir of 1860.

PASTEUR'S MEMOIR OF 1860 ON ALCOHOLIC FERMENTATION

As we have seen, Pasteur was initially drawn to the problem of fermentation because optically active amyl alcohol could not be coaxed into conformity with his law of hemihedral correlation. But amyl alcohol was only one among a host of optically active products of fermentation, and it is hardly astonishing that he would take up the problem in general—not only because of what he later called the "inflexible logic" of his successive research programs, and not only because fermentation was a subject of immense economic significance. There was yet a third enticement, this one of a sort that was bound to appeal to Pasteur's scientific ambitions and to the thrill he got from the challenge of engaging in scientific combat. For the fact is that fermentation had become a leading arena of theoretical dispute between biological and chemical conceptions of the process—and thus, more grandly, between "vitalistic" and "mechanistic" approaches to the explanation of scientific problems.

Thanks in no small part to Pasteur's own self-serving "histories" of the problem, the consensus has been that the chemical theory of fermentation thoroughly dominated the field until Pasteur arrived on the scene. He, of course, came to the arena already predisposed toward the biological theory of fermentation because its products are so often optically active and thus, for Pasteur, linked with life, with the activity of living microbes. In his pursuit of this point of view, he challenged some of the leading chemists of the day, notably Justus von Liebig and Jacob Berzelius, but he was by no means alone. His basic position was neither obscure nor novel. Since at least 1837, several distinguished observers—including notably Charles Cagniard de Latour and Theodore Schwann of cell theory fame—had insisted that alcoholic fermentation depended on the vital activity of brewer's yeast.[45]

This view had been challenged, even ridiculed, by Liebig and Berzelius, who insisted that the process was chemical rather than vital or biological. Their position drew its most impressive support from indisputably chemical processes that were widely considered to be analogous to fermentation—most notably the action of the soluble digestive "ferments" (for us, enzymes) diastase and pepsin. But the alternative, biological theory was also based on a range of highly suggestive evidence that must have given Pasteur enormous comfort when he launched his campaign against the chemical theory, and especially Liebig, its most famous and leading representive.

As we have seen, Pasteur's intial salvo on behalf of the biological theory of fermentation came in the form of his Lille paper of August 1857 on lactic fermentation. That effort attracted considerable attention, but the central arena of dispute was always alcoholic fermentation, the prototypical and most significant example of the phenomenon. Pasteur did not tarry long before marching directly onto this field. In December 1857, within months of his paper on lactic fermentation, Pasteur published the first in a series of abstracts, notes, and letters on alcoholic fermentation that culminated in a long and classic memoir of 1860. Divided into two major sections, dealing respectively with the fate of sugar and of yeast in alcoholic fermentation, it inflicted on the chemical theory what Emile Duclaux rightly called "a series of blows straight from the shoulder, delivered with agility and assurance."[46]

Among other things, Pasteur established that alcoholic fermentation invariably produces not only carbonic acid and ethyl alcohol, as was well known, but also appreciable quantities of glycerin and succinic acid as well as trace amounts of cellulose, "fatty matters," and "indeterminate products." On the basis of these results, Pasteur emphasized the complexity of alcoholic fermentation and attacked the tendency of chemists since Lavoisier to depict it as the simple conversion of sugar into carbonic acid and alcohol. If the alleged simplicity of the process had once been seen as evidence of its strictly chemical nature, then its actual complexity, which he had now established, ought to be seen as evidence of its dependence on the activity of a living organism. The full complexity of alcholic fermentation, he insisted, was such as to prevent the writing of a complete equation for it, since chemistry was "too little advanced to hope to put into a rigorous equation a chemical act correlative with a vital phenomenon."[47]

An even more impressive line of attack against the chemical theory derived from Pasteur's ability to produce yeast and alcoholic fermentation in a medium free of organic nitrogen. To a pure solution of cane sugar he added only an ammonium salt and the minerals obtained by incineration of yeast, then sprinkled in a trace of pure brewer's yeast. Although the experiment was difficult and not always successful, this method could produce an alcoholic fermentation accompanied by growth and reproduction in the yeast with the evolution of all the usual products. If any one constituent of this medium were eliminated, no alcoholic fermentation took place. Obviously, argued Pasteur, the yeast must grow and develop in this mineral medium by assimilating its nitrogen from the ammonium salt, its mineral constituents from the yeast ash, and its carbon from the sugar. In fact, it was precisely the capacity of yeast to assimilate combined carbon from sugar that explained how it could decompose sugar into carbonic acid and alcohol. Most important, there was in this medium none of the "unstable organic matter" required by Liebig's theory.[48]

Pasteur's 1860 memoir on alcoholic fermentation marked a watershed in the debate over biological vs. chemical explanations of the phenomenon. It did not, of course, end the debate once and forever, though it is easy enough to declare Pasteur the victor in his long and rancorous dispute with Liebig, which clearly drew part of its heat from nationalism. We need not pursue the later history of the debate any further here, except to say that Pasteur's commitment to the biological theory of fermentation was so pronounced that he even produced a memoir, unpublished during his lifetime, in which he tried (unsuccessfully) to find a microbe responsible for the relatively modest chemical process known as the inversion of sugar. In fact, Pasteur's biological theory of fermentation amounted to a virtual tautology, since he limited the range of "fermentations properly so called" precisely to those in which he could establish the role of a living microbe. As in the case of his definition of "naturally occurring organic products," properly so called, Pasteur again limited and *defined* the pertinent cases in a way guaranteed to favor his biological theory.[49]

Eventually, of course, the debate over the theory of fermentation was resolved in a way that allowed a sort of *via media* between the chemical and biological theories. But this resolution had not yet begun when Pasteur died in 1895, the first step being Buchner's discovery of 1897 that yeast could be made to yield a cell-free, "dead" alcoholic enzyme (zymase) that produced fermentation independently of the yeast from which it had been isolated.[50]

In the end, Pasteur's victory in the fermentation debate depended on his skillful use of "semantic stratagems."[51] But we should also recognize that his definition of fermentation applied to the most traditional and familiar examples of the phenomenon. Even if Pasteur's theory was, strictly speaking, a virtual tautology, not all circles are vicious. Pasteur's work on fermentation was immensely fruitful, both scientifically and practically. And not the least of its fruits was that it led Pasteur into the study of spontaneous generation, in yet another apparent confirmation of that "inflexible [internal] logic" on which Pasteur insisted so strongly.

In fact, Pasteur traced his interest in spontaneous generation directly to his work on fermentation, and more specifically to his recognition that the ferments were living organisms. As he put it at the end of the "historical" introduction to his prize-winning memoir of 1861 on "The Organized Corpuscles that Exist in the Atmosphere . . .":

> Then I said to myself, one of two things must be true. The true ferments being living organisms, if they are produced by the contact of albuminous materials with oxygen alone, considered merely as oxygen, then they are spontaneously generated. But if these living ferments are not of spontaneous origin, then it is not just the oxygen as such that intervenes in their production—the gas acts as

> a stimulant to a germ carried with it or already existing in the nitrogenous or fermentable materials. At this point, to which my study of fermentation had brought me, I was thus obliged to form an opinion on the question of spontaneous generation. I thought I might find here a powerful support for my ideas on those fermentations which are properly called fermentations.[52]

As this passage suggests, it is perhaps artificial to separate Pasteur's study of fermentation from his work on spontaneous generation. And certainly Pasteur entered into his next great arena of debate armed with the cluster of preconceived ideas that had guided his research almost from the outset. In the case of spontaneous generation, as we shall see, he was also armed with a rather different set of "prejudices," including his philosophical, religious, and political views.

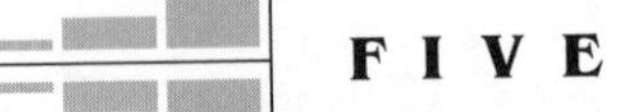

FIVE

Creating Life in Nineteenth-Century France: Science, Politics, and Religion in the Pasteur-Pouchet Debate over Spontaneous Generation

ON THE EVENING of 7 April 1864, Pasteur took the stage at the large amphitheater of the Sorbonne to give a wide-ranging public lecture on spontaneous generation and its religio-philosophical implications.[1] It was the second in a glittering new series of "scientific soirées" at the Sorbonne, and *tout Paris* was there, including the writers Alexandre Dumas and George Sand, the minister of public instruction Victor Duruy, and Princess Mathilde Bonaparte. They were expecting a grand performance; Pasteur did not disappoint them. He opened the lecture with a list of the great problems then agitating and dominating all minds: "The unity or multiplicity of human races; the creation of man several thousand years or several thousand centuries ago; the fixity of species or the slow and progressive transformation of one species into another; the reputed eternity of matter . . . ; and the notion of a useless God (*Deux inutile*)." In addition to these questions—indeed transcending them all, since it impinged on the others and since it alone could be subjected to experimental inquiry—was the question of spontaneous generation: "Can matter organize itself? In other words, can organisms come into the world without parents, without ancestors? That's the question to be resolved."

After a brief historical sketch of the controversy—in which his aim was to show that the doctrine of spontaneous generation "has followed the developmental pattern of all false ideas"—Pasteur struck to the heart of the matter:

> Very animated controversies arose between scientists, then [in the late eighteenth century] as now—controversies the more lively and passionate because

> they have their counterpart in public opinion, divided always, as you know, between two great intellectual currents, as old as the world, which in our day are called *materialism* and *spiritualism*. What a triumph, gentlemen, it would be for materialism if it could affirm that it rests on the established fact of matter organizing itself, taking on life of itself; matter which already has in it all known forces! . . . Ah! If we could add to it this other force which is called life . . . what would be more natural than to deify such matter? What good then would it be to resort to the idea of a primordial creation, before which mystery it is necessary to bow? Of what use then would be the idea of a Creator-God? . . .
>
> Thus, gentlemen, admit the doctrine of spontaneous generation, and the history of creation and the origin of the organic world is no more complicated than this. Take a drop of sea water . . . that contains some nitrogenous material, some sea mucus, some "fertile jelly" as it is called, and in the midst of this inanimate matter, the first beings of creation take birth spontaneously, then little by little are transformed and climb from rung to rung—for example, to insects in 10,000 years and no doubt to monkeys and man at the end of 100,000 years.
>
> Do you now understand the link that exists between the question of spontaneous generations and those great problems I listed at the outset?

But if Pasteur thus explicitly recognized the religio-philosophical implications of the spontaneous generation controversy, he hastened to deny that his scientific work had been motivated or influenced by such concerns:

> But, gentlemen, in such a question, enough of poetry . . . , enough of fantasy and instinctive solutions. It is time that science, the true method, reclaims its rights and exercises them. Neither religion, nor philosophy, nor atheism, nor materialism, nor spiritualism has any place here. I may even add: as a scientist, I don't much care. It is a question of fact. I have approached it without preconceived idea, equally ready to declare—if experiment had imposed the view on me—that spontaneous generations exist as I am now persuaded that those who affirm them have a blindfold over their eyes.

After this grandiloquent introduction, Pasteur turned to the humble facts of the matter. Specifically, he reviewed what he took to be the most important experiments on both sides of the controversy. His survey contained little or nothing that was new to informed scientists in the audience. To put ourselves in their shoes, we need to break away from Pasteur's lecture at this point and describe in some detail the experiments that he was about to summarize. In effect, Pasteur was about to reprise and, he hoped, to conclude his debate over spontaneous generation with Félix Pouchet, which had much agitated the scientific and popular press since 1859. Because Pouchet is now almost forgotten, we need first to learn a bit about him and his place in the spontaneous generation debate.

FÉLIX-ARCHIMÈDE POUCHET (1800–1872) AND SPONTANEOUS GENERATION

Pouchet was a respected naturalist from Rouen, directer of the Museum d'Histoire Naturelle in that city and a corresponding member of the Académie des sciences in Paris.[2] When the debate with Pasteur began, Pouchet was nearly sixty years old, a full generation older than Pasteur, who was then thirty-seven. Pasteur had only recently turned to the study of biological problems, before which his training and expertise lay in the fields of crystallography and chemistry. Pouchet, by contrast, entered the debate after a long career in traditional biology, with a special interest in embryology and reproductive biology, about which he had published two books in the 1840s. He was best known for his theory of "spontaneous ovulation," which challenged the once widely accepted belief that the formation of eggs in the ovary was preceded by and dependent upon fertilization with sperm. Pouchet showed that ovulation occurred in female animals "spontaneously," that is, independently of any contact with male sperm. Among the "fundamental laws of Physiology," he insisted, was that "in all the animal kingdom, generation occurs by means of eggs that preexist at fertilization."[3] In 1845, for his work on "spontaneous" ovulation, Pouchet was awarded the prestigious Montyon prize in physiology by the Académie des sciences, which named him a corresponding member in 1849. In 1853, Pouchet published a massive book on Albertus Magnus and the history of science in the Middle Ages.[4] Then, in 1859, he brought out his long and immediately notorious *Hétérogenie, ou traité de la génération spontanée* (*Heterogenesis: A Treatise on Spontaneous Generation*), in which he presented all the evidence he could marshall—whether embryological, experimental, philosophical, or theological—in favor of spontaneous generation.[5]

One striking feature of *Hétérogenie* was Pouchet's insistence that his version of spontaneous generation had nothing in common with any atheistic and dangerous versions of the doctrine from the past. Indeed, *Hétérogenie* began with a 137-page historical and metaphysical justification for a belief in spontaneous generation, and Pouchet emphasized throughout that his version of the doctrine was in complete accord with orthodox biological, geological, and religious beliefs. Heterogenesis, he argued, was not the "chance" doctrine of the ancient atomists. According to his theory, new organisms arose from the effects of a mysterious and unknowable "plastic force" that could be found in plant and animal debris as well as in living organisms themselves. But, Pouchet argued—and this is the most distinctive feature of his version of spontaneous generation—it is not adult organisms that are thereby spontaneously generated, but rather their eggs: "Spontaneous generation does not produce an adult being. It proceeds in the same manner as sexual generation, which, as we will show, is initially a completely spontaneous act by which the plastic force brings together in a special organ [the egg] the primitive elements of the organism."[6]

In the second chapter of *Hétérogenie*, Pouchet sought to reconcile his version of spontaneous generation with traditional notions of the Creator-God. He agreed, for example, that the first appearance of life was "a true spontaneous generation operating under divine inspiration," but he saw no reason to deny that other spontaneous generations had occurred since that first moment. To deny the existence of subsequent spontaneous generations was to yield to an "illegitimate fear, for if the phenomenon exists, it is because God has wished to use it in his design." Indeed, "the laws of heterogenesis, far from weakening the attributes of the Creator, can only augment the Divine Majesty."[7] In keeping with his vitalistic conception of spontaneous generation, Pouchet denied the abiogenetic production of life. For him, only "organic molecules" and not inorganic matter was endowed with this mysterious *force plastique*.[8] In the end, however, Pouchet's efforts to reconcile his version of spontaneous generation with traditional beliefs fell on deaf ears in the religious and political climate of the Second Empire. Like most French scientists, Pasteur paid no public attention to Pouchet's Christian "apologetics" and focused instead on experimental evidence in the debate that soon flared between them.

EXPERIMENTAL ISSUES IN THE PASTEUR-POUCHET DEBATE

The Pasteur-Pouchet debate began in private and quite politely. Pouchet's experimental efforts to demonstrate spontaneous generation appeared just as Pasteur was reaching the conviction that fermentation depended on the "germs" of living organisms that could not arise spontaneously. In 1858 Pouchet sent the Académie des sciences a brief but widely noticed paper that claimed to offer experimental proof of spontaneous generation.[9] This paper described the appearance of microorganisms in boiled hay infusions under mercury after exposure to artificially produced air or oxygen. In February 1859, in a note on lactic acid fermentation, Pasteur asserted that the "lactic yeast" in his experiments always came "uniquely by way of the atmospheric air." On this point, he wrote, "the question of spontaneous generation has made an advance."[10] This note prompted a letter from Pouchet, which apparently has not survived, but Pasteur's reply has: "The experiments I have made on this subject," he began, "are too few and, I am obliged to say, too inconsistent in results . . . for me to have an opinion worth communicating to you." Nevertheless, he repeated the conclusion he had just announced in his published note and advised Pouchet that if he repeated his experiments with the proper precautions, he would see that "in your recent experiments you have unwittingly introduced [contaminated] common air, so that the conclusions to which you have come are not founded on facts of irreproachable exactitude." Thus, wrote Pasteur, "I think . . . you are mistaken—not for believing in spontaneous generation, for it is difficult in such a question not to have a preconceived idea—but rather for affirming its existence." He concluded by apologizing for "taking the liberty of telling you what I think on so delicate a subject which has taken only an incidental and very small part in the direction of my studies."[11]

Within a year, however, the question of spontaneous generation had taken a central—indeed dominant—place in Pasteur's research. Beginning in February 1860, Pasteur presented the Académie des sciences with a series of five papers on the topic, the results of which were eventually brought together in his prize-winning essay, *Mémoire sur les corpuscles organisés qui existent dans l'atmosphère*, published in the *Annales des sciences naturelles* in 1861. Recognizing that the existence of atmospheric germs was not yet demonstrated, Pasteur set out to show that ordinary air did contain living organisms and to deny that "there exists in the air a more or less mysterious principle, gas, fluid, ozone, etc., having the property of arousing life in infusions."[12]

In the first and most important of these five papers to the Académie des sciences,[13] Pasteur began by examining the solid particles of the air, which he collected by aspirating atmospheric air through a tube plugged with guncotton. When this guncotton was dissolved in a sedimentation tube containing an alcohol-ether mixture, the solid particles trapped by it settled at the bottom. Although this method killed any germs or microorganisms in the trapped particles, microscopic examination always revealed a variable number of corpuscles, the form and structure of which closely resembled those of living organisms. But were these "organized corpuscles" in fact the "fecund germs" of the microorganisms that appeared in alterable materials exposed to the air? In search of an answer, Pasteur deployed three methods. With the first, involving the use of a pneumatic trough filled with mercury—*Pouchet's own method, it should be emphasized*—he obtained inconsistent results and soon abandoned it in favor of a second method, which he

1. Pasteur's mother. From a portrait drawn in pastel by Pasteur at the age of thirteen. (Musée Pasteur, Paris)

2. Pasteur's father. From a portrait drawn in pastel by Pasteur at the age of fifteen.(Musée Pasteur, Paris)

3. Pasteur's birthplace in Dole. (Musée Pasteur, Paris)

4. Pasteur in 1846, while a student at the Ecole Normale Supérieure. From a drawing by Lebayle, based on a daguerreotype. (Musée Pasteur, Paris)

5. Pasteur in 1857, while dean of the Faculté des sciences at Lille. (Musée Pasteur, Paris)

6. Pasteur's first laboratory at the Ecole Normale Supérieure. (Musée Pasteur, Paris)

7. Pasteur and Madame Pasteur in 1884. (Musée Pasteur, Paris)

8. Emile Roux. (Musée Pasteur, Paris)

9. Pasteur in 1884. (Musée Pasteur, Paris)

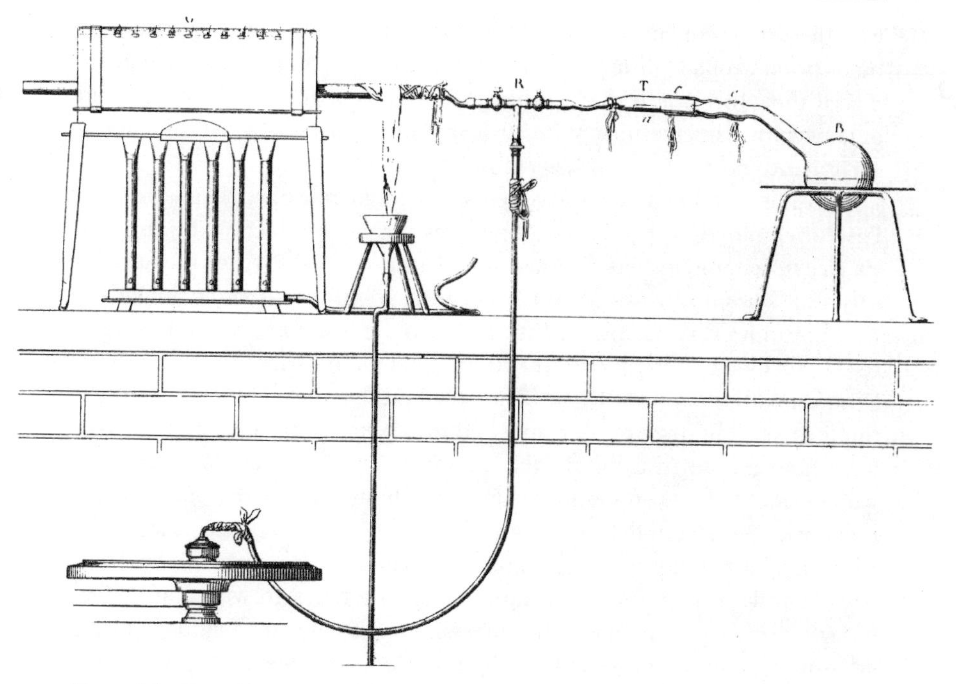

Figure 5.1. Experiment against spontaneous generation: Pasteur's apparatus for collecting solid particles from atmospheric air and then introducing them into a previously sterile flask. (From Pasteur, *Oeuvres*, II, p. 239)

characterized as "unassailable and decisive." In a flask of 300 cubic centimeters, he placed 100 to 150 cubic centimeters of sugared yeast-water, which he boiled for a few minutes. After the flask had cooled, he filled it with calcined air (by means of a neck connected to a red-hot platinum tube) and then sealed it in a flame. The liquid in such a flask, deposited in a stove at 28°–32° C, could remain there indefinitely without alteration.

Thus far, Pasteur had basically repeated the well known earlier experiments of Schwann and others on boiled yeast infusions. But he now introduced an important modification. After a month to six weeks he removed the flask from the stove and connected it to an elaborate apparatus so arranged that a small wad of guncotton previously charged with atmospheric dust could be made to slide into the hitherto sterile liquid in the flask (see fig. 5.1). In twenty-four to thirty-six hours, the once limpid fluid swarmed with familiar microorganisms. Thus, Pasteur concluded, the dust of the air, sown in an otherwise sterile medium, could produce organisms of the same sort and in the same period of time as would have appeared if the liquid had simply been freely exposed to ordinary air. Finally, to counter the objection that these microorganisms arose not from germs in the atmospheric dust but "spontaneously" from the organic matter in the guncotton, Pasteur replaced the guncotton with dust-charged asbestos, a mineral substance, and obtained the same results. With dust-free or precalcined asbestos, on the other hand, no microorganisms appeared in the flask.

To confirm and extend these conclusions on the role of atmospheric dust, Pasteur employed a third method, perhaps the most influential by virtue of its elegant simplicity: his famous "swan-necked" flasks. After preparing a series of flasks in the same manner as in the second method, he drew their necks out into very narrow extensions, curved in various ways and exposed to the air by an opening one to two millimeters in diameter (see fig. 5.2). Without sealing these flasks, he boiled the liquid in most of them for several minutes, leaving three or four unboiled to serve as controls. If all the flasks were then placed in calm air, the unboiled liquids became covered with various molds in twenty-four to thirty-six hours, while the boiled flasks remained unaltered indefinitely despite their exposure to the same air. More than that, Pasteur continued, if the curved necks were snapped off the swan-necked flasks and dipped upright into them, vegetative growths appeared in a day or two. He concluded that the "sinuosities and inclinations" of his swan-necked flasks protected the liquids from growths by capturing the dusts that entered with the air. In fact, he insisted, nothing in the air—whether oxygen or other gases, fluids, electricity, magnetism, ozone, or some unknown or occult agent—was required for microbial life except the "germs" carried by atmospheric dusts.[14]

According to Emile Duclaux, Pasteur's disciple and biographer, the swan-necked flask method was suggested to him by Professor Jérôme Balard; and Pasteur openly admitted that Michel Eugène Chevreul had already done "similar experiments" in his chemistry lectures.[15] Thus here, as in his experiments with calcined air, Pasteur borrowed importantly from the techniques of his predecessors, but he developed and exploited them with much greater effect. By the confident forcefulness of its conclusions and the variety and ingenuity of its experimental techniques, this paper of 6 February 1860 propelled Pasteur into the forefront of the opponents of spontaneous generation. All of his subsequent work on the topic can be seen as an extension, elaboration, and defense of the principles and methods set forth here. By May 1860 he had extended his conclusions to media other than albuminous sugar water—namely, to urine and milk, two highly alterable fluids that could nonetheless be kept sterile by using his techniques, although milk had to be heated above the boiling point, to 110° or 112° C to prevent the appearance of the microorganisms found in spoiled milk.[16]

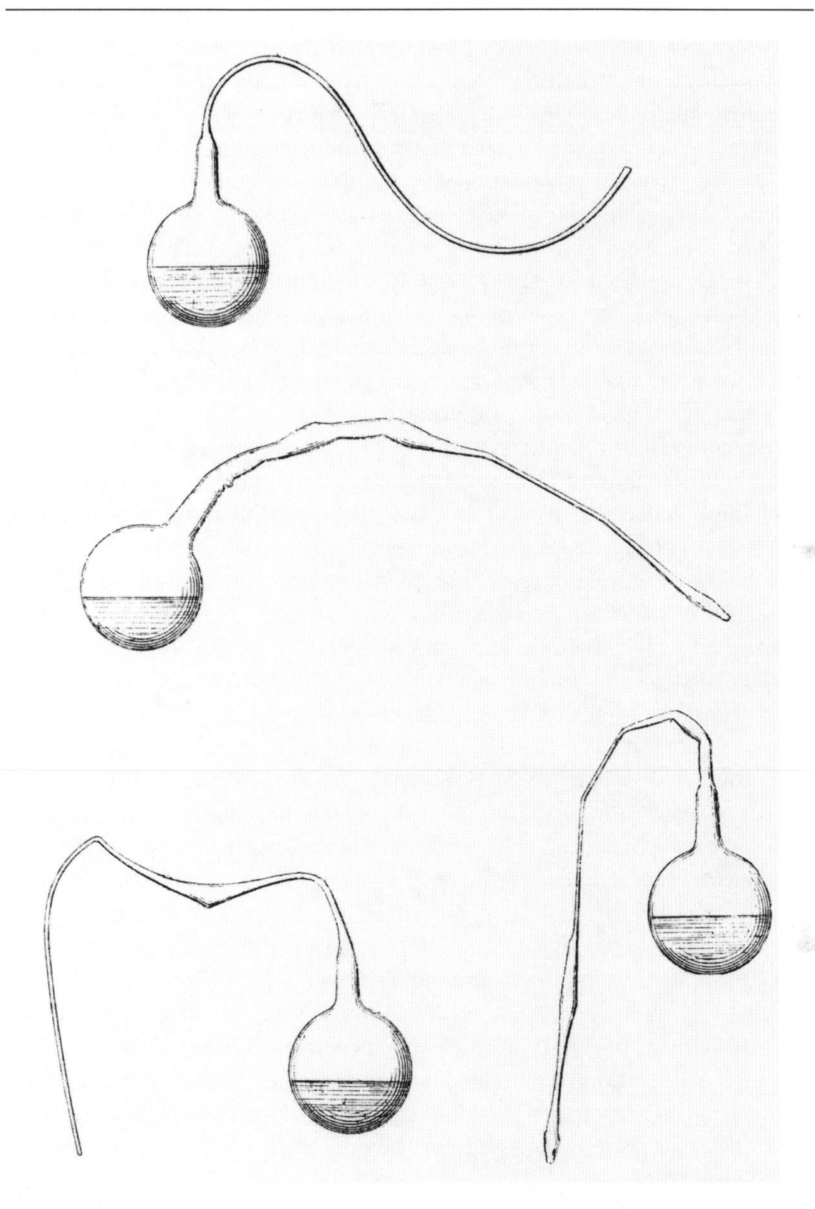

Figure 5.2. The "swan-necked" flasks used in Pasteur's most elegant experiments against spontaneous generation. (From Pasteur, *Oeuvres*, II, p. 260)

In September and November 1860, Pasteur described another famous set of experiments in which he exposed alterable liquids to the natural atmosphere of different locales and altitudes, with the aim of discrediting the belief that any quantity of air, however minute, sufficed for the production of organized growths in any kind of infusion. For scientists of that time, this belief was based mainly on Joseph Louis Gay-Lussac's analysis of Appert's "canned preserves" and his experiments with grapes crushed under mercury—both of which led Gay-Lussac to conclude that fermentation or putrefaction could be set in motion by the presence of even minuscule amounts of oxygen. Using this conclusion as their point of departure, the partisans of spontaneous generation had elaborated a seemingly impressive argument against the notion of airborne germs. For if, as Pasteur claimed, each decomposition resulted from a specific germ carried in the air, and if even the tiniest amount of atmospheric oxygen invariably sufficed to induce all of these varied decompositions, then the atmosphere must be so thick with a variety of germs as to appear foggy, if not as dense as iron.[17]

Pasteur's experimental response to this argument was ingenious and wonderfully theatrical. Basically, he boiled sugared yeast-water in sealed flasks, then broke their necks to admit the surrounding air, immediately resealed the flasks in a flame, and then stored them in a stove at a temperature favorable to the development of microorganisms. The percentage of such flasks that fermented or putrified depended on the locale and altitude at which they had been exposed to the surrounding air. In the vaults of the Paris Observatory, for example, only a few of the flasks became cloudy with microbes. In the air that surrounded Pasteur's laboratory at the Ecole Normale on the busy rue d'Ulm, many of the flasks supported vegetative growths. But Pasteur, with his typical flair, was not content to remain in the basement or at ground level. He launched elaborate expeditions to the foothills of the Jura Mountains, where eight of twenty flasks eventually showed vegetative growths, and most spectacularly to the glacier (Mer de Glace) near Montanvert in the Alps, 2,000 meters above sea level, where only one of twenty flasks underwent subsequent alterations. Thus did Pasteur refute the heterogenicists who claimed that the air must everywhere be dense with germs if germs were responsible for the appearance of microorganisms in boiled and sealed flasks.[18]

In May 1861, at a meeting of the Société chimique de Paris, Pasteur gathered together all of this evidence, as well as new experiments on the heat resistance of fungal spores, in a lecture that was later expanded into his prize-winning *Mémoire sur les corpuscles organisés qui existent dans l'atmosphère*. Although this memoir is mainly a summary of his earlier papers on the topic, it is richer in detail and contains some new material, including a "historical" introduction that has had a powerful effect on standard histories of the spontaneous generation debate ever since.[19] Pasteur also described more fully his microscopic observations of atmospheric dust and the specific organisms found in different infusions, and he gave new attention to the role of alkalinity in the heat resistance of microorganisms; in effect, he claimed that acidic infusions could be sterilized by boiling at 100°C, whereas alkaline infusions or fluids (notably milk) required higher temperatures or a more prolonged period of boiling in order to be sterilized. Last but far from least, he emphasized the role of contaminated mercury as a source of error in the experiments of Pouchet and others. In his first major paper of February 1860, Pasteur had barely touched on this issue. In September 1860, he made his concerns more explicit. Pouchet's experimental case for spontaneous generation rested mainly on his ability to produce microbial life by adding germ-free air to boiled hay infusions in a mercury trough. Pasteur conceded that Pouchet's precautions seemed to eliminate every possible source of contamination except one—the mercury. But, Pasteur insisted, ordinary laboratory mercury is often contaminated with germs. As evidence he cited the following comparative experiments: If a globule of ordinary mercury is dropped into an alterable liquid in an atmosphere of calcined (and hence germ-free) air, microbial life appears within two days. But if the mercury, too, is previously calcined, not a single living organism will appear.[20]

At this point, we can circle back to the beginning of our story and return to Pasteur's famous Sorbonne lecture of 7 April 1864. Now that we have placed ourselves in the same position as the well-informed scientists in his audience, we are not surprised to hear him focus on contaminated mercury as *the* cause of Pouchet's alleged cases of spontaneous generations. Recall that Pasteur considered Pouchet's experiments by far the most important evidence yet produced on behalf of spontaneous generation. And so, he asked, what objections can we make to Pouchet's experiments? If his microorganisms are not generated "spontaneously," where do they come from? What can be the source of the germs, of the "contamination"? Can it be that the oxygen contains germs? No, for it has been prepared artificially, under purely chemical conditions. Can it be the water? No, for it has been boiled and any germs it may have harbored would have lost their power to generate life. *Can it be the hay?* No, for it comes from a stove heated to 100° C. But we know that some germs can survive that temperature. Is Pouchet's stove hot enough? No problem, Pouchet answers, and he heats the hay to 200° or 300° C . . . even to the point of carbonization.

Thus far, Pasteur continued, Pouchet's experiment is irreproachable. But he has overlooked one cause of error: the mercury. Let's make the amphitheater dark. A beam of light from the stage will allow you to see that this room is full of dusts, which should not be trivialized, for sometimes they carry the germs of disease and death: typhus, cholera, yellow fever, and many other plagues. As for Pouchet, he has apparently eliminated the dusts by using oxygen gas and artificial air, and destroyed, by boiling, any germs that may be in the water or the hay. But he has *not* removed the dusts at the surface of the mercury, which carries the germs of the atmosphere into the flasks. In fact, in all such experiments, "*it is absolutely necessary to banish the use of the mercury bath* [*il faut absolument proscrire l'emploi de la cuve à mercure*]."[21] That was, of course, tantamount to banishing Pouchet himself from the arena, as Pasteur knew full well.

Having thus disposed of the most important evidence in favor of heterogenesis, Pasteur rehearsed his own famous experiments in which yeast-water was prevented from alteration by denying the access of any atmospheric dusts. By thus depriving the sugary yeast-water of germs from the air, Pasteur said with mounting excitement, "I have removed from it the only thing that it has not been given to man to produce . . . I have removed life, for life is the germ, and the germ is life. Never will the doctrine of spontaneous generation recover from the fatal blow that this simple experiment delivers to it."[22] But there is, Pasteur informed his Sorbonne audience, yet another, more recent lethal blow that deserves your attention. Until 1863 he had relied solely on experiments in which organic infusions had been vigorously heated, leaving him vulnerable to the charge that such high temperatures might destroy any "vegetative force" in the infusion and thereby render it incapable of generating life.[23] But in March 1863 he had finally succeeded in preserving two highly alterable natural liquids—blood and urine—without heating them at all, but merely by collecting them directly and hermetically from the veins or bladder of healthy animals and then exposing them only to germ-free air. Here, then, was another powerful, indeed decisive, proof that "spontaneous generation is a chimera."[24] Even we fully informed scientists are now convinced, and we join the rest of the audience in a standing ovation as Pasteur concludes his lecture by saying he hopes in the future to shed light on "the immense, marvelous, truly moving" role of microorganisms in "the general economy of creation," and to do so before an audience as brilliant as ours.[25]

But what of the rest of Pasteur's audience? Why should Princess Mathilde care about this Pouchet and his contaminated mercury? What was the point of all those technical details? What larger issues are at stake in this "merely" scientific controversy. At this point we take our permanent leave of Pasteur's "scientific soirée" of 7 April 1864, but only to join other privileged guests at Princess Mathilde's famous salon, where we learn at some length of the past and present cultural and political significance of the magnificent lecture we have just heard from M. Pasteur.

THE RELIGIOUS AND POLITICAL BACKGROUND TO THE PASTEUR-POUCHET DEBATE

Although advanced in a variety of more or less sophisticated forms, the doctrine of spontaneous generation rests ultimately on the notion that living organisms can arise independently of any parent, whether from inorganic matter (abiogenesis) or organic debris (heterogenesis). Following an erratic historical career in which it long enjoyed the support both of natural philosophers and of Christian theology, only to be declared heretical by both in later eras, this doctrine reached its zenith of popularity during the first three decades of the nineteenth century, particularly in Germany where the early parasitologists and the *Naturphilosophen* argued forcefully in its favor.[26] In France, too, spontaneous generation received support through the writings of the materialist Georges Cabanis; the transformist Jean-Baptiste de Monet, *chevalier* de Lamarck; and the putative *Naturphilosophen* Etienne Geoffroy Saint-Hilaire and his student Antoine Dugès. But the popularity of spontaneous generation was short-lived in France. There, by its presumed association with the doctrines of materialism and transformism, it became not only scientifically discredited, but also politically, socially, and theologically suspect.

This French tendency to associate spontaneous generation with the transmutation of species derived in large measure from the eventual commitment of Lamarck and Geoffroy to both notions. In the full-fledged version of his theory of transformism, Lamarck insisted that continuous spontaneous generations were necessary in order to replenish the lowest forms on the escalator of life as they moved upward to become more complex organisms. Without such continuous spontaneous generations, he argued, the earth would be devoid of primitive life. Especially after Geoffroy revealed his allegiance to similar ideas, the French tended to associate spontaneous generation with any evolutionary theory.

Beginning about 1802, Georges Cuvier launched a vigorous campaign against the doctrines of Lamarck and Geoffroy, culminating in his celebrated debate with Geoffroy in 1830. Most witnesses to this debate awarded the palm of victory to Cuvier. The scientific evidence that he marshalled against transformism is too well known to need elaboration here, but among its central features were his emphasis on discontinuities in the known fossil record; his widely acclaimed taxonomic scheme, which denied unity of type in favor of four independent *embranchements*; and his concept of the "correlation of parts," which restricted variation within narrow bounds and on the basis of which he produced virtuoso reconstructions of extinct organisms from one or very few surviving fragments.[27]

But Cuvier's attack gained additional force from less directly pertinent sources. As Toby Appel's excellent analysis of the debate shows, the outcome also turned on the question of scientific style, with Geoffroy seen as defender of the broadly philosophical aims of traditional natural history, while Cuvier represented the sober, cautious, "professional" position that science should deal only with strictly limited problems and "positive facts." In spite of this posture, Cuvier did not hesitate to buttress his scientific arguments against Geoffroy with religio-philosophical and political supports forged for him by his influential post in the Académie des sciences and by events in the national arena. With the rise to power of Napoleon Bonaparte, followed by the Restoration under Charles X, Cuvier hastened to associate his opponents and their doctrines with the speculative and supposedly pantheistic *Naturphilosophie* of the German enemy and with the materialism of the late eighteenth-century *philosophes* and *idéologues*, who were considered responsible for much of the chaos and terror of the French Revolution.[28] It scarcely helped Geoffroy or spontaneous generation that Cabanis, a known advocate of the doctrine, had also been a major figure in the educational program initiated by the National Assembly during the Revolution.[29] And it seems to have made no difference that Geoffroy repeatedly and explicitly tried to dissociate himself from *Naturphilosophie*, materialism, and impiety.[30] Whether consciously or not, Cuvier and much of the French public displayed a convenient disregard for the complexity of the relationships among spontaneous generation, transformism, pantheism, *Naturphilosophie*, and materialism. What mattered was the public perception that spontaneous generation somehow belonged with these politically and religiously dangerous doctrines, and ought therefore to receive its full share of blame for the turmoil of the Revolution.

A generation later, when Pasteur launched his famous battle against spontaneous generation, the scientific and political situation bore a striking resemblance to that which had obtained during the Geoffroy-Cuvier debate. In the scientific arena, the similarities reflected in part the continuing influence of Cuvier, dead since 1832. Many French biologists long paid obeisance to his principles and precepts, including his cautious attitude toward theory. Although by 1860 belief in universal providential catastrophes had been replaced by a naturalistic concept of localized mountain erogeny, French geologists remained convinced that little or no continuity could be established between organisms in different geological strata.[31] Unable to explain the sudden appearance of distinct new fossil species, most French biologists and geologists ascribed the phenomenon to Divine Will, to an unknown natural cause, or avoided the question entirely. Any suggestion that these fossil species or the earliest ones known could have arisen spontaneously from non-living substances was considered absurd because of their complexity. In the wake of Cuvier's work, the doctrine of the transmutation of species still seemed ridiculous at worst, or an unprovable philosophical speculation at best.

In the political arena, France had again entered a period of conservatism following the republican experiments of the 1840s.[32] As Cuvier had waged his campaign against transformism and spontaneous generation during the First Empire, so did Pasteur wage his—more strictly against spontaneous generation—during the Second. Napoleon Bonaparte's nephew, Louis Napoleon, had been elected president of the Republic in 1848, thanks in part to the support of the Catholic Church, which effectively controlled the votes of the newly enfranchised French peasants. In 1850 the new president had signed the notorious Falloux Law which allowed religious teaching in the state schools. In December 1852, his power newly legitimized by a second plebiscite, Louis Napoleon declared himself emperor, once again with the general support of the Catholic Church. Thus, from the outset of the Second Empire, religious issues were simultaneously political issues. The forces of church and state united in the face of the common enemy—republicanism and atheism. For opposition to church and state came not only from republican or liberal ranks but also from positivists, materialists, and atheists, all of whom associated themselves with the scientific movement of the nineteenth century. Indeed, for many the new scientific movement became a sort of religion in its own right; the historian, philosopher, and critic Hippolyte Taine looked "forward to the time when it will reign supreme over the whole of thought and over all man's actions."[33]

In response to this liberal undercurrent, the church became increasingly authoritarian and reactionary, culminating in the papal encyclical of Pope Pius IX in 1864, which emphasized the dangers of religious tolerance and of accommodation with the forces of liberalism and republicanism. In Albert Guérard's words, "God, the Pope, property, law and order were all attacked by the same enemies; practically all Catholics became reactionaries and all reactionaries . . . stood as defenders of the Pope and the Church."[34] In 1860 Ernst Faivre remarked with passion that the problem of spontaneous generation "excites at the moment the best minds, for it touches science, philosophy, and religious beliefs." The destruction of spontaneous generation, he concluded, "is capable of lifting us from the consideration of physical laws to that of general truths, which enlighten our reason and confirm our religious beliefs."[35] Even such a Protestant as François Guizot, the historian and politician, joined in the defense of the Catholic Church against the materialist attack, which he regarded as an attack upon the whole of Christian religion. In a book of 1862, he insisted that "under the blows that [the materialists] bring against Christian dogma, the entire religious edifice collapses and the entire social edifice shakes; the Empire, the essence of religion itself, vanishes."[36]

This climate was further exacerbated by the appearance of Clémence Royer's translation of Darwin's *On the Origin of Species* in 1862 and of Ernst Renan's *Vie de Jesus* in 1864. The latter attempted to rewrite the life of Christ on the basis of historical criticism and scientifically verifiable events. The former was even more incendiary since Royer adhered simultaneously to every doctrine the conservative forces loathed: atheism, materialism, and republicanism. Her preface to the *Origin* was an extended diatribe against the Catholic Church, which she described as a "religion spread by an ignorant, domineering and corrupt priesthood" and which she identified as the major cause of all social ills. It is hardly surprising, then, that Darwinian evolution was regarded in France as a politico-theological doctrine allied with the forces that threatened church and state. Nor is it surprising that so many French critics of Darwinian evolution focused on the issue of spontaneous generation. For besides its historical association in France with evolutionary theories, spontaneous generation was perceived as a threat to the belief in a providential Creator.[37]

Against this background, the outcome of the Pasteur-Pouchet debate carried implications of enormous importance to the political culture of the Second Empire, as had the Cuvier-Geoffroy debate for the First Empire and the Restoration that succeeded it. The great British anatomist Richard Owen, who lived through both debates, long ago emphasized their similarity. "The analogy of the discussion between Pasteur and Pouchet, and that between Cuvier and Geoffroy, is curiously close," he wrote in 1868. In part, this analogy rested on the circumstance that "Pasteur, like Cuvier, had the advantage of subserving the prepossessions of the 'party of order' and the needs of theology." More than that, Owen suggested, Pouchet might soon win for his position on the "origin of monads" the sort of vindication that Geoffroy had already won for his position on the origin of species—"a suggestive and instructive fact in the philosophy of mind and the history of progress."[38] That Owen misrepresented Geoffroy's "vindication" and misjudged Pouchet's ultimate fate is of little concern to us here. What does matter is that even a foreigner like Owen could clearly see that, in nineteenth-century France, the debate over spontaneous generation had profound religio-political implications—so much so that it aroused passion even among scientists and even within their official institutions, above all the Académie des sciences in Paris.

THE ACADÉMIE DES SCIENCES AND THE PASTEUR-POUCHET DEBATE

In the highly centralized structure of French science, the outcome of any scientific controversy depended crucially on the reaction of the Académie des sciences. During the nineteenth century, the Académie often responded to controversies by appointing formal commissions to adjudicate between the conflicting parties in order to arrive at a presumably objective decision, which thereby became the quasi-official position of the French scientific community. To no small extent, Pasteur's victory over Pouchet was determined by the response of the two commissions that the Académie appointed in the 1860s to examine the question of spontaneous generation.

The controversy aroused by the appearance of Pouchet's *Hétérogenie* in 1859 almost surely stimulated the Académie to propose a prize of 2,500 francs to be awarded in 1862, "to him who, by well-conducted experiments, throws new light on the question of so-called spontaneous generations." The commission appointed to award the prize consisted initially of Geoffroy Saint-Hilaire, Antoine Serres, Henri Milne-Edwards, Adolphe-Thèodore Brongniart, and Pierre Flourens. But before a judgment could be rendered Geoffroy died and Serres was dropped from the panel. Their places were taken by Claude Bernard and Jacques Coste, thereby producing a panel unanimously unsympathetic to Pouchet from the outset.[39] Milne-Edwards and Bernard had already responded critically to Pouchet's initial experimental paper of 1858; Brongniart and Flourens were disciples of Cuvier; and Coste opposed Pouchet's embryological views on the origin of infusoria in hay infusions. In addition, all of them, with the possible exception of Coste, were Catholics. Nonetheless, according to Georges Pennetier, Pouchet and his two collaborators, Nicolas Joly and Charles Musset, entered the competition, only to withdraw when some members of the commission announced their decision before even examining the entries. Pasteur was awarded the prize on the strength of his 1861 memoir.[40]

But Pouchet and his allies withdrew only temporarily, much to the chagrin of leading members of the Académie des sciences but to the delight of the anti-establishment popular press.[41] In 1863, Pouchet, Joly, and Musset climbed high in the Pyrenees to repeat Pasteur's famous experiments on the glacier at Montanvert in the French Alps—with one crucial difference: their flasks did not contain Pasteur's yeast-water, but rather hay infusions of the sort that Pouchet had already used in his experiments with the mercury trough. They reported that all eight of the flasks they opened in the Pyrenees underwent subsequent alteration, as would be expected if their organic infusions required only oxygen to generate life.[42] In the face of Pasteur's contemptuous response to these experiments, Pouchet and his collaborators issued a challenge that ended in the appointment of a second Académie commission on spontaneous generation in 1864, just two years after the first commission had completed its work by awarding Pasteur its prize.

Pierre Flourens probably spoke for many of his colleagues when he insisted in 1863 that "the experiments of M. Pasteur are decisive."[43] In fact, the "new" five-member Académie commission of 1864 included three holdovers from the 1862 commission—including, incredibly enough, Flourens himself. Milne-Edwards was also back, with no sign that he had changed his mind since 1859, when he asserted, in response to Pouchet's first published experimental claim for spontaneous generation, that "brute matter cannot organize itself in such a way as to form an animal or plant," and that the "life force has been passed on successively through an uninterrupted chain of being since creation."[44] Nor were the two new members of the 1864 commission—Jérôme Balard and Jean-Baptiste Dumas—likely to tip the scale in Pouchet's favor. Balard was not only Pasteur's mentor in chemistry; he had even played a direct role in Pasteur's work against spontaneous generation by suggesting to him the famous swan-necked flask experiments. Dumas was, if anything, even more predisposed toward Pasteur, whose career he had long and actively promoted, not least by introducing him to Emperor Louis Napoleon, who had named Dumas a senator and minister of agriculture.

Faced with this patently biased commission, Pouchet and his collaborators displayed a precipitous loss of nerve, dragging the commission through a long and complicated dispute about the timing and nature of the experiments they would be allowed to present before it. In general, Pouchet and his collaborators sought to expand the scope of the inquiry and of the experimental program, while Pasteur and the commission continued to insist that the inquiry was to be confined to the single original question: *Does the least quantity of air invariably suffice to induce fermentation in fermentable media?*[45] In the end, Pouchet and his collaborators once again withdrew in the belief that they would be denied a fair hearing.[46]

The biased composition of these two commissions and the uncritical acclaim they heaped upon Pasteur's experiments were only part of the "official" position of the French scientific community on spontaneous generation. Concurrently the French scientific elite invested considerable energy in a campaign against Darwinian evolution based precisely on Pasteur's experiments against spontaneous generation. In fact, Flourens—who had succeeded Cuvier as perpetual secretary of the Académie at the latter's own request—published his *Examen du livre de M. Darwin sur l'origine des espèces* in 1864, the very year that the second commission was constituted. The central theme of Flourens's book was that Darwinian evolution depended on the occurrence of spontaneous generations and therefore could no longer be maintained because "spontaneous generation is no more. M. Pasteur has not only illuminated the question, he has resolved it."[47] Other leading French scientists rallied to the cause and did so in terms that left no doubt as to the political and religious danger of evolutionary ideas.[48]

In this politically charged climate, many members of the French scientific elite surely preferred Pasteur over Pouchet on political grounds alone, especially since many who joined the two-pronged attack against Darwinism and spontaneous generation were dubiously qualified to do so. No one seemed to pay any attention to Pouchet's insistence on the orthodoxy of his version of spontaneous generation; like Geoffroy before him in the First Empire, Pouchet found himself associated with the forces of materialism, transformism, and atheism—all heresies that he explicitly repudiated. However decked out, the doctrine of spontaneous generation was simply too dangerous to the established order of things. At the Académie des sciences, not only were the two commissions appointed to adjudicate the dispute clearly biased, but none of the other academicians seemed to notice how superficially the commissioners carried out their charge.

In short, the Académie des sciences, and much of the rest of the French scientific establishment, was predisposed against the very possibility of spontaneous generation. At least one famous Parisian scientific institution, the Muséum d'Histoire Naturelle, actually banned discussion of spontaneous generation by a professorial decree of 1869. At the Muséum, as Anna Diara recently put it, "spontaneous generation was institutionally eliminated."[49] And when George Pouchet, son of Félix Pouchet, vehemently attacked the decree of the professors of the Muséum in an article in *L'Avenir national*, he was deprived of his position there as *aide-naturalist*. True, a few brave or stubborn souls, even at the Académie des sciences and the Académie de médecine, and certainly in the "scientific" press, continued to advocate spontaneous generation. And Pasteur—despite the entreaties of his friends and colleagues that he was wasting his precious time—could not resist returning to the battleground. For the vast majority of French scientists, however, spontaneous generation was a dead issue by about 1870.

Happily for the sake of freedom of speech and of thought, and for the future of bacteriology and medicine, scientists outside of France continued to investigate the possibility of spontaneous generation. Perhaps the most important was H. Charlton Bastian, an English doctor-scientist who published prolifically on the subject. This is not the place to discuss Bastian's complex and curious career, nor even the details of his position on spontaneous generation, which are described at some length in John Farley's admirable general survey of the history of the controversy.[50] For current purposes, it will suffice to point out that in early 1877 Bastian claimed, in opposition to Pasteur, that he could produce microorganisms "spontaneously" in neutral or alkaline urine. In part, Pasteur responded to this challenge simply by pointing to his own earlier experiments on the heat resistance of germs in alkaline infusions. As usual, however, he also took the occasion to impugn his opponent's technical skill, insisting that Bastian, like Pouchet before him, must have unwittingly "contaminated" his flasks with germs. In Pouchet's case, the source of contamination was the mercury trough; in Bastian's case, Pasteur surmised, the source of "error" must be the solution of potash that Bastian added to boiled urine in order to render it neutral or alkaline. In response, Bastian sent a letter to *Nature*, declaring himself "perfectly ready to reproduce before competent witnesses the results of which I have above spoken." Writing also to *Nature*, Pasteur immediately leapt to the challenge, proposing a face-to-face encounter with Bastian "in presence of competent judges," by which he meant yet another (third) commission on spontaneous generation to be appointed by the Académie des sciences:

> I defy Dr. Bastian to obtain . . . the result to which I have referred, with sterile urine, on the sole condition that the solution of potash which he employs be pure, i.e., made with pure water and pure potash, both free from organic matter. If Dr. Bastian wishes to use a solution of impure potash, I freely authorize him to take any . . . on the sole condition that that solution shall be raised to 110° for twenty minutes or 130° for five minutes.[51]

Bastian accepted Pasteur's challenge, but this time it was he, rather than Pasteur, who sought to define the terrain, insisting that the commission limit its inquiry to this one "mere question of fact": "*Whether previously boiled urine, protected from contamination, can or cannot be made to ferment and swarm with certain organisms by the addition of some quantity of liquor potassae which had been heated to 110°C, for twenty minutes at least.*" At the February 1877 meeting of the Académie des sciences a three-member commission was duly appointed, of whom two were once again holdovers from the Pouchet commissions of the 1860s: Dumas and the thoroughly rigid Milne-Edwards. The third member was Joseph Boussingault, a distinguished agricultural chemist. This commission, Bastian complained, included not "a single member who could be considered as representing my views, or even as holding a neutral position between me and my scientific opponents."[52] After a long and confusing exchange of letters between Bastian and Dumas, Bastian went to Paris on 15 July 1877 and met with Dumas and Milne-Edwards, the latter of whom quickly made it plain that he would not participate in the commission if the scope of its inquiry was to be limited to Bastian's single "question of fact." Bastian tried to arrange a compromise, and a meeting of the commission was scheduled for 18 July. There was, it seems, great confusion about the scheduled time of the meeting, as well as the exact scope of the inquiry, and in the end the commissioners and the two disputants never did get together at the same time and place. Bastian returned to London without the commission ever having witnessed an experiment or rendering a judgment.[53]

None of this would be especially interesting or important except for two things: (1) It suggests the extent to which the Académie des sciences continued to insist on controlling the terms of the debate and even its likely outcome by repeatedly appointing biased commissions; and (2) despite the aborted third commission, Emile Duclaux, who was in a position to know, testified that of the many debates over spontaneous generation, it was the discussion with Bastian that bore the most fruit. In his 1896 biography of Pasteur, Duclaux reported that Bastian, despite some flaws in his own experiments and interpretations, nonetheless pushed Pasteur and his collaborators Jules Joubert and Charles Chamberland toward a firmer grasp of the relative distribution of germs in the air, in water, and on solid objects, and—more important—toward a new appreciation for the heat resistance of many microorganisms, one result of which was Chamberland's autoclave for sterilization at high temperatures. Indeed, Duclaux went so far as to say that "all our present [bacteriological] technique has arisen from the objections made by Bastian to the work of Pasteur on spontaneous generations."[54]

PASTEUR, SPONTANEOUS GENERATION, AND THE SCIENTIFIC METHOD

In his famous Sorbonne lecture of 7 April 1864, Pasteur insisted that the question of spontaneous generation was a matter of fact, which he had approached "without preconceived idea." Let us pretend for a while that we believe that statement and that we also believe in the so-called Scientific

Method. If so, we can only be dismayed by some surprising lapses in Pasteur's modus operandi. Thus, at one point in his prize-winning memoir of 1861, Pasteur admitted that his own repeated attempts to prevent the appearance of microbial life in infusions under mercury succeeded only rarely, perhaps less than 10 percent of the time. But rather than draw the seemingly obvious conclusion that this microbial life had originated spontaneously, Pasteur refused to accept this experimental evidence at face value and pressed relentlessly toward an alternative explanation. "I did not publish these experiments," Pasteur wrote, "for the consequences it was necessary to draw from them were too grave for me not to suspect some hidden cause of error in spite of the care I had taken to make them irreproachable."[55] Although Pasteur failed to specify what "grave consequences" he feared, it seems likely that the very possibility of spontaneous generation was chief among them. As a matter of fact, throughout the spontaneous generation controversy, Pasteur *defined* as "unsuccessful" any experiments—including his own—in which life mysteriously appeared and as "successful" any experiments which gave an opposite result. Happily for him, he managed to indict contaminated mercury as the source of the microbial life that appeared in the many "unsuccessful" experiments conducted with the mercury trough.

If this achievement seems to justify Pasteur's approach—if indeed it might even seem in keeping with the precept that the scientist should suspend judgment until "all the facts are in"—no such interpretation can be applied to other aspects of his scientific conduct. Most strikingly, Pasteur failed to repeat Pouchet's disputed experiments in the Pyrenees. The distinctive feature of those experiments was the *absence* of mercury from Pouchet's flasks of boiled hay infusions. In his 1864 Sorbonne lecture, Pasteur entirely ignored this problem, choosing to discuss only Pouchet's early experiment on boiled hay infusions in the mercury trough. Only once did Pasteur attempt directly to challenge Pouchet's experiments in the Pyrenees. In a note of November 1863, he criticized Pouchet and his collaborators for limiting their flasks to so small a number as eight (thereby introducing the possibility that their results were due to mere "chance") and on the quite desperate ground that they had broken their sealed flasks in the Pyrenees with a heated file rather than with a pair of pincers, as Pasteur had done in the Alps.[56] Not even in the benevolent presence of the Académie commissioners did Pasteur repeat or directly refute Pouchet's experiments. Instead, he chose merely to repeat his own secure experiments with *yeast* infusions, in spite of which the commission praised his exactitude in a report that scarcely veiled its contempt for the opposite side.[57]

That Pasteur should thus have violated one of the presumably fundamental precepts of the Scientific Method—namely, to "falsify" his opponents' experiments—is no less remarkable than the failure of any member of the commission to perceive the violation. In the case of spontaneous generation, moreover, this violation was particularly serious since a single uncontroverted experiment in support of the doctrine automatically carried greater weight than any number of experiments against it. Advocates of spontaneous generation did not need to show that they could produce life artificially under a variety of circumstances, nor even that they could do so consistently. They needed only to show that the feat was possible—a situation that Pasteur sometimes turned to his advantage by emphasizing how difficult his task was compared to the heterogenesists and by noting that "in the observational sciences, unlike mathematics, the absolutely rigorous demonstration of a negation [i.e., that spontaneous generation does not exist] is impossible."[58]

And in fact, as Emile Duclaux pointed out a century ago, the Pasteur-Pouchet debate might have ended quite differently had Pasteur carefully repeated Pouchet's experiments, or had Pouchet and his collaborators maintained their nerve in the face of Pasteur's self-assurance and the commission's contempt.[59] Thanks mainly to continued experimental work outside of France, where scientists were relatively isolated from the presumed political dangers of spontaneous generation and from judiciary commissions of the Parisian Académie des sciences, it became clear by the early 1870s that microbial life did in fact often appear in boiled infusions of hay (as well as cheese, among other materials) even in experiments conducted with "irreproachable exactitude" and Pasteur-perfect technique. In 1876, the German botanist Ferdinand Cohn and the English physicist John Tyndall were able to offer an explanation for many such cases of putative spontaneous generations. In separate works, they showed that the life cycle of the hay bacillus (Cohn's *Bacillus subtilis*) included a highly resistant endospore phase which could survive boiling and develop into the usual form of the bacillus upon the introduction of oxygen.[60] For this reason, Pouchet's flasks of boiled hay infusions might well have produced life even in Pasteur's sterile hands upon exposure to the atmosphere, and might therefore have lent crucial support to spontaneous generation during the 1860s.

As noted above, Pasteur himself had argued as early as 1861 that the heat resistance of certain microbes increased in alkaline media[61] (including hay infusions as well as milk), and in his 1864 Sorbonne lecture he briefly raised the possibility that Pouchet's hay infusions might contain some unknown heat-resistant microorganism.[62] But he mentioned this possibility only in passing and seemed fully satisfied that Pouchet's precautions were sufficient to preclude it. For Pouchet's early experiments, this posed no serious problem for Pasteur, since he was able to indict contaminated mercury as the cause of the supposedly spontaneous appearance of microbial life in Pouchet's flasks. But this explanation could not be applied to the mercury-free Pyrenees experiments of Pouchet and his collaborators. Perhaps because he remained satisfied with Pouchet's precautions, Pasteur did not now even mention the possibility that his opponents' hay infusions might have contained some unknown heat-resistant microorganism from the outset. Once again, but now on highly dubious grounds, he preferred instead to accuse Pouchet of having contaminated his flasks through sloppy technique.[63] If Pasteur ever did repeat Pouchet's mercury-free experiments with hay infusions, he kept the results to himself.

None of this is to say that Pouchet was robbed of a victory that rightly belonged to him. For the fact remains that Pasteur was a more ingenious and more skillful experimentalist as well as a more effective rhetorician—all in all, a more persuasive advocate for his point of view. In drawing his analogy between the Cuvier-Geoffroy and Pasteur-Pouchet debates, Richard Owen admitted that the similarities extended beyond the fact that Cuvier and Pasteur "had the advantage of subserving the prepossessions of the 'party of order' and the needs of theology." For Pasteur, like Cuvier, also had "the superiority in fact and argument," and "the justice of . . . awarding to [Pasteur] the palm of superior care and skill both in devising and performing the experiments, and exposing the inferiority of [Pouchet] in polemical ability and coolness of argumentation, cannot be denied."[64]

Now that we have completed our little methodological exercise, we can drop the pretense that we believe, at face value, Pasteur's statement that spontaneous generation was for him a simple matter of fact, which he had approached "without preconceived idea." But we should also stop pretending that we believe in the Scientific Method. That has a liberating effect, for Pasteur as for us. For what sounded like criticisms of Pasteur just paragraphs ago are really criticisms of a simplistic and passé notion of the Scientific Method. It is not Pasteur who has fallen short; it is this Scientific Method. As Bruno Latour has archly suggested, Pasteur was a subtle philosopher of science and a shrewd sociologist of knowledge.[65] He knew how and when to draw on his rhetorical talents and other resources. When lecturing to Princess Mathilde, it was a good move to act as if the question of spontaneous generation were simply a matter of fact. To tell the princess only that spontaneous generation was a dangerous doctrine would have had little effect; it would have been preaching to the converted. But to show her that objective science could prove that spontaneous generation was not only dangerous but wrong—now *that* was to provide her with a new resource and to forge a new alliance.

In his recent discussion of Pasteur's 1864 Sorbonne lecture, Latour rightly emphasizes that this *conférence* was much more than a "talk." It was also, and more importantly, a "demonstration," a sort of scientific *mise en scène* that made dramatic use of light and darkness, beams, shades, and shadows; of instructive sounds like aspiration and inrushing air; of tangible "props" such as projecters, micrographs, cotton balls, swan-necked flasks, bubbling chemicals, metallic tubes, etc.; and, far from least for Latour, a host of microbial "actors" ready to perform for the crowd. In manipulating, orchestrating, and directing all of these elements, instruments, and actors, Pasteur was an awesome prestidigitator, who could produce the desired results "at will" (*à volonté*).[66] A century ago Emile Duclaux wrote eloquently of Pasteur's "mastery" over microbes, of his ability to sow, cultivate, and domesticate these creatures from the world of the infinitely small so that they would, through their effects and acts, become visible and tangible in the ordinary world.[67] That sort of feat is impossible without rigorous techniques, executed with "irreproachable exactitude." But it is not accomplished by the routine application of some mechanical Scientific Method. It is more than that. It is a gift, a talent, a skill, an art—and Pasteur was most decidedly an artist of the invisible world.

PASTEUR, "PRECONCEIVED IDEAS," AND HIS CAMPAIGN AGAINST SPONTANEOUS GENERATION

Even as we justly admire Pasteur as masterful craftsman and tactician, we should not forget that he was also a thinker. And his ideas, especially his "preconceived ideas," were also "actors" for him: they shifted and twisted and changed, and they had effects on him and other human actors. Often enough, in the right contexts, Pasteur was perfectly willing, even proud, to acknowledge that he operated under the sway of preconceived ideas. He explicitly traced his discovery of optical isomers in the tartrates to his preconceived idea of a correlation between molecular asymmetry, crystalline asymmetry, and optical activity. He was equally explicit in tracing his interest in fermentation and his biological or germ theory of the process to a related preconception: the correlation between optical activity and life.[68] It is therefore noteworthy, but in the end not surprising, to hear Pasteur say that he approached the question of spontaneous generation as a matter of fact, "without preconceived idea." We already know that Pasteur's political beliefs would have predisposed him to deny the existence of spontaneous generation. In fact, it was precisely the question of spontaneous generation that most fully engaged his preconceptions, his prejudices, his ideology, his faith; and, for strategic reasons, precisely the question where it was most important for him to deny the effects of any such commitments on his own scientific work.

At heart, which is where he located his "philosophy," Pasteur was a sincere believer in a Creator-God, with no particular doctrinal passion, an exemplar of the nineteenth-century French bourgeoisie and a fervent patriot, Bonapartist, and political conservative, who once ran for the Senate on a pledge "never to enter into any combinations whose goal is to upset the order of things." In short, Pasteur's values and political beliefs conformed precisely to the reigning orthodoxies of the Second Empire, and he knew full well that his campaign against spontaneous generation was a sort of gift to the emperor, who returned the favor in several symbolic and tangible ways. Especially during the 1860s, when evolutionary theory and spontaneous generation were seen as part of a broader threat to the established order of things, Pasteur was clearly eager to destroy the doctrine on political grounds alone.

But there was more to it than that. For Pasteur had the good fortune to be predisposed against spontaneous generation on other grounds as well. His "political" campaign against the doctrine was fortified by a set of "merely scientific" preconceptions that pointed in the same direction. No wonder Pasteur fought so hard and so well. In fact, his set of scientific preconceptions were themselves mutually reinforcing. Much has been written, both for and against, the role of preconceived ideas in Pasteur's research, and the division reflects an ambiguity that can be found in his own explicit statements about his scientific modus operandi. He sometimes spoke of the fertility of *idées préconçues*, but at other times (as in his blistering attack on the posthumous laboratory notes of his colleague Claude Bernard) he drew attention to the "tyranny of systems" and the danger of *idées fixes*. The tone shifted according to the immediate audience and context. Yet from roughly 1860 until his death in 1895, there was a remarkable consistency in his most fundamental ideas about fermentation, spontaneous generation, disease, and life in general. Pasteur had an uncommonly coherent and wide-ranging vision of the natural world.

Some of these links have already been discussed. In Chapter Four, we have seen how a stubborn exception to Pasteur's supposed "law of hemihederal correlation"—that is, amyl alcohol—led directly to his interest in fermentation. And once drawn to the study of fermentation, with its many optically active products, his mind was "prepared" to associate it with life in the form of microorganisms. The link between fermentation and spontaneous generation was twofold. First, in order to maintain his biological theory of fermentation, he had to show that his "germs" came from outside the fermenting material, for if microbes arose "spontaneously" within a medium already undergoing fermentation, it would be easy enough to see them as products rather than as causes of the phenomenon. Second, as Emile Duclaux emphasized, Pasteur's notion of the specificity of each fermentation implies an ordinary sort of generation for them. Only through ordinary generation, it would seem, could they retain the specific hereditary properties that must account for the specificity of their actions during fermentation.[69]

What is perhaps less obvious is that there was one preconceived idea that served as the glue for the rest of the cluster: the sharp distinction Pasteur drew between the world of "natural" and "artificial" substances—between the world of simply symmetrical, optically inactive, "dead" substances, on the one hand, and on the other the world of molecularly asymmetric, optically active substances, "natural organic products properly so-called," that made up "the immediate, essential principles of life." As discussed at some length in Chapter Four, Pasteur first drew full and explicit attention to this division between "living" and "nonliving" compounds in a pair of famous lectures in 1860. But we know from his laboratory notebooks and other evidence that he was already committed to this doctrine by 1851, and there is good reason to believe that he held it almost from the moment he began to study with Jean-Baptiste Biot and Auguste Laurent. Both of these mentors, and especially Biot, instilled in him from the outset the maxim that was the most fundamental and enduring of his preconceived ideas, namely, that *optically active compounds, which are associated with the organic world, cannot be produced artificially from optically inactive starting materials*.

As early as February 1851, Pasteur inscribed the following version of the maxim in his laboratory notebook: "The power to rotate polarized light [i.e., optical activity] has never been found in a compound [*corps*] artificially prepared from other compounds not possessing this power."[70] In August of that year the maxim appears, in very slightly different form, in the extract of Pasteur's memoir on Dessaignes's aspartic and malic acids.[71] In Chapter Four I argued that it was precisely this "preconceived idea" that prompted Pasteur to rush by train to Dessaignes's laboratory in Vendôme to test the optical activity of his artificially produced aspartic acid. What deserves emphasis here is just how firmly and how long Pasteur embraced this idea—in fact he did so to the end of his life. Thanks to the editorial labors of Pasteur's grandson, Pasteur Vallery-Radot, we can find all the evidence conveniently brought together in one place in the collected works.[72]

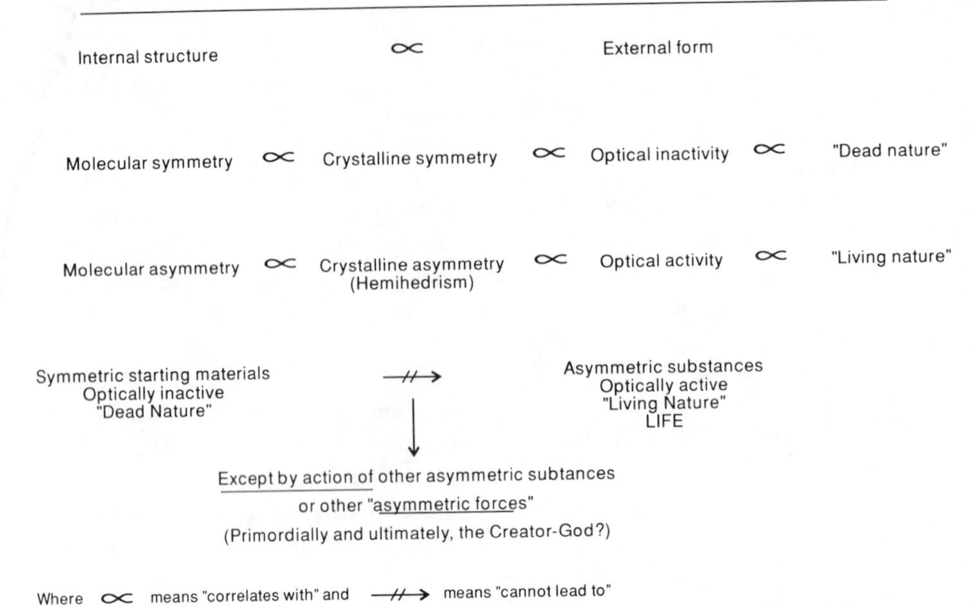

Figure 5.3. Pasteur on the correlations between internal structure, external form, optical activity, and life.

There, arranged chronologically, we find five brief notes (one each from 1861, 1862, 1866, 1873, and 1875) and one extensive lecture (in 1883) in which Pasteur doggedly and cleverly defended his preconceived idea in the face of a series of apparent challenges to it. He repeatedly and explicitly referred to Dessaignes's aspartic acid as the archetypical example of these more recent would-be "exceptions." Thereafter, whenever Pasteur came across an example of the allegedly artificial production of an optically active substance, he raced into print with objections or suggestions for further experiments, few if any of which he carried out himself. Several different compounds were subjected to scrutiny, but repeated attention was given to the alleged production of tartaric acid (ordinarily optically active) by several different methods of "total synthesis" from succinic acid—which is to say, beginning from optically inactive chemical elements. In every case but one, it seems, Pasteur managed to persuade the chemists who had apparently managed this feat that they had been mistaken—whether because their starting materials were not really inactive, or because the compounds produced from inactive materials were not really active.

The exception was Emile Jungfleisch, who in 1873, beginning with indisputably inactive succinic acid, produced a paratartrate that then immediately and "spontaneously" resolved itself into ordinary (right-handed, optically active) tartaric acid and its left-handed isomer without the intervention of any of the "asymmetric influences" that Pasteur claimed were required for such resolutions. In short, Jungfleisch believed that he had created optical activity without the intervention of life, and many of his peers agreed. But Pasteur never flinched. Even assuming that all of the details of Jungfleisch's experiments were correct, it was still the case that "to tranform *an inactive compound into another inactive compound* that has the power of resolving itself simultaneously into a right-[handed] compound and its opposite, is in no way comparable to the possibility of transforming *an inactive compound into a single [simple] active compound.*"[73] Saying that only "nature" could accomplish that feat, he described it as the last barrier between organic and inorganic phenomena. Jungfleisch and others did not agree. But in what was apparently his last published word on the subject, in a lecture of 22 December 1883, Pasteur ended by repeating his claim: "No. Chemistry has never made an active compound from inactive products. A paratartrate is an inactive, non-hemihedral compound. It has no asymmetry. In order to resolve it one must introduce asymmetric actions. Chemistry will remain powerless to make sugar, quinine, [and other immediate principles of life] so long as it continues on the erroneous path [*errements*] of its current procedures, which are exclusive of the use and exercise of asymmetric forces. That's what M. Jungfleisch does not understand."[74]

On the surface, this conclusion would seem to fit perfectly with Pasteur's campaign against spontaneous generation. Optically active compounds, and thus "the immediate principles of life," cannot be created artificially. Is that not a simple and forceful objection to spontaneous generation? In one sense, yes, and it is fascinating to watch Pasteur plunge simultaneously into both arenas, attacking the advocates of spontaneous generation in one forum and then the chemists who had claimed to create optical activity in another. But there is a paradox here that needs to be addressed. For Pasteur did not always and forever deny the possibility that life or at least "the immediate principles of life" might be created artificially—indeed he tried to do so himself, although no one except his inner circle would have had any way of knowing that during the Second Empire. Especially during the 1860s, the most politically sensitive phase in the spontaneous generation debate, Pasteur's public image was that of fearless crusader against the dangerous doctrine.

During those years, surely, outsiders would have been surprised to hear that Pasteur considered the artificial creation of asymmetry and life a theoretical possibility. They would have been astonished to learn that he had actually pursued the problem experimentally in the early 1850s. For thirty years, Pasteur said nothing in public about these remarkable experiments and very little about the "preconceived ideas" that had encouraged him to undertake them in the first place. Not until his lecture of 22 December 1883 did Pasteur publicly disclose his early attempts to "imitate nature" and to "introduce asymmetry into chemical phenomena."[75] He said that he had been silent until now because nothing had come of those early experiments, which rather fails to explain why he decided to talk about them now since nothing more had come of them in the long interval since. I have deliberately ignored the existence of these experiments until now in an effort to keep us in the same state of mind as an average well-informed scientist of the Second Empire, whose only source of information about Pasteur's work was the published record. Such a scientist, it seems fair to say, would have been amazed to learn that Pasteur had managed to wage a vocal public campaign against spontaneous generation even as he speculated about the creation of asymmetry (and thus life) in his own special version of this dangerous doctrine. For René Dubos, it was "a striking fact, perhaps worthy of the attention of psychoanalysts, that Pasteur devoted much of his later life demonstrating that nature operates as if it were impossible to achieve what he—Pasteur—had failed to do."[76]

So there does indeed seem to be a paradox: at one and the same time—indeed at least once in the very same lecture (of 22 December 1883)—Pasteur insisted that all chemists who had claimed they had created optical activity from inactive materials were mistaken, even as he disclosed his own belief that the creation of artificial asymmetry might be possible through the sorts of experiments he had briefly pursued in the early 1850s. But there is a way out of the paradox, and here is the key: for Pasteur, there was a profound difference between the "ordinary chemical procedures" used by chemists in their laboratories and the "asymmetric forces" whose origins Pasteur sought in physical forces at work in the cosmos at large. Unless and until they found a way to bring these physical asymmetric influences to bear in their laboratories, chemists would never be able to create optically active substances from optically inactive starting materials. "That," as Pasteur put it at the end of his lecture of 1883, "is what M. Jungfleisch does not understand."

PRIVATE THOUGHTS AND EXPERIMENTS: ASYMMETRIC FORCES, GOD, AND THE CREATION OF LIFE

On 12 December 1851, in a letter to his best friend and former schoolmate Charles Chappius, Pasteur wrote that he was "on the trail of some mysteries, and the veil that covers them is getting thinner and thinner." He reported that his lectures at the University of Strasbourg, where he was now serving as acting professor of chemistry, took so little of his time that he was able to devote five full days a week to this exciting new research. He was often scolded by his bride for working too hard. But, wrote the youthful Pasteur, still shy of his thirtieth birthday, "I console [her] by telling her that I will lead her to posterity."[77] Two years later, in November 1853, Pasteur's posterity-bound wife informed his father that Louis was well enough, if perhaps "always a little too preoccupied with his experiments." But, she continued, "the experiments he is undertaking this year should give us a Newton or Galileo if they succeed."[78] Just a few weeks later, however, Pasteur himself wrote his father that his experiments were not going well. He continued to hope for the best, but also admitted that "one must be a bit mad to undertake what I've undertaken."[79]

In none of these letters did Pasteur or his wife give any hint as to what sort of experiments had aroused so much hope and excitement. Nor did Pasteur's published papers from this period provide any more information. But from other sources, including Pasteur's unpublished correspondence, manuscripts, and laboratory notebooks, we can very briefly reconstruct the central features of these few and tentative experiments, and recapture some sense of the theoretical concerns that lay behind them. In doing so, we will come to appreciate why young Louis could briefly dream that this research might bring him into the sublime company of Galileo and Newton. For Pasteur, too, was in quest of a new and fundamental force in nature—a "cosmic asymmetric force" that was ultimately responsible for life itself.

From Pasteur's correspondence and laboratory notebooks of the early 1850s, we know that his research was followed almost every step of the way by his old mentor and patron, Biot, though not always with his full blessing. No one knew better that Biot what Pasteur had already accomplished and what a promising future lay ahead of him in the fields of crystallography and molecular chemistry. He was therefore disappointed to hear that his protégé wanted to undertake such a bold and, so Biot thought, unpromising research program. He tried to discourage Pasteur but ultimately relented, even securing a modest research grant that ultimately allowed Pasteur to buy a "Ruhmkorff apparatus," a new instrument "designed to facilitate the exhibition of optical phenomena produced by transparent bodies when they are placed between the opposite poles of a magnet of great power," as Biot himself described it in a report of 1846 to the Académie des sciences.[80]

In fact, the Ruhmkorff apparatus was the main instrument Pasteur used in his asymmetry experiments, as one would expect given his intention to focus his search for the cosmic asymmetric force on electromagnetic phenomena as well as polarized light. The earliest asymmetric experiments recorded in his laboratory notes, dating from July 1853, investigated the effects of polarized light on the crystallization of the tartrates of cinchonine and quinine, with no apparent effects. Beginning in October 1853 Pasteur began to crystallize various salts under the influence of the Ruhmkorff apparatus.[81] By December he was focusing on formiate of strontium, presumably because it had such close analogies in its physical and crystallographic properties to quartz, the most famous of the optically active substances.[82] For a heady few days or weeks in the winter of 1853, Pasteur was beginning to believe that the application of the Ruhmkorff electromagnet was consistently producing asymmetrical forms in formiate of strontium: "The fact indicated is constant . . . I have repeated it many times."[83] But by February 1854 it was clear that these experiments, too, would not fulfill his hopes for them. After his transfer from Strasbourg to Lille in 1854, Pasteur tried to modify the normal character of optically active substances by using a large clockwork mechanism to rotate a plant continuously in alternate directions and by using a reflector-and-heliostat arrangement to reverse the natural movement of solar rays directed on a plant from its moment of germination. These experiments, too, failed to yield any striking results, and Pasteur abandoned his experimental search for the cosmic asymmetric force. There was to be no Newton or Galileo for now.

Even so, Pasteur continued to speculate about asymmetric forces and the origin of life. In manuscript notes he wrote at Arbois in autumn 1870, while Paris was embroiled in the Commune, he jotted down some thoughts about the origin of life and projected a new series of experiments designed to create or modify life by means of magnets and other asymmetric influences—experiments very similar in conception to those he had already tried almost two decades before, though he did plan to focus this time on the application of asymmetric forces to simple inorganic compounds such as sulphur, potassium, copper, hydrogen, oxygen, cholorine, and carbon (in the form of diamond).[84] Pasteur apparently never carried out these projected experiments, and no published results emerged from them.

Beginning in the mid-1870s Pasteur began to develop and articulate his previously tentative notion of asymmetric forces—forces about which he had been publicly silent since his famous but ambiguous pair of 1860 lectures, "On the Asymmetry of Naturally Occurring Organic Compounds." Now in the 1870s he made it clear, as he had not done before, that he considered these asymmetric forces to be within the bounds of experimental inquiry and began to speak of them as the means by which the barrier between asymmetric (living) and symmetric (dead) nature might actually be breached. Such an achievement, he wrote in 1874, "would give admittance to a new world of substances and reactions and probably also of organic transformations." This was the direction from which one should attack "the problem not only of the transformation of species but also of the creation of new ones."[85] In a note of July 1875, he stated this position in essentially identical terms.[86]

In his familiar "confessional" lecture of 22 December 1883, Pasteur reached the oratorical peak of his efforts to describe the asymmetric forces and their relation to life. In order to create life, he said, it is necessary to manufacture some asymmetric forces, to resort to the actions of a solenoid, of magnetism, of the asymmetric movements of light:

> The line of demarcation of which we speak is not a question of pure chemistry or of the obtaining of such or such products. It is a question of forces. Life is dominated by asymmetric forces that present themselves to us in their enveloping and cosmic existence. I would even urge that all living species are primordially, in their structure, in their external form, functions of cosmic asymmetry. Life is the germ and the germ life. Now who can say what the *destiny* of germs would be if one could replace the immediate principles of those germs—albumin, cellulose, etc., etc.—by their inverse asymmetric principles? *The solution would consist in part in the discovery of spontaneous generation, if such is within our power;* on the other hand, in the formation of asymmetric products with the aid of the elements carbon, hydrogen, nitrogen, sulphur, phosphorus, if in their movements these simple bodies may be dominated at the moment of their combination by asymmetric forces. Were I to try some asymmetric combinations from simple bodies, I would make them react under the influence of magnets, solenoids, elliptically polarized light—finally, under the influence of everything which I could imagine to be asymmetric actions.[87]

There is something at once materialist and spiritual about this conception of life as being "dominated by asymmetric forces that present themselves to us in their enveloping and cosmic existence," and according to which "all living species are primordially . . . functions of cosmic asymmetry." In fact, I fully believe, though the evidence is scattered and thin, that Pasteur ultimately saw the Creator-God as the source of the original cosmic asymmetric force. It was God who had set the world of the living on its asymmetric path, an asymmetry that had been handed down from generation to generation under the influence of the sun's light and heat. There was some danger of heresy here, for in trying to capture and deploy aspects of this cosmic asymmetric force, Pasteur ran the risk of trying to play God. But one could look at it just the other way around: to "capture" those asymmetric forces and display their powers on earth was to provide evidence of the existence of God in our world. In February 1875, in the heat of Académie debates over fermentation and spontaneous generation, Pasteur once said that "in good philosophy, the word cause ought to be reserved to the single divine impulse that has formed the universe."[88] A month later, in a poetic image of the cosmic cycle from life to death, and then from death to life again, "as oxygen, hydrogen, and carbon, now in suspension in the gaseous state, [are] ready to be borne by the winds to all the parts of the globe where they are able to re-enter into the cycle of life under the benificent influence of the heat of the sun," Pasteur continued that "it is here that I would love to place the providential idea, not by sentiment alone, this time, but by serious and true scientific deduction and because it seems to me that we have just satisfied one of the great laws of nature!"[89] It is not hard to see Pasteur finding evidence of "the providential idea" in the great law of asymmetric forces.

Vaccines, Ethics, and Scientific vs. Medical Mentalities: Anthrax and Rabies

SIX

The Secret of Pouilly-le-Fort: Competition and Deception in the Race for the Anthrax Vaccine

ON THE AFTERNOON of Thursday, 2 June 1881, Pasteur stepped off a train in Melun, 40 kilometers southeast of Paris. Escorted by his three leading collaborators and various dignitaries, he made his way to the nearby commune of Pouilly-le-Fort and to the large farm of Hippolyte Rossignol, a local veterinary surgeon. Rossignol's large farmyard easily accommodated an expectant crowd of more than two hundred government officials, local politicians, veterinarians, farmers, agriculturists, even calvary officers and newspaper reporters. Among the latter was the Paris correspondent for the London *Times*, who had been invited to attend. His route to Rossignol's farm took him along one of the "splendid roads, lined with limes and acacias," that crisscrossed the fertile agricultural region around Melun. It brought to his mind the close connection between politics and agriculture in the region, "the peasants often being influenced in their votes by a good or bad harvest, voting, according as it turns out, for or against the Government." But the next "electoral harvest" was some time off and attention was focused for now on the results of a public trial of a vaccine that Pasteur and his collaborators had developed in hopes of combatting the disease anthrax.[1] A major killer of sheep, anthrax had become a source of grave concern to French agriculturists, whose annual losses from the disease in recent years were estimated at 20–30 million francs.[2] The size and composition of the crowd in Rossignol's farmyard was a reflection of the economic significance of the disease—and of Pasteur's efforts to combat it.

The crowd had gathered to observe the fortunes of fifty sheep, half of which had been marked with a hole in their ears and "vaccinated" by Pasteur's collaborators in two stages. The first "protective injection" had been made on 5 May; the second, on 17 May. The other twenty-five sheep had received no injections until 31 May, when both they and their twenty-five vaccinated counterparts were injected with a culture of virulent anthrax bacilli. In a bold prophecy given wide public circulation, Pasteur had predicted that the vaccinated sheep would all survive, while the unvaccinated sheep would all succumb to anthrax. He had set today, 2 June, as the date by which it should have become clear whether or not his vaccine had been a success. Quite apart from the economic significance of the outcome, Pasteur had aroused great excitement by predicting such decisive results in what was, after all, the world's first public trial of a laboratory vaccine.

As Pasteur and his collaborators entered the farmyard at 2 P.M., the crowd burst into applause and congratulations. All of the vaccinated sheep were alive and all but one ewe were seemingly healthy. Most of the unvaccinated sheep were already dead and the survivors were obviously not long for this world. It was a moment of high drama in an uncommonly dramatic scientific career.[3]

For Pasteur himself, much of the drama had already been played out in private. He knew before he boarded the train in Paris that he would find a triumphant reception at Pouilly-le-Fort that afternoon. A telegram in the morning from Rossignol had already assured him of a "stunning success."[4] Until that telegram arrived, the outcome had seemed less certain. During the prior two days, some of the vaccinated sheep had become alarmingly feverish, and Pasteur briefly feared that his bold prophecy might end in public ridicule. At one point, it has even been said, Pasteur accused his devoted collaborators of carelessness and thought of sending one of them, Emile Roux, to face alone the embarrassment he dreaded.[5] Certainly Pasteur's friends and associates were puzzled by his atypical loss of confidence. "As if," Roux later wrote, "the experimental method might fail him."[6]

But if Pasteur's collaborators did not fully share his transient fear of failure, he and they did share an important secret: the method by which immunity had been achieved in the animals that survived the Pouilly-le-Fort experiments. Pasteur himself never disclosed in print the real nature of the vaccine deployed at Pouilly-le-Fort. Indeed, his published accounts conveyed the impression that the Pouilly-le-Fort vaccine had been prepared by a method entirely and significantly different from the one actually used.

THE PUBLIC VERSION OF THE TRIAL AT POUILLY-LE-FORT

On 13 June 1881, less than two weeks after his triumphant reception at Pouilly-le-Fort, Pasteur came before the Académie des sciences to summarize the results of his already famous experiments. He spoke almost at once of the precise program of experiments as set forth in a signed agreement with the official sponsor of the Pouilly-le-Fort trial, the Agricultural Society of Melun. The signed protocol actually differed in several details from the commonly repeated story of experiments on fifty sheep, but this simplified version of the Pouilly-le-Fort trial accurately captures its basic thrust and boldly prophetic character.[7] Pasteur himself now drew attention to his audacity in taking on the challenge of Pouilly-le-Fort:

> This program, I admit, had a boldness of prophecy that only a striking success could excuse. Several people were good enough to point this out to me, not without adding some reproach as to my scientific imprudence. But the Académie [des sciences] ought to realize that we did not draw up such a program without having solid support from prior experiments, although none of these had been of the magnitude of the one which was now prepared. Besides, chance favors the prepared mind, and it is in this sense, I think, that one should understand the poet's [i.e., Virgil's] inspired phrase: *Audentes fortuna juvat* [luck comes to the bold].[8]

Pasteur's account of the trial itself was lean but dramatic. Emphasizing its stunning success, he did concede that one of the vaccinated sheep died a day after the crowd had left Pouilly-le-Fort. But armed with the results of an autopsy by Rossignol and another veterinarian, he dismissed the death of this pregnant ewe by linking it with the prior death of the fetus she carried. Pasteur reported with pride that the skeptical veterinarians who had come to Pouilly-le-Fort to follow his experiments—once "very far from accepting as true the artificial preparation of virus-vaccines"—had become instant converts to his point of view in the wake of the decisive results they had now seen with their own eyes. These veterinarians would soon serve as "propagators of the anthrax vaccination." But for some time at least, Pasteur insisted, it was crucial that "the vaccinal cultures . . . be prepared and controlled in my laboratory." Otherwise, "a bad application of the method might compromise the future of a practice which is called upon to render great services to agriculture."[9]

Conspicuously absent from this triumphant address of 13 June was any specific description of the vaccine responsible for the success at Pouilly-le-Fort. Pasteur merely noted that each of the vaccinated animals had been inoculated on 5 May with "five drops of an attenuated anthrax virus" and then again on 17 May with "a second anthrax virus, also attenuated but more virulent than the preceding." On 31 May, all of the animals—vaccinated and unvaccinated—had been injected with a "very virulent virus . . . regenerated from some spores of the anthrax parasite conserved in my laboratory since 21 March 1877."[10] In at least two ways, however, Pasteur led curious and well-informed members of his audience to believe that the vaccine used at Pouilly-le-Fort had been prepared by a method in which exposure to atmospheric oxygen played a crucial role.

In the last paragraph of his address of 13 June, Pasteur drew an explicit link between the *modus fasciendi* of his new anthrax vaccine and the method by which he had already produced a vaccine against chicken cholera:

> In sum, we now possess some virus-vaccines of anthrax, capable of providing protection against the fatal disease without ever being fatal themselves, living vaccines, cultivatable at will, transportable anywhere without alteration, and, lastly, *prepared by a method that one may consider capable of generalization since it served a previous time for the discovery of the chicken cholera vaccine*. By virtue of the conditions I have enumerated here, and to look at things solely from the scientific point of view, the discovery of these anthrax vaccines constitutes a considerable advance over the Jennerian vaccine against smallpox, for the latter has never been obtained experimentally.[11]

The chicken cholera vaccine, as Pasteur had disclosed nine months earlier, had been prepared by exposing cultures of the implicated microbe to atmospheric air for prolonged periods of time. In the paper of October 1880 that described this vaccine, Pasteur reported that no attenuation (and thus no vaccine) resulted when the chicken cholera microbe was cultivated in sealed tubes, however long the intervals between cultures might be, and therefore ascribed attenuation to the action of atmospheric oxygen. He further suggested that oxygen might have a similar effect on other microbes and might even be responsible for the observed behavior of natural epidemics, in which an initially virulent contagious disease becomes progressively less lethal and ultimately burns itself out.[12]

Besides directing attention to his earlier work on the chicken cholera vaccine, Pasteur's address of 13 June 1881 on the Pouilly-le-Fort trial also referred to a paper of the previous February in which he had first announced his discovery of a vaccine against anthrax and had described its *modus fasciendi* in some detail. The method disclosed there, just two months before the Pouilly-le-Fort trial began, involved two basic steps. The more delicate task was to produce a spore-free culture of anthrax bacilli. By careful manipulation of ambient conditions, the culture was maintained at a temperature of 42°–43° C. Within that very narrow temperature range, the anthrax bacillus could be cultivated without forming spores that resisted the action of external agents. Then, in the absence of such resistant spores, the anthrax culture underwent steady and fairly rapid attenuation when maintained at 42°–43° in the presence of pure air. It soon lost its lethal effects when injected into susceptible adult animals and gained instead the capacity to protect them against virulent cultures of anthrax bacilli. It had become a vaccine against anthrax. Here, as with the earlier chicken cholera vaccine, atmospheric oxygen was presumably essential to attenuation and thus to Pasteur's new anthrax vaccine.[13]

On 22 June 1881, nine days after his triumphant address at the Académie des sciences, Pasteur spoke again of the Pouilly-le-Fort experiments in a lecture at Versailles before the International Congress of Directors of Agronomic Stations.[14] He offered this second public account of the Pouilly-le-Fort trial in the context of a general disquisition on the virtues and promise of oxygen-attenuated vaccines. At one point, Pasteur claimed that he had now extended the method of oxygen attenuation beyond the chicken cholera microbe and anthrax bacillus to a previously unknown "microbe of saliva" that he had first detected in a child who had died of rabies.[15]

In public, then, Pasteur spoke of the Pouilly-le-Fort trial as if it were part and parcel of his more general quest for oxygen-attenuated vaccines against microbial diseases. He never published a different—or more explicit—account of the *modus fasciendi* of the anthrax vaccine used at Pouilly-le-Fort. Small wonder that the best informed and most interested scientists of Pasteur's time assumed that the Pouilly-le-Fort vaccine had been prepared by the method of oxygen attenuation. Small wonder that virtually all subsequent studies of Pasteur have adopted the same assumption.[16] And small wonder, too, that these studies ignored or dismissed the very different, indeed opposing, testimony of one sympathetic and firsthand observer of Pasteur's work on anthrax vaccines.

In 1937, forty years after Pasteur's death, his nephew and sometime research assistant, Adrien Loir, published a series of recollective essays under the general (and apt) title, "In the Shadow of Pasteur." In one of these anecdotal but revealing essays, Loir gave passing attention to Pasteur's search for an anthrax vaccine and the famous Pouilly-le-Fort trial. Although vague and sometimes mistaken about the precise details and sequence of events, Loir's account is perfectly clear in its claim that the vaccine used at Pouilly-le-Fort had been prepared *not* by atmospheric attenuation, but rather by the "antiseptic" action of potassium bichromate:

> At the same time that [Pasteur] sought attenuation of the anthrax bacillus by atmospheric oxygen, Chamberland and Roux tried the action of different antiseptics on this microbe. They had obtained an obvious attenuation with potassium bichromate. . . . Pasteur, at this time, pursued the attenuation of viruses by atmospheric oxygen. It was a theory that he had conceived. Oxygen

> destroyed the virulence of all microbes. This immense role of oxygen was an idea he had long held. He pursued its demonstration. It was the agent responsible for the disappearance of diseases, so he said to Chamberland and Roux: "So long as I am alive, you will not publish the results of this experiment on potassium bichromate before having found attenuation by oxygen." It was, in fact, only a long time later that they obtained from Pasteur the authorization to publish a note on this subject.
>
> But, at the time, Pasteur was enticed by the Académie de médecine into making the celebrated experiment of Pouilly-le-Fort. His enemies made him sign the protocol of an experiment that they judged impossible of being realized. Pasteur, in the heat of passion, signed the protocol. . . . On returning to the laboratory, where he announced the thing, his collaborators—in the course of making some objections—asked him what vaccine he was going to use. He answered, "The potassium bichromate one." It was, in fact, the one that was used.[17]

Until very recently, this remarkable passage attracted little attention from students of Pasteur. In the wealth of literature on Pasteur and his disciples, exceedingly few sources even mention Loir's version of the Pouilly-le-Fort episode.[18] Even Rene Dubos, whose celebrated 1950 biography of Pasteur used other parts of Loir's reminiscences to good effect, simply ignored Loir's account of the secret of Pouilly-le-Fort.[19]

The reasons for this neglect are not far to seek. Loir's essays were initially published in scattered issues of an obscure journal. More important, he offered no documentary evidence whatever for his version of the affair of Pouilly-le-Fort, and no other member of Pasteur's inner circle ever corroborated it in print. That applies even to Emile Roux, who was sometimes sharply at odds with Pasteur and who played a central role in the Pouilly-le-Fort trial. Even after Pasteur's death, Roux never challenged the standard account of the events at Pouilly-le-Fort. Quite the contrary. In an essay of 1896 on Pasteur's veterinary and medical research, Roux repeated the official line that the vaccine used at Pouilly-le-Fort had been an atmosphere-attenuated and spore-free culture of the anthrax bacillus.[20] Thus, to accept Loir's characterization of the Pouilly-le-Fort vaccine would be to deny the public testimony of Emile Roux as well as Pasteur himself. Finally, as we shall see, Loir does little to help us understand what meaning or significance should be attached to his version of the affair of Pouilly-le-Fort. He claims that the vaccine used there had been prepared by exposure to potassium bichromate rather than atmospheric oxygen, but he does not tell us exactly what difference that makes or what motives might lie behind the secret of Pouilly-le-Fort.

THE SECRET OF POUILLY-LE-FORT: THE LABORATORY NOTES

It is only by turning to Pasteur's laboratory notebooks that we are able to establish conclusively the nature of the vaccine actually used at Pouilly-le-Fort. Working independently, Antonio Cadeddu and I have analyzed the pertinent notebook, and our interpretations agree on this central point: Pasteur deliberately deceived the public and the scientific community about the nature of the vaccine actually used at Pouilly-le-Fort.[21]

The crucial pages, reproduced in fig. 6 (a, b, c), partly transcribed and translated in Appendixes A and B at the back of this book, come from the notebook that Pasteur labeled "10ème cahier. Du 20 novembre 1880 au 10 avril 1882." First, at the bottom of a page entitled "Charbon. Vaccination à Melun" and dated 26 April [1881], Pasteur added a footnote indicating that he and the Agricultural Society of Melun reached an agreement on 28 April as to the experimental protocol to be followed at Pouilly-le-Fort. He then goes on to specify his projected modus operandi in a way that finds no echo whatever in any of his published work. On 5 May, he writes in the footnote, the twenty-five sheep to be vaccinated will be injected with an anthrax culture already so "weakened by potassium bichromate" that it has become harmless to mice and then further weakened by three successive passages through mice.[22]

That this projected modus operandi was in fact followed at Pouilly-le-Fort is established by Pasteur's notes on page 113 of the same notebook. There Pasteur records that "the anthrax culture [bactéridie] employed for the first vaccine, this 5th of May . . . was an anthrax culture attenuated by Ch[amberland] with bichromate and which, no longer being lethal at all, had been reinforced by three successive passages in three mice." On the remainder of this page, Pasteur further informs us that the *second* culture injected into the vaccinated sheep (on 17 May) had also been attenuated by potassium bichromate, but this time exposure to the antiseptic had been limited to just a few days and the resulting attenuation had not been reinforced by passage through mice. This second culture, considerably more virulent than the first, had killed two of the four unvaccinated sheep into which it had been injected in Pasteur's laboratory. On 28 May and then again on 29 May, in a departure from the signed protocol, Pasteur's collaborators injected a virulent anthrax culture into one vaccinated and one unvaccinated sheep. By the morning of 31 May, the day originally scheduled for all virulent injections, the two unvaccinated sheep already injected were dead, while their two vaccinated counterparts had suffered only a slight elevation in temperature. The rest of the sheep—vaccinated

Charbon. Vaccination à Melun. 106 108

26 avril. Projet de convention avec la Société d'agriculture de Melun (1):

La Société d'agriculture de Melun ayant proposé à M. Pasteur par l'organe de son président M. de la Rochette de se rendre compte par elle même, sous le rapport pratique, des résultats des expériences faites par M. Pasteur & MM Chamberland & Roux au sujet de l'affection charbonneuse, il a été convenu ce qui suit:

1° La Société d'Agriculture de Melun met à la disposition de M. Pasteur 60 moutons.

2° 10 de ces moutons ne subiront aucun traitement & serviront comme témoins

3° 25 de ces moutons subiront deux inoculations vaccinales à 12 ou 15 jours d'intervalle par le virus charbonneux atténué

4° Ces 25 moutons seront en même temps que les 25 restant inoculés 12 ou 15 jours après par le charbon très virulent. Les 25 moutons non vaccinés périront tous, les 25 vaccinés résisteront & on les comparera ultérieurement avec les 10 témoins réservés ci dessus afin de montrer que les vaccinations n'ont pas empêché les moutons de revenir après un certain temps, à un état normal.

5° Après l'inoculation du virus très virulent aux deux séries de 25 moutons vaccinés & non vaccinés, les 50 moutons resteront réunis dans la même étable; on distinguera une des séries de l'autre en faisant avec un emporte pièce un trou à l'oreille des 25 moutons vaccinés.

6° Les 10 moutons témoins resteront toujours dans une bergerie à part afin qu'ils ne soient pas exposés à la contagion des moutons malades.

7° Tous les moutons qui mourront seront enfouis un à un dans des fosses distinctes, voisines les unes des autres & situées dans un enclos palissadé.

8° Au mois de mai 1882 on fera parquer dans l'enclos dont il vient d'être

(1) Ce projet a été accepté par M. le Président le Baron de La Rochette [illegible] le 28 avril [illegible] 25 moutons [illegible]

Figure 6.1 (a,b,c). Pasteur's handwritten record of the agreed upon protocol for the trial of the anthrax vaccine at Pouilly-le-Fort. Dated 26 April [1881]. From Pasteur, *Cahier 91*, fols. 106, 106v, 107 (using Pasteur's handwritten pagination; the stamped pagination 108, 108v, 109 was added by the staff of the Bibliothèque Nationale). It is on fol. 106, in footnote (1) in Pasteur's microscopic and hard-to-decipher

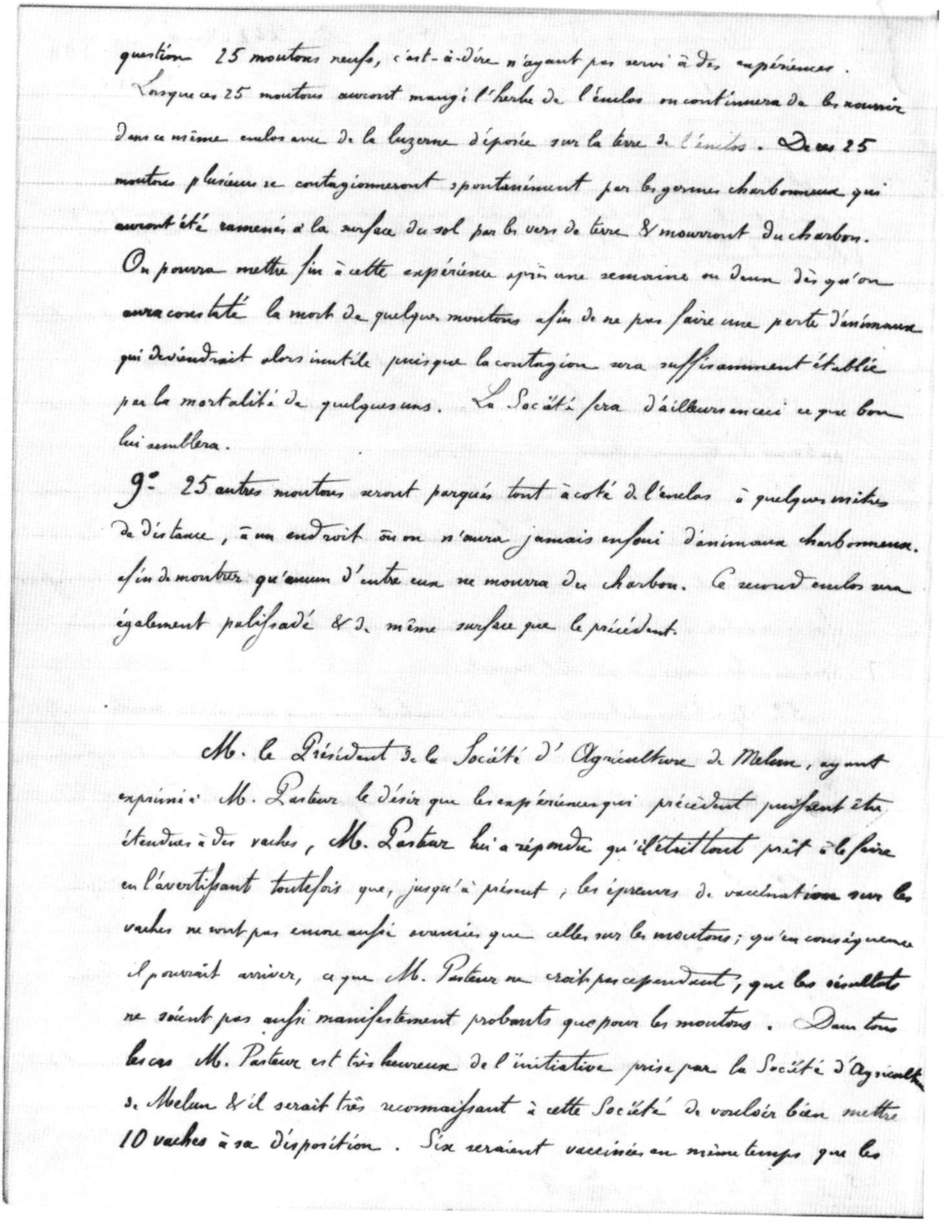

question 25 moutons neufs, c'est-à-dire n'ayant pas servi à des expériences.
Lorsque ces 25 moutons auront mangé l'herbe de l'enclos on continuera de les nourrir
dans ce même enclos avec de la luzerne déposée sur la terre de l'enclos. De ces 25
moutons plusieurs se contagionneront spontanément par les germes charbonneux qui
auront été ramenés à la surface du sol par les vers de terre & mourront du charbon.
On pourra mettre fin à cette expérience après une semaine ou deux dès qu'on
aura constaté la mort de quelques moutons afin de ne pas faire une perte d'animaux
qui deviendrait alors inutile puisque la contagion sera suffisamment établie
par la mortalité de quelques uns. La Société fera d'ailleurs en ceci ce que bon
lui semblera.

9° 25 autres moutons seront parqués tout à côté de l'enclos à quelques mètres
de distance, à un endroit où on n'aura jamais enfoui d'animaux charbonneux.
afin de montrer qu'aucun d'entre eux ne mourra du charbon. Ce second enclos sera
également palissadé & de même surface que le précédent.

M. le Président de la Société d'Agriculture de Melun, ayant
exprimé à M. Pasteur le désir que les expériences qui précèdent puissent être
étendues à des vaches, M. Pasteur lui a répondu qu'il était tout prêt à le faire
en l'avertissant toutefois que, jusqu'à présent, les épreuves de vaccination sur les
vaches ne sont pas encore aussi avancées que celles sur les moutons; qu'en conséquence
il pourrait arriver, ce que M. Pasteur ne croit pas cependant, que les résultats
ne soient pas aussi manifestement probants que pour les moutons. Dans tous
les cas M. Pasteur est très heureux de l'initiative prise par la Société d'Agriculture
de Melun & il serait très reconnaissant à cette Société de vouloir bien mettre
10 vaches à sa disposition. Six seraient vaccinées en même temps que les

Figure 6.1b.

handwriting, that he records his intention to use the vaccine that Chamberland (here abbreviated as Ch[d]) had produced by attenuating the anthrax bacillus with potassium bichromate. (Papiers Pasteur, Bibliothèque Nationale, Paris)

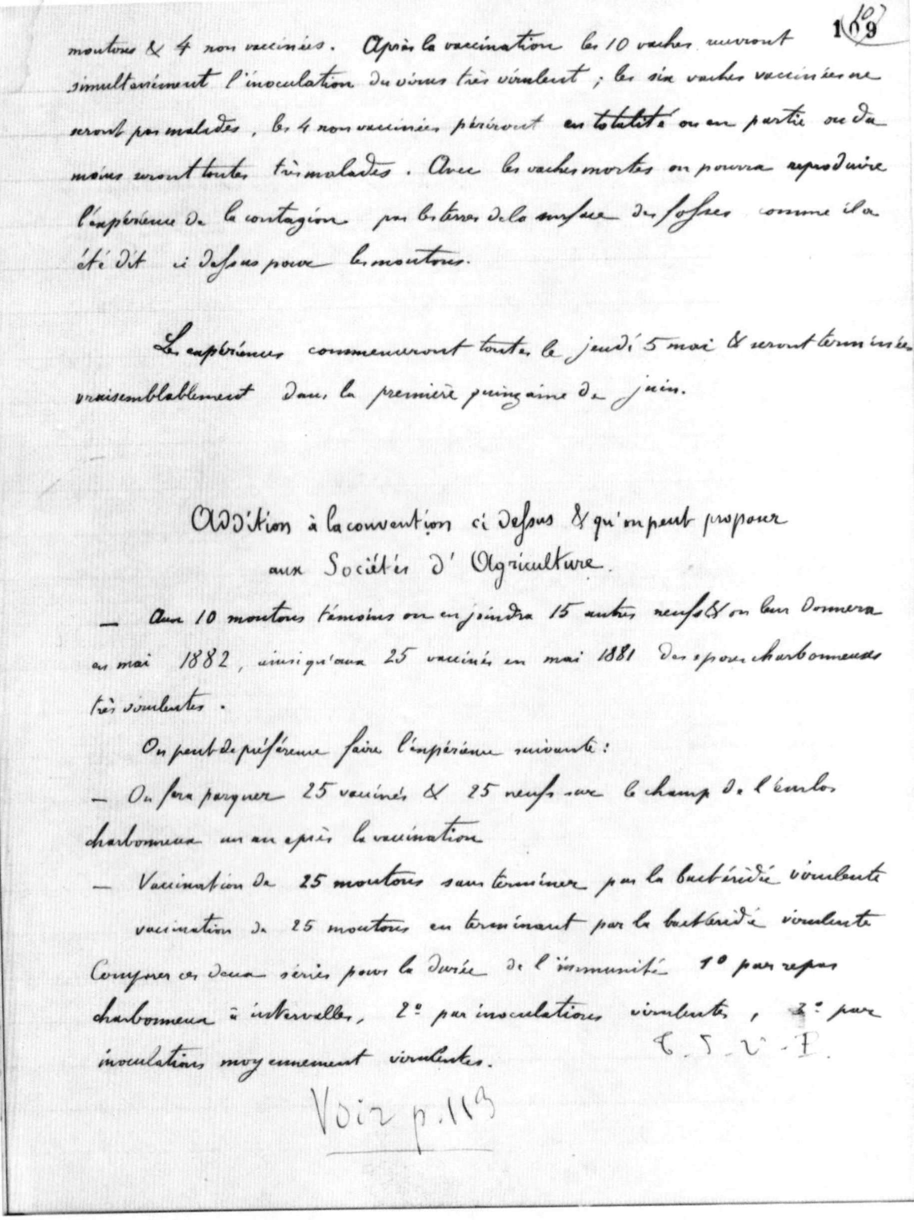

109

moutons & 4 non vaccinées. Après la vaccination les 10 vaches recevront simultanément l'inoculation du virus très virulent ; les six vaches vaccinées ne seront pas malades, les 4 non vaccinées périront en totalité ou en partie ou du moins seront toutes très malades. Avec les vaches mortes on pourra reproduire l'expérience de la contagion par les terres de la surface des fosses comme il a été dit ci dessus pour les moutons.

Les expériences commenceront toutes le jeudi 5 mai & seront terminées vraisemblablement dans la première quinzaine de juin.

Addition à la convention ci dessus & qu'on peut proposer aux Sociétés d'Agriculture.

— Aux 10 moutons témoins on en joindra 15 autres neufs & on leur donnera en mai 1882, ainsi qu'aux 25 vaccinés en mai 1881 des spores charbonneuses très virulentes.

On peut de préférence faire l'expérience suivante:

— On fera parquer 25 vaccinés & 25 neufs sur le champ de l'enclos charbonneux un an après la vaccination

— Vaccination de 25 moutons sans terminer par la bactéridie virulente

vaccination de 25 moutons en terminant par la bactéridie virulente

Comparer ces deux séries pour la durée de l'immunité 1° par repas charbonneux à intervalles, 2° par inoculations virulentes, 3° par inoculations moyennement virulentes. E. S. V. F.

Voir p. 113

Figure 6.1c.

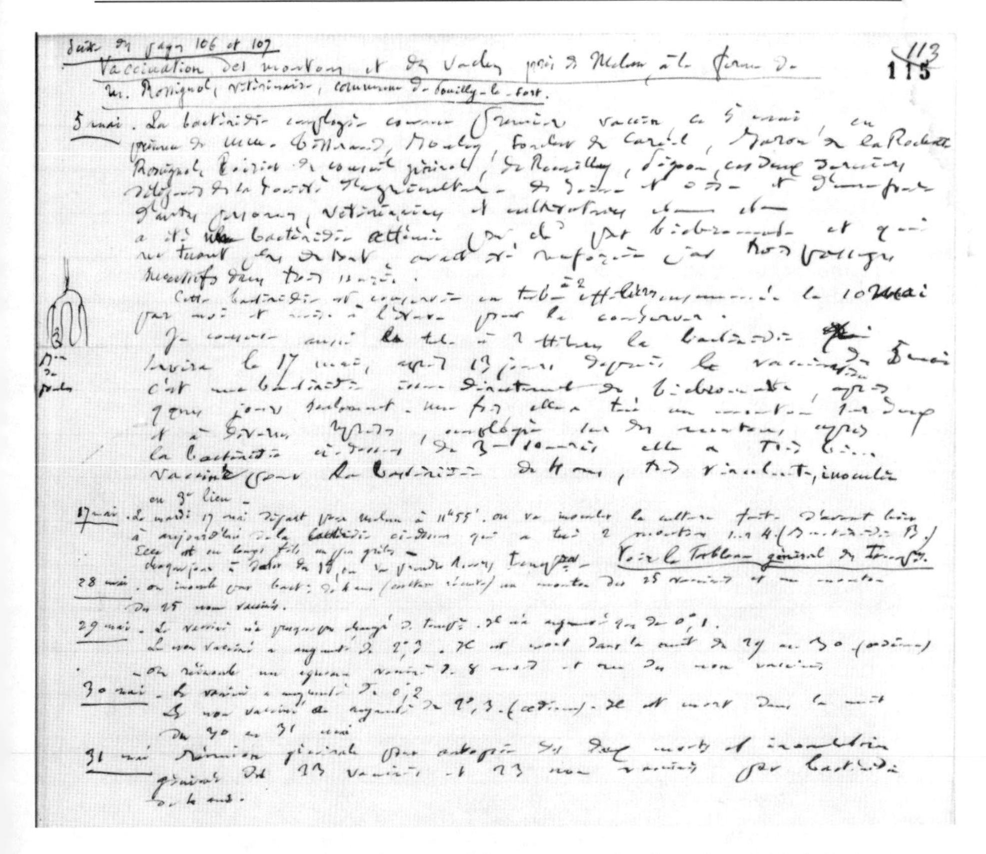

Figure 6.2. This page, from the same laboratory notebook as figure 6.1 (a,b,c), establishes that Pasteur did in fact use the potassium bichromate vaccine at Pouilly-le-Fort. Pasteur, *Cahier 91*, fol. 113 (using Pasteur's handwritten pagination). (Papiers Pasteur, Bibliothèque Nationale, Paris)

and unvaccinated—were then inoculated as scheduled on 31 May with a highly virulent culture that had been preserved in Pasteur's laboratory for four years. The results on the four prematurely injected sheep gave Pasteur what he called a "foretaste" of the spectacular success he would find on 2 June 1881, despite a few anxious moments in the meantime.[23]

In his public accounts of the Pouilly-le-Fort experiments, Pasteur described only the last, virulent injection with any degree of accuracy and specificity. It is only because of the evidence recorded in Pasteur's own carefully preserved laboratory notebooks that we can now insist, beyond any shadow of doubt, that Loir's memory had not failed him: the vaccine used at Pouilly-le-Fort had in fact been prepared by exposure to potassium bichromate. Nor did Pasteur merely suppress that fact. Rather, as we have seen, his public accounts portrayed the Pouilly-le-Fort trial as a striking demonstration of the virtues and potential of oxygen-attenuated vaccines. The conclusion is unavoidable: Pasteur deliberately deceived the public, including especially those scientists most familiar with his published work, about the nature of the vaccine actually used at Pouilly-le-Fort.

* * *

It is one thing to expose Pasteur's deception, and quite another to explain it. Why did he do it and what difference does it make? In his reminiscences of 1937, Adrien Loir offered a benign assessment of his uncle's conduct in the affair of Pouilly-le-Fort. Indeed, Loir virtually dismissed any question of impropriety by noting that the experimental protocol agreed upon by Pasteur and the Agricultural Society of Melun did not specify what method was to be used to produce the Pouilly-le-Fort vaccine. Loir further minimized the significance of the episode by insisting that Pasteur's search for an effective oxygen-attenuated vaccine did, after all, soon succeed. Despite his temporary resort to an antiseptic vaccine at Pouilly-le-Fort, Pasteur would "return to the role of oxygen," wrote Loir. "He was tenacious."[24]

Here, too, Loir's account has a substantial basis in fact. It is true that the signed protocol of the Pouilly-le-Fort experiments made no reference to the *modus fasciendi* of the vaccine to be used there. It is also true that Pasteur did soon develop effective anthrax vaccines by exposing cultures of the bacillus to the atmosphere at a temperature of 42° to 43° C. The same laboratory notebook that reveals the secret of Pouilly-le-Fort also shows that Pasteur had begun to achieve increasingly secure results with his oxygen-attenuated vaccines even as the Pouilly-le-Fort trial was underway. Within a month of his triumphant reception at Pouilly-le-Fort on 2 June 1881, Pasteur felt sufficiently confident of his new vaccines to test them on a flock of seventy-five sheep. The results of these and subsequent trials were overwhelmingly positive, although it did eventually become clear that the oxygen-attenuated vaccines lost some of their immunizing powers over time.[25]

Meanwhile, in the wake of the celebrated trial of Pouilly-le-Fort, Pasteur's laboratory was flooded with requests for supplies of his anthrax vaccines. The available notebooks do not seem to provide any definitive indication as to what method of attenuation was employed in meeting this demand. Pasteur delegated day-to-day responsibility for the manufacture, sale, and distribution of these vaccines to Chamberland, who now worked in a separate annex on the rue Vauquelin, two blocks away from Pasteur's main laboratory on the rue d'Ulm. According to Loir, Pasteur did not pay close or regular attention to the work at the new annex and was at least briefly unaware of certain important details in Chamberland's method of producing the new

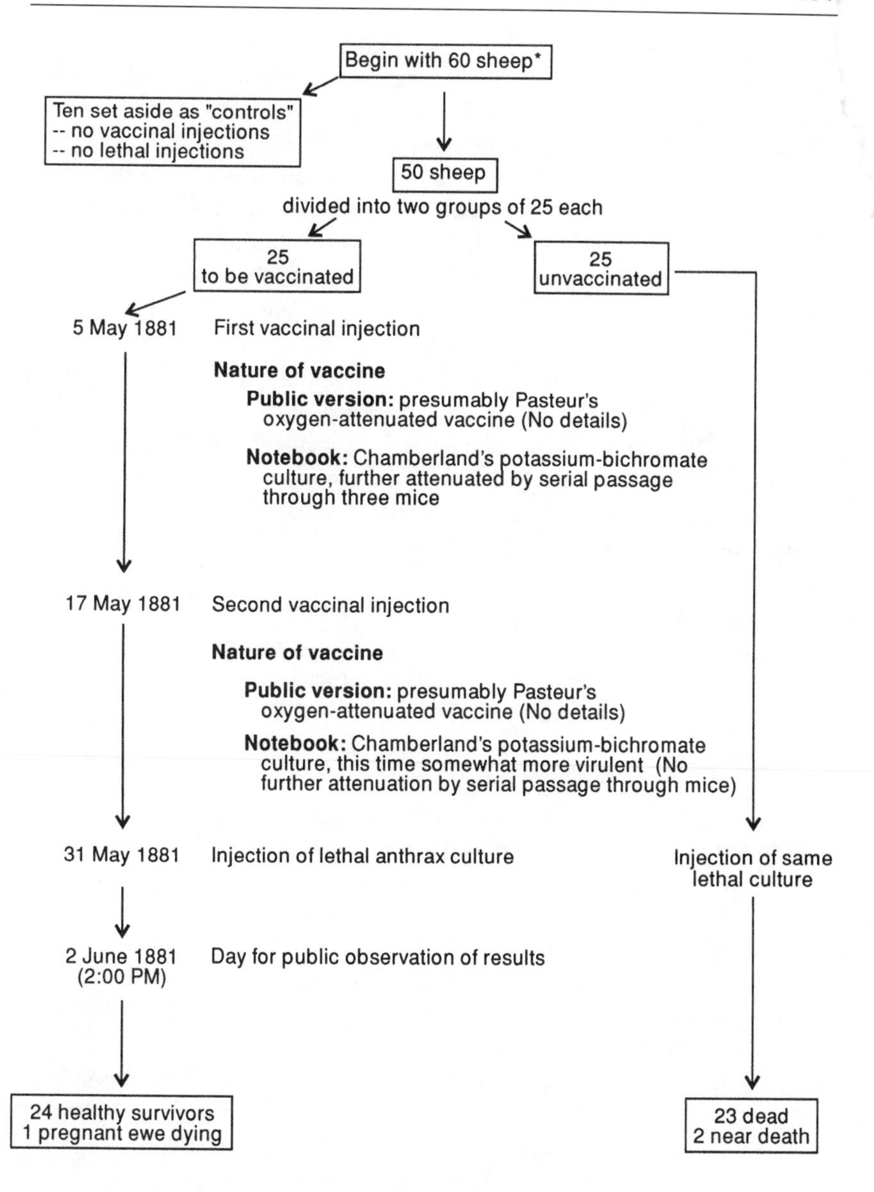

Figure 6.3. Schematic diagram of "The Secret of Pouilly-le-Fort."

vaccines. Yet there seems no reason to doubt that oxygen-attenuation had now become a central feature of Chamberland's vaccines. At the least, Pasteur believed that to be so.[26] And if these commercial vaccines did sometimes fail and did attract some sharp criticism, notably from Robert Koch in Germany, there is nonetheless ample evidence that by and large they were very successful.[27] In this respect, too, Loir's account deserves our credence.

Extrapolating slightly from Loir's account, it could further be argued that the specific method used to produce the Pouilly-le-Fort vaccine was really only a minor matter of detail, so long as it was consistent with Pasteur's biological theory of immunity. As we shall see more fully below, Pasteur had committed himself to a "biological" theory of immunity, according to which vaccines were living but attenuated microbial strains. And the vaccine actually used at Pouilly-le-Fort was such a "live" vaccine that had "merely" been attenuated with potassium bichromate rather than oxygen. It is even conceivable—despite the absence of any documentary evidence for it—that Pasteur traced the attenuating action of bichromate to its properties as an oxidizing agent, thus confirming his conception of attenuation as some (unspecified) sort of oxidation process.[28] By this point, it might begin to seem as if the secret of Pouilly-le-Fort is hardly worth revealing: the potassium bichromate vaccine breached no explicit agreement with the Agricultural Society of Melun and was arguably in keeping with Pasteur's central theoretical commitments.

Yet several nagging questions remain. Why did Pasteur not simply, even eagerly, publish this interesting version of the story? Even if the signed agreement did not require him to reveal the nature of the vaccine, why did he choose to conceal it? If Pasteur had nothing to lose by disclosing the secret, why did he bother to keep it? Why did Roux and Chamberland follow suit? When they published the results of their work on the attenuation of the anthrax bacillus by antiseptics—including notably carbolic acid and potassium bichromate—why did they conspicuously avoid any reference to the Pouilly-le-Fort trial in which they had participated just two years before?[29] And most important, why was Pasteur not content merely to keep the secret of Pouilly-le-Fort? Why did he actively purvey the impression that the vaccine had been prepared by the method of oxygen attenuation?

These questions are left unanswered in Adrien Loir's sketchy account of the affair of Pouilly-le-Fort. In fact, Loir does not even raise them. His account is not inaccurate, strictly speaking, but it is woefully incomplete. We cannot begin to make sense of the secret of Pouilly-le-Fort unless and until we focus on part of the drama that finds no place whatever in Loir's script—Pasteur's competition with a now obscure young veterinarian, Jean-Joseph Henri Toussaint (1847–1890), a professor at the Toulouse Veterinary

School who was also in quest of an anthrax vaccine.[30] The story of their race to produce the first effective anthrax vaccine begins with Pasteur's earlier discovery of a vaccine against chicken cholera.

THE COMPETITION BETWEEN PASTEUR AND TOUSSAINT

In February 1880, after a year of research on chicken cholera and the microbe that caused it, Pasteur announced that he had discovered a vaccine against the disease.[31] His work on chicken cholera had begun in December 1878, when Toussaint sent him some blood from a cock dead of the disease.[32] Like a few others before him, Toussaint linked the disease with a microbe, which he claimed to have found in the blood of all hens having the disease. Beginning with the blood sent to him by Toussaint, Pasteur immediately sought to isolate the microbe in a state of purity and to demonstrate that it was the true and sole cause of chicken cholera. He soon found that this microbe developed much more readily in neutral chicken broth than in the neutral urine that Toussaint had used as his culture medium. Pasteur thanked Toussaint for sending him the blood with which his research on chicken cholera had begun, but he left little doubt that he considered Toussaint's work and techniques decidedly inferior to his own.[33]

In announcing the discovery of a vaccine against chicken cholera, Pasteur declined to reveal the details of the method by which he had produced it. He disclosed only that his new vaccine was an attenuated form of the chicken cholera microbe itself, which he had obtained "by certain changes in the mode of culture." He justified this reticence on the grounds that he wished to assure temporary independence in his ongoing research.[34] For nine more months, until October 1880, Pasteur continued to keep private the method by which he had produced his chicken cholera vaccine.

Pasteur's reticence may have owed something to his initial uncertainty about the precise immunizing power of his vaccine.[35] But a second consideration was probably much more important. As we shall see in more detail below, the method by which the vaccine had been prepared was remarkably simple, involving little more than allowing a culture of the microbe to sit exposed to ordinary air for a prolonged period of time. Given the simplicity of his method, Pasteur clearly hoped and expected that it could quickly be extended to other (and more important) microbial diseases. Had he revealed immediately the method of attenuation through exposure to ordinary air, Pasteur might have faced a host of competitors in the search for other vaccines. Instead, while others waited for him to disclose the *modus fasciendi* of his chicken cholera vaccine, Pasteur was already seeking to extend the method of atmospheric attenuation to another microbial disease of far greater economic significance to animal husbandry—anthrax, a disease on which he had already been working for three years. The pace of his search for an anthrax vaccine was sharply accelerated by competition from Toussaint, who announced in July 1880 that he had already discovered an effective anthrax vaccine.

The announcement came slightly earlier than Toussaint himself had intended or foreseen. In the first few days of July 1880, he received a visit at the veterinary school in Toulouse from Henri Bouley (1814–1899), formerly a professor at the Alfort Veterinary School who was now serving as inspector general of French veterinary schools. Touissaint told Bouley of some promising initial results in his search for an effective anthrax vaccine. With evident pride that a fellow veterinarian may have made so momentous a discovery, Bouley encouraged Toussaint to make his work public and served ever after as a defender of his achievements and honor.[36]

When Bouley returned from Toulouse to Paris, where he was a member of the Académie des sciences and the Académie de médecine, he carried with him a note in which Toussaint briefly described the results to date of his efforts to find an effective anthrax vaccine. Bouley had also been entrusted with a sealed envelope containing a separate note in which Toussaint outlined the method used to produce his vaccine. He had asked Bouley to wait until 12 July to read the first note to the Académie des sciences and to deposit the sealed envelope with the secretariat of the Institut de France. In France, the official deposit of such a sealed note (*pli cacheté*) had long been an established mechanism for securing or protecting one's priority for a scientific discovery. It did not, however, establish one's right to the commercial exploitation of a discovery. For that, an official patent (*brevet d'invention*) was required.[37]

In the event, Bouley found himself unable to respect Toussaint's deadline. A week early, on 6 July 1880, he alluded to Toussaint's new vaccine in a meeting of the Académie de médecine. He did so in the context of a discussion of a paper on "malignant pustule," the name commonly given to anthrax in humans. The author of that paper, a veterinarian named Gabriel Colin, was a bitter opponent both of Bouley and of Pasteur and his microbiological doctrines. In his acerbic response to Bouley's comments, Colin complained of this premature announcement of Toussaint's "secret" vaccine and claimed priority himself for preventive inoculations against anthrax.[38]

Colin's outburst eventually drew a reply from Toussaint, who did not belong to the Académie de médicine and therefore could not reply on the spot. In the meantime, Bouley set about fulfilling Toussaint's original instructions to him. On 12 July, as planned, he passed on to the secretariat of the Institut de France the sealed note in which Toussaint outlined the method by which he had produced his vaccine. On the same day, also in keeping with Toussaint's instructions, Bouley read before the Académie des sciences the note in which Toussaint described the initial results of his experimental trials. As published in the Académie's *Comptes rendus*, this note reported that Toussaint had conducted trials of his vaccine on eight dogs and eleven sheep. Of the eight dogs, four had been injected with his vaccine and had then survived a series of four successive injections of virulent anthrax blood. By contrast, all four unvaccinated dogs succumbed to the first injection. Of the eleven sheep, six were vaccinated and five served as unvaccinated controls. When injected with virulent anthrax, all five unvaccinated sheep died. One of the six vaccinated sheep also died when injected with virulent anthrax blood, but the other five survived. After a second injection of Toussaint's vaccine, these five vaccinated sheep proved immune to three further injections of virulent anthrax blood or spores.[39] The method by which Toussaint had produced his vaccine remained sealed up in the envelope entrusted to the secretariat of the Institut de France.

Two weeks later, on 27 July 1880, the names of Pasteur and Toussaint became linked in the course of a stormy session of the Académie de médecine. In a letter read to the Académie by Bouley, Toussaint now replied to Colin's outburst at the meeting of 6 July, saying in effect that Colin should have published an account of his results if he had in fact produced immunity against anthrax. In further defense of his honor, Toussaint also asked the Académie de médecine to publish in its *Bulletin* his note of 12 July, which had already been published in the *Comptes rendus* of the Académie des sciences. That request provoked an animated discussion and considerable opposition. Several members of the Académie de médecine objected on the grounds that the *Bulletin* should not publish accounts of "secret remedies," which label could be applied to Toussaint's vaccine so long as its *modus fasciendi* remained under seal. Bouley, eager as always to defend Toussiant's interests and reputation, then invoked the precedent established in February, when the *Bulletin* had published Pasteur's announcement of his discovery of a *chicken cholera vaccine* by a method he had yet to reveal. Ultimately, the Académie de médecine decided to publish Toussiant's note, but only on condition that the *Bulletin* would omit his strictures against their fellow member, Colin, while also recording the debate over the decision to publish. Because of this latter proviso, the published account of the meeting of 27 July 1880 lumped Pasteur's name with Toussaint's as an alleged purveyor of "secret remedies."[40]

Pasteur and Toussaint shared a concern that their personal integrity had been impugned at the 27 July 1880 meeting of the Académie de médecine. But they responded in strikingly different ways. Pasteur, a member of the Académie since 1873, threatened to resign. He changed his mind when he was allowed to publish a note defending his reticence about the *modus fasciendi* of his chicken cholera vaccine and after he had been assured by the president of the Académie that there had been no intention of impugning his personal honor.[41] Toussaint, who was not a member of the Académie de médecine, responded more quickly to the substance of the charges brought against him. To remove any suspicion that he intended to exploit a secret remedy, he immediately directed that his sealed note be opened and its contents revealed, initially to the Académie des sciences and then to the Académie de médecine.[42] Little more than a week after they had complained of Toussaint's "secrecy," his critics could read a published account of the method by which he had produced his anthrax vaccine. Yet the *modus fasciendi* of Pasteur's chicken cholera vaccine remained private for some time to come.

In the sealed note that was now made public, Toussaint took but a few lines to describe the method by which he had prepared his anthrax vaccine. Initially, he had simply defibrinated and filtered the blood of animals dead of anthrax. But injections of the resulting liquid sometimes killed the animals it was meant to protect. Toussaint assumed that these "accidents" occurred when anthrax bacilli slipped through his paper filters. He therefore resorted to heat "in order to kill the bacilli," and he presumed that heating defibrinated anthrax blood for ten minutes at 55° C was sufficient to accomplish that goal. After being injected with 3 cubic centimeters of his heated blood, five sheep had proved immune to subsequent injections of untreated virulent anthrax blood. Toussaint hoped that few difficulties would be encountered in the task of making his procedure suitable for large-scale vaccinations, at which point he had planned to disclose the contents of the sealed note that was now being read prematurely.[43]

When Toussaint was revealing the *modus fasciendi* of his anthrax vaccine, Pasteur was vacationing at the familial home in Arbois. On 10 August 1880, in a letter to Bouley, Pasteur responded as follows to Toussaint's announcement:

> My very good colleague,
>
> Since yesterday morning, when I received your letter, the extracts of the journals, and the Compte rendu [of the Académie des sciences]—all at the same time—I have been in astonishment and admiration over the discovery of M. Toussaint—in admiration that it exists, in astonishment that it can be. It

overturns all the ideas I had on viruses, vaccines, etc. I no longer understand anything. Ten times yesterday, I had the idea of taking the train to Paris. I really cannot believe this surprising fact until I've seen it, seen it with my own eyes, though the observations that establish it seem irrefutable to me. It is the importance of the fact that makes me want to confirm it to my own satisfaction.

The Académie de médecine has thus received a severe lesson. It will surely have grasped that one does not deal lightly with facts of this order in public, that contemplation is appropriate in the face of such solutions to such problems.

I am too moved to write you more fully. I have dreamed about it, both asleep and awake, all through the night.

Best to you and thanks.

L. Pasteur[44]

Pasteur's expression of surprise and agitation make sense only in the context of his general theoretical views on disease and immunity. He had come to the study of disease after a remarkably successful campaign on behalf of the biological (or "germ") theory of fermentation. In opposition to the theory that fermentation could result from ordinary chemical processes, Pasteur insisted that fermentation depended on the activity of living microbes. Given the long-standing analogy between fermentation and disease, he was therefore predisposed to believe that infectious disease also resulted from the activity of microbes. In explanation of his chicken cholera vaccine, he had proposed a similarly biological theory of immunity. Linking immunity with the biological, and specifically the nutritional, requirements of the pathogenic microbe, he suggested that the tissues of the invaded animal might contain only trace amounts of substances required for the nutrition of the invading microbe. If so, the invading microbe might soon exhaust the supply of these trace substances, rendering the host an unsuitable medium for the microbe's subsequent cultivation. When and if the host survived the initial invasion, it would be henceforth more or less immune to that pathogenic organism. And so a living but weakened (attenuated) pathogen, one which exhausted trace nutrients without killing the host, could provide protection (that is to say, could act as a "vaccine") against future invasions of its unaltered and more virulent relatives. An attenuated strain of the chicken cholera microbe, for example, could provide protection against virulent strains of the microbe and thus against the disease itself.[45]

Central to Pasteur's conception of immunity, then, was the biological activity of a living, if attenuated, microbe. Toussaint, by contrast, had a chemical conception of immunity, as is clear from his assumption that the anthrax bacilli were *killed* in the course of preparing his vaccine. Like his eminent mentor Auguste Chauveau, Toussaint supposed that the developing anthrax bacillus released a soluble substance into the bloodstream that was toxic to the microbe itself.[46] And this opened the possibility that such a soluble substance, once produced and captured, might act as a vaccine independently of living anthrax bacilli. It was Toussaint's claim that he had in fact produced such a "dead" vaccine against anthrax that moved Pasteur to say "it overturns all the ideas I had on viruses, vaccines, etc." In the public critique that Pasteur was soon to issue against Toussaint's work, the central theoretical concern was precisely this question of "live" vs. "dead" vaccines.

For a few days in early August 1880, after he had revealed the *modus fasciendi* of his anthrax vaccine, Toussaint could bask in the applause his work received at the Académie des sciences and the Académie de médecine. But he was very soon to regret the speed with which he had released the contents of his sealed envelope. His own experimental trials of the vaccine quickly aroused his concern. In fact, he had modified his method of producing the vaccine within days of sending his sealed note to Paris. At some point before 8 August, he had already decided that heating anthrax blood at 55° C did not consistently yield an effective vaccine. He had switched to procedures in which the application of heat was either supplanted or supplemented by other agents. In particular, he had begun to subject anthrax blood to the action of carbolic acid,[47] which had long been used as a disinfectant and had more recently become famous as Joseph Lister's "antiseptic" of choice in the treatment of surgical patients.

On Sunday, 8 August 1880, Toussaint and Bouley undertook a relatively large-scale trial of this new "antiseptic" vaccine against anthrax, injecting it into twenty healthy sheep at the Alfort Veterinary School. By Thursday, 12 August, four of the twenty sheep had died of anthrax, and the remaining sixteen were seriously ill. Toussaint and Bouley then feared a total disaster, but the sixteen surviving sheep recovered and eventually proved immune to injections of virulent anthrax. A week later, on Thursday morning, 19 August, Bouley went to Pasteur's laboratory on the rue d'Ulm in Paris. He spoke with Emile Roux, who remained at work in the Paris laboratory while Pasteur and Chamberland were conducting experiments in the field near Arbois. Bouley disclosed, in confidence, the partial failure of Toussaint's vaccine in the trial at Alfort and sought Roux's opinion as to what had gone wrong.[48]

That same day, Roux sent Pasteur a detailed report of his meeting with Bouley. On Roux's account, he informed Bouley that he was not surprised by the disappointing results of Toussaint's Alfort trial. During the previous several days, Roux had already found that anthrax bacilli were not uniformly killed by ten minutes of heating at 55° C, as Toussaint had presumed in the sealed note that had since been made public. More specifically, Roux continued, Toussaint's allegedly "dead" vaccine sometimes killed experimental animals when injected into them and sometimes gave rise to fertile cultures of anthrax bacilli. At that point, Bouley informed Roux that the Alfort vaccine had been prepared not by heating, but rather by the "measured action of carbolic acid." On the surface, that revelation could have been taken to undermine the pertinence, if not the validity, of Roux's experiments with heated anthrax bacilli or blood. But Roux did not see it that way. He conveyed to Pasteur his a priori conviction that Toussaint's new "antiseptic" vaccine would also prove to be a "live" rather than a "dead" vaccine. In both cases, Roux presumed, successful vaccinations could be ascribed to the unwitting attenuation of still living anthrax bacilli, while Toussaint's failures or "accidents" showed only that his methods were too crude to produce such a result consistently. Roux did not yet have any experimental evidence of his own to support this interpretation. But he and Pasteur had approached Toussaint's heat-produced vaccine with a similar predisposition toward a biological interpretation of its effects. That predisposition had since been vindicated by Roux's recent experiments, and he saw every reason to expect that the outcome would be the same in the case of Toussaint's new antiseptic vaccine.[49]

In the face of his partial failure at Alfort, Toussaint moved quickly to abandon the theoretical interpretation he had initially given to his vaccines. Pasteur, Toussaint, and Bouley attached great importance to the precise date on which this shift in interpretation took place. Already in a race to produce a safe and effective anthrax vaccine, Pasteur and Toussaint were now to engage in a dispute over which of them deserved recognition for establishing the biological, as opposed to chemical, interpretation of Toussaint's results. Toussaint, with the customary support of Bouley, claimed that he had abandoned his chemical interpretation in a paper delivered at Reims on 19 August 1880, the very day that Bouley informed Roux in Paris of the outcome of the trial at Alfort.[50] If so, Toussaint must have changed his mind independently of any knowledge of Roux's ideas or results.

There can be no doubt that Toussaint spoke of anthrax vaccines at Reims on 19 August on the occasion of the annual meeting of the French Association for the Advancement of Science. And in the published version of that address, Toussaint definitely did state that his vaccines produced their effects not by virtue of a soluble vaccinal substance (his initial chemical theory), but rather by virtue of "an *attenuated state* of the parasite."[51] In other words, the published version of Toussaint's address of 19 August 1880 endorsed the same biological interpretation of his results that Pasteur and Roux favored a priori. But several months intervened between Toussaint's oral address and its printed version, and the rules of the French Association allowed speakers to revise their oral communications before they were published in the Association's annual *Compte rendu*.[52] Since the printed text of Toussaint's address refers to experiments undertaken *after* the Reims meeting of 19 August,[53] it is obvious that he had revised his oral communication before publishing it. Pasteur could therefore plausibly suspect that Toussaint had abandoned his chemical interpretation only after the Reims meeting—and thus only after he had learned of Roux's results from Bouley. Eventually, Pasteur said as much in print.[54]

For our purposes, it is not important to settle this priority dispute, though the available evidence does seem unfavorable to Toussaint's claim. Nor is it necessary to discuss in any detail several other instances of competition between Pasteur and Toussaint—including notably their earlier parallel work on chicken cholera and its microbe.[55] But it is crucial to appreciate the nature and extent of the competition between them. For it is only in that context that we can begin to understand Pasteur's conduct in his quest for an effective anthrax vaccine, including especially the discrepancy between his public and private accounts of the experiments at Pouilly-le-Fort.

Pasteur finally revealed the *modus fasciendi* of his chicken cholera vaccine in late October 1880, nine months after announcing its discovery. The first step in the preparation of this vaccine, Pasteur now disclosed, was to procure the chicken cholera microbe in its most virulent form by taking it from a chicken dead of the chronic form of the disease. In successive cultures made at brief intervals, this virulence remained constant, but attenuation set in when the intervals reached two or three months. To explain how this attenuation was achieved, Pasteur invoked the effect on the microbe of prolonged exposure to atmospheric oxygen.[56] Neither here nor elsewhere did Pasteur say exactly why oxygen should weaken microbes, especially the aerobic microbes (including the chicken cholera microbe) that ordinarily depended on it for life.

At one point in this paper of October 1880, Pasteur alluded to his prior silence on the method of attenuation, seeking once again to deflect repeated complaints about his "secrecy" from some members of the Académie de médecine. The "true reason" for his prior reticence, he said, ought now to be clear. "Time was an element in my researches."[57] The pace of his research had been slowed by the long intervals required for attenuation of the chicken cholera microbe and the variable results to which these long intervals contributed. What Pasteur did not reveal even now was the complex, fitful, and, in fact, still inconclusive program of research that had produced his oxygen-attentuated anthrax vaccine. As briefly noted above, in Chapter Two, Antonio Cadeddu has recently shown that Pasteur's laboratory notebooks are sharply at odds with the appealing legend, originating with Emile Duclaux, that the discovery of the vaccine was an "accidental" result due to a sudden moment of illumination and a single "crucial experiment" devised by Pasteur's "intuitive genius." It was instead the still-imperfect outcome of an extensive and twisting program of research in which an independent set of experiments by Roux played a crucial role. From Cadeddu's account, it is clear that Pasteur had exceptionally little experimental basis for announcing the "discovery" of an anthrax vaccine in January 1880.[58] More than that, as we shall soon see, even the oxygen-attentuated vaccine that Pasteur described for the first time in October 1880 was by no means yet fully established through decisive experiments.

By October 1880, when Pasteur finally disclosed his method of producing the chicken cholera vaccine, Toussaint had already announced the discovery of his anthrax vaccine and in August had already published the sealed note in which he described the *modus fasciendi* of that vaccine. Not until Feburary 1881 did Pasteur announce the discovery of his own anthrax vaccine. As noted early in this chapter, Pasteur's account of this new vaccine linked it with his earlier chicken cholera vaccine by ascribing attenuation in both cases to the action of atmospheric oxygen. There were, to be sure, important differences between the *modi fasciendi* of the two vaccines. Unlike the chicken cholera microbe, the anthrax bacillus formed spores that resisted the attenuating action of atmospheric oxygen. It had taken much time and effort to ascertain that a spore-free culture of anthrax bacilli could be produced at a temperature of 42°–43° C and would only then undergo attenuation. Although this procedure raised the possibility that the elevated temperature of 42°–43° C might itself play some role in attenuation, Pasteur stressed the role of atmospheric oxygen and thus the link between the new anthrax vaccine and his earlier chicken cholera vaccine.[59]

In announcing the discovery and *modus fasciendi* of his new anthrax vaccine, Pasteur asserted in passing that it was superior to Toussaint's "uncertain" method of heating anthrax blood.[60] A month later, on 21 March 1881, Pasteur delivered his most extended critique of Toussaint's work. He now unveiled in print the biological interpretation of Toussaint's results that he (and Roux) had maintained in private since mid-August and referred to Toussaint's rapid switch from a chemical to a biological interpretation—from a "dead" vaccine to an "attenuated" one—in such a way as to make himself responsible for it. He also insisted again on the practical deficiencies of Toussaint's "artificial procedure" of heating anthrax blood, which might lead to "great losses" if applied to sheep on a large scale. Even when Toussaint's heated anthrax blood did work, it failed to maintain its vaccinal properties in subsequent cultures. By contrast, stressed Pasteur, his own atmosphere-attenuated cultures of anthrax bacilli could be produced and maintained at any desired degree of attenuation in successive cultures.[61] In Pasteur's cultures, as Roux later put it, attenuation was *hereditary*.[62]

By the time he delivered this critique of Toussaint's work, Pasteur was well aware that his competitor had switched from heat to carbolic acid as the chief agent in his search for an effective anthrax vaccine. Yet his critique referred only to Toussaint's initial heat-produced vaccine. He said nothing here or elsewhere in print about the more recent antiseptic (carbolic acid) vaccine that Toussaint had used to produce immunity in the sheep that survived the trial at Alfort in August 1880. Nor did Pasteur ever refer in print to the related earlier work of his friend Casimir-Joseph Davaine (1812–1882), discoverer of the anthrax microbe, who had created something of a stir in 1873 by claiming that a wide range of antiseptics could render virulent anthrax blood inoffensive and could even be used to treat active anthrax.[63] In 1877, as only Pasteur's laboratory notebooks from that year reveal, he and his collaborators had themselves subjected the anthrax bacillus to the action of antiseptics, including notably carbolic acid, and had concluded that prolonged exposure to this antiseptic destroyed the bacillus.[64] It was almost surely Toussaint's novel claim of August 1880 that carbolic acid could be used to produce a *vaccine* against anthrax that now quickened the Pastorians' interest in the effects of antiseptics and other chemical agents on the anthrax bacillus.[65]

The most important experiments of this sort were conducted by Charles Chamberland, whose results are only sketchily recorded in the laboratory notebooks and other manuscripts now deposited in the Pasteur collection at the Bibliothèque Nationale in Paris. But the few details that these sources do record are crucial to the task of penetrating the secret of Pouilly-le-Fort. From two letters sent to Pasteur by Roux in late August 1880, for example, we know that Chamberland's experiments became intertwined with those of Toussaint in a curious and significant way that deserves some elaboration here.[66]

On 21 August 1880, just two days after informing Roux of Toussaint's Alfort trial with the carbolic acid vaccine, Bouley returned to Pasteur's laboratory on the rue d'Ulm. This time, Toussaint came with him. They told Roux of their plans to inject virulent anthrax into sheep that had survived the Alfort trial to see if these sheep had been rendered immune to the disease. They asked Roux to supply them with a virulent anthrax culture prepared according to the usual standards of Pasteur's laboratory. Roux gave them what he took to be such a virulent culture. But in a letter of 27 August, Bouley informed Pasteur that this allegedly virulent culture had failed to kill a rabbit used as a control, so that their first effort to test the immunity of the surviving sheep at Alfort had come to naught. They would need to try again with virulent anthrax blood rather than the inactive culture given them by Roux.[67]

In a letter of 30 August 1880, Roux now recalled that he had shown two different cultures to Toussaint and Bouley during their visit of 21 August. The first culture consisted of ordinary fresh anthrax blood. The second culture, prepared in July by Chamberland, had seemed upon microscopic examination to be a pure, "beautiful" culture containing a large number of anthrax bacilli in the form of spores. What Roux now suspected was that Toussaint and Bouley had taken with them a tube of Chamberland's culture, leaving the fresh anthrax blood behind. The results of their injection of this second culture into the control rabbit showed that it had for some reason lost its virulence. In a letter of 19 August, Roux had already drawn a tentative comparison between Toussaint's vaccines and some "enfeebled" anthrax cultures that Chamberland had produced by exposure to gasoline vapors.[68] Now in his letter of 30 August, Roux returned to the possibility that Chamberland's cultures might produce immunity in sheep. In a passage that Pasteur conspicuously underlined, Roux wrote as follows: "Why this anthrax culture proved to be inactive, I have no idea. Could it be that the anthrax cultures in many of the flasks we have at the laboratory would show themselves to be benign for sheep and give them immunity—notably the cultures of the spores that Chamberland has left exposed for some time to gasoline vapors? Perhaps it was a culture of this sort that I gave Toussaint without realizing it."[69]

Here Roux was suggesting that agents other than atmospheric oxygen, including at the least gasoline vapors, might have the capacity to attenuate cultures of the anthrax bacillus. He had been alerted to this possibility by the results of Toussaint's experiments, as disclosed to him by Bouley and Toussaint in their meetings of 19 and 21 August. It may very well have been these private disclosures that inspired Roux and Chamberland to undertake a focused search for an antiseptic anthrax vaccine.[70] At the same time, however, Pasteur pursued his quest for an effective oxygen-attenuated vaccine against anthrax, similar in principle to the chicken cholera vaccine he had already produced. And as this search continued, neither he nor his collaborators said anything in public about gasoline vapors or any other attenuating agents except oxygen.[71]

THE BACKGROUND TO THE TRIAL OF POUILLY-LE-FORT

When Pasteur announced the discovery of a new anthrax vaccine on 28 February 1881, he created great excitement among agriculturists and veterinarians. Whatever they thought of Toussaint's ongoing attempts to produce a vaccine against anthrax, they surely welcomed the news that the celebrated scientist from the Ecole Normale now had his own remedy to offer against this economically destructive livestock disease. The excitement only increased after 21 March, when Pasteur reported successful results in preliminary tests of his new vaccine on sheep and projected a full-scale field trial in the Beauce district when the sheep-penning season arrived there.[72] But these plans were forestalled by Hippolyte Rossignol, the veterinary surgeon from Pouilly-le-Fort whose farm was to serve as the site of the famous public trial. Up to this point, Rossignol had been profoundly skeptical of the germ theory of disease and of its "high priest" and "prophet," Louis Pasteur. Perhaps hoping that he now had an opportunity to embarrass Pasteur and his germ theory, Rossignol immediately challenged him to undertake an independent public trial of the newly announced vaccine and organized a campaign to secure funding for such a trial.[73]

Rossignol's challenge came at an awkward moment for Pasteur. He had now twice claimed in public that his laboratory already possessed an effective anthrax vaccine, produced by the method of oxygen attenuation and decidedly superior to Toussaint's heat-produced vaccine. But the boldly confident tone of Pasteur's public reports exaggerated the actual results to date of his experiments with the new vaccine. His laboratory notebook from this period tells a rather different story. Since mid-January, to be sure, Pasteur had been accumulating solid (if not remarkably extensive) evidence that anthrax cultures could be attenuated by exposing them to air at 42°–43° C.[74] In a notebook entry of 1 February, just a month before he publicly announced the discovery of his anthrax vaccine, Pasteur described one of these attenuated cultures as "very probably a vaccine" against anthrax in sheep.[75] Yet not until early March did Pasteur's notebook record the results of attempts to test his oxygen-attenuated vaccines on sheep. Well into April, by which time Rossignol's campaign for a public trial of Pasteur's new anthrax vaccine was in full swing, the notebook shows that the results of such tests remained decidedly inconclusive, if promising, and encompassed only a small number of sheep. One index of Pasteur's uncertainty is that he continued to test several different strains of his oxygen-attenuated vaccines.[76]

On 13 April 1881, just two weeks before Pasteur signed the exacting protocol of the experiments to be performed at Pouilly-le-Fort, his notebook records the results of a small-scale comparative test of one of his oxygen-attenuated cultures and one of Chamberland's potassium-bichromate vaccines. Of two sheep "vaccinated" with Pasteur's strain, one died when injected with virulent anthrax, while Chamberland's potassium-bichromate vaccine preserved both of the sheep into which it had been injected. Chamberland's vaccine was thus more secure (*sûr*), as Pasteur wrote in his notebook at the end of this modest trial.[77] Clearly, to borrow Adrien Loir's phrase, Pasteur's oxygen-attenuated vaccine was not yet "au point."[78]

What to do? Had Pasteur now confessed his private uncertainty about the efficacy of his oxygen-attenuated vaccines, he surely would have exposed himself to charges that he, like Toussaint, had made his announcement prematurely, in the absence of adequate evidence. But the same conclusion would very likely have been drawn had he now simply declined to accept Rossignol's well-publicized challenge. Worse yet, Pasteur's refusal to participate in an independent public trial of his new vaccine might have revived suspicions that he really was seeking to profit from "secret remedies" against livestock diseases. To deflect such criticism—and surely also to preserve his richly deserved reputation for taking on any scientific challenge in a public arena—Pasteur "impulsively" accepted the challenge of Pouilly-le-Fort. He signed the detailed and demanding protocol of experiments on 28 April 1881, just a week before his collaborators made the first of their preventive inoculations into the twenty-five earmarked sheep at Pouilly-le-Fort.[79]

If Pasteur's collaborators were at first dismayed by his decision to undertake the boldly uncompromising trial of Pouilly-le-Fort, their concerns apparently abated when the master told them that he would use Chamberland's more fully developed potassium-bichromate vaccine rather than his own still-uncertain oxygen-attenuated vaccine. As the trial neared its end, anxiety seems to have shifted from his collaborators to Pasteur himself.[80] But none of these concerns ever found their way into the public eye. In his published accounts of the Pouilly-le-Fort trial, Pasteur wrote with the casual assurance of one who had never for a moment doubted the safety or efficacy of the vaccine used there. Those same public accounts not only failed to disclose but actively misrepresented the nature of the vaccine actually used at Pouilly-le-Fort. From Pasteur's public statements, one could never have guessed that the triumphant results at Pouilly-le-Fort had been achieved through the use of an antiseptic vaccine, very similar in principle to the studiously ignored carbolic acid vaccine of Jean-Joseph Henri Toussaint.

CONCLUSION

It should be abundantly clear by now that Pasteur's research on anthrax vaccines was accelerated and otherwise greatly influenced by his competition with Toussaint. Toward a focused summary of this influence, let us begin by recalling Pasteur's surprise and agitation when he learned of Toussaint's claim that he had produced the first effective anthrax vaccine by injecting experimental animals with heated anthrax blood. When this news reached Pasteur in early August 1880, he immediately suspended his annual late-summer holiday at the familial home in Arbois and asked Roux and Chamberland to forego their own summer holidays in order to undertake tests of Toussaint's new vaccine.[81] In a previously quoted letter of 10 August 1880 to Henri Bouley, Pasteur ascribed his "astonishment" to the theoretical implications of Toussaint's allegedly "dead" vaccine against anthrax, which "overturns all the ideas I had on viruses, vaccines, etc."[82] But surely Pasteur's swift and agitated response also owed something to his concern about the distribution of credit and priority between him and Toussaint. That concern may well have been exacerbated by the fact that Pasteur had yet to reveal the *modus fasciendi* of his new chicken cholera vaccine. As Pasteur emphasized, this was the first alleged example of an artificial vaccine, produced in the laboratory rather than taken from nature in the matter of Jenner's famous "vaccine" (cowpox) against smallpox.[83] But its mode of production remained a secret, and Toussaint's new anthrax vaccine had therefore become the first artificial vaccine whose *modus fasciendi* was known.

Worse yet for Pasteur, Toussaint's new vaccine was directed against the very disease, anthrax, on which the Pastorians had lavished much of their time and energy for the past three or four years. They had long been in search of an effective vaccine. It is easy to imagine Pasteur's disappointment upon learning that all of these efforts might yield nothing better than second place in the race with Toussaint. Happily for him, neither Toussaint's initial heat-produced vaccine nor its later variants—including the carbolic acid vaccine he used in the trial at Alfort—fully met the great expectations they aroused. Yet despite Toussaint's "accidents" (that is to say, failures) and despite his rapid switch from a chemical to a biological interpretation of his results, he had in fact produced immunity in most of his experimental animals. The Pastorians' own preliminary and very private tests of Toussaint's carbolic acid vaccine yielded successful results—depending, to be sure, on the precise dosage employed.[84] One of Toussaint's experiments also focused Roux's attention on the possibility that Chamberland may already have produced vaccinal cultures by exposing the anthrax bacillus to agents other than atmospheric action—including, at the least, gasoline vapors. In these and other ways, Toussaint's work almost surely stimulated Chamberland and Roux to undertake a focused and quite rapidly successful search for an effective antiseptic vaccine against anthrax.

When "trapped" by Hippolyte Rossignol's challenge to undertake the public trial of Pouilly-le-Fort, Pasteur momentarily shelved his oxygen-attenuated vaccines and resorted instead to Chamberland's more fully developed potassium-bichromate vaccine. And though the Pastorians could have claimed priority for subjecting the anthrax bacillus to the action of potassium bichromate—which neither Toussaint nor his acknowledged precursor Davaine had done[85]—Pasteur was more immediately concerned with diverting attention from antiseptic vaccines of any sort. Already shaken by the recognition Toussaint had received since announcing his discovery of the first artificial vaccine against anthrax, Pasteur clearly wished to avoid any suggestion that the Pouilly-le-Fort vaccine had been prepared by a method so analogous to Toussaint's use of carbolic acid. He therefore suppressed the fact that an antiseptic vaccine had been responsible for the triumphant results at Pouilly-le-Fort and presented those results instead as if they were another demonstration of the virtues and promise of oxygen-attenuated vaccines. In the process, he very effectively diverted attention from the work and claims of his rival, Toussaint. In the wake of Pouilly-le-Fort, Toussaint effectively disappeared from the field of competition.

It would be easy to conclude by drawing a pathos-filled portrait of Pasteur's hapless competitor Toussaint: a lonely worker in Toulouse who did not belong to the prestigious Parisian institutions (notably the Académie des sciences and the Académie de médecine) through which Pasteur exerted much of his influence; a "mere" veterinarian who lacked the extensive laboratory facilities, talented assistants, and munificent state support that Pasteur enjoyed; a political *naif* who was repeatedly frustrated in his attempts to gain recognition for work that seemed similar in its central thrust to Pasteur's; a "true scientist" who published the contents of his sealed envelope as soon as members of the Académie de médecine complained about his "secret remedy" against anthrax and who never sought to patent or license his vaccine; finally, and most pathetically of all, a sensitive young man whose "mind gave way under the weight of the great thoughts it carried" within a year of Pasteur's triumph at Pouilly-le-Fort. Although Toussaint lingered on until 1890, when he died at the age of forty-three, he published only two papers after 1881, the last in 1882.[86] Two years before, Roux had written Pasteur of Toussaint's extremely nervous and agitated response to the news that the Pastorians were aware of deficiencies in his anthrax vaccines—perhaps already exhibiting signs of the mental derangement that became fully manifest within a year and that ultimately carried him off.[87]

This moving finale would only gain in pathos if we now considered the possibility that Pasteur sought to discredit and divert attention from Toussaint's work in order to corner the market on commercial anthrax vaccines. For it would be remarkably easy—some might find it appallingly easy—to build a case that Pasteur's conduct vis-à-vis Toussaint and Pouilly-le-Fort was motivated by the prospect of large fiscal returns on the sale of vaccines. From Pasteur's published work, we know that he sought exclusive control over the production of anthrax vaccines at least briefly, lest "a bad application of the method compromise the future of a practice that is called upon to render great services to agriculture."[88] His correspondence includes a letter of Christmas Day 1881 to the president of the Council of Ministers proposing the creation of a state factory for the manufacture of anthrax vaccines, of which he would be named director and in return for which he asked "only" that his family be "freed of material preoccupations."[89] The French government rejected this proposal, but Pasteur's laboratory nonetheless soon acquired a de facto monopoly over the manufacture of commercial anthrax vaccines. Even Adrien Loir, Pasteur's indulgent nephew, conceded that the Pastorians were sometimes sorely tempted to profit from this de facto monopoly, especially in the case of foreign sales of their anthrax vaccines.[90] More than that, there exists solid evidence that this temptation was not always resisted. By one perhaps biased but apparently well-documented estimate, Pasteur and his laboratory enjoyed a net annual profit of 130,000 francs from the sale of anthrax vaccines in the mid-1880s.[91]

In fact, Pasteur's unpublished correspondence offers additional and even more compelling evidence of his interest in profiting from vaccines against anthrax and other livestock diseases. Yet whatever may be said about Pasteur's pecuniary interest in vaccines, his conduct vis-à-vis Toussaint was in keeping with his treatment of other competitors who encroached on what he considered his territory, whether or not they represented a threat to his fiscal interests. Indeed, if only out of personal regard for Henri Bouley, Toussaint's steadfast advocate and defender, Pasteur displayed a rather more conciliatory tone toward Toussaint than he did toward most of his other rivals. Even as he criticized Toussaint's work or denied his claims to priority for this or that contribution, Pasteur did acknowledge some indebtedness to him and some grudging admiration for him. He even recommended that his sometime competitor be awarded an important prize by the Académie des sciences.[92]

Nor, in the end, was Toussaint—any more than Pouchet in the spontaneous generation controversy—somehow robbed of a victory that rightly belonged to him. Here, too, Pasteur's criticisms were persuasive. We need only recall Toussaint's own testimony (as recorded by Roux) that he was unable to produce consistently pure cultures of his vaccines.[93] His methods and procedures, consisting mainly of injecting variably virulent fluids into experimental animals, were indeed relatively crude and inadequate to the task at hand. He apparently never did find a way to reproduce in successive cultures whatever vaccinal properties his treated anthrax blood did acquire. Insofar as Pasteur's anthrax vaccines did have this property—insofar as their attenuation was hereditary—they enjoyed undeniable practical advantages over Toussaint's vaccines, which lacked such relatively stable levels of attenuation and would have had to be produced in immense quantities to immunize sheep on a large scale.[94] Pasteur could thus insist on the technical superiority of the anthrax vaccines produced in his laboratory.

And if Toussaint was strategically inept in advancing his claims, if he was less skillful than Pasteur at persuading others of the value and importance of his work, that is hardly Pasteur's fault. The ability to persuade, the effective use of "rhetoric" in the classical sense of the word, is one crucial index of talent in science as in any discursive field of creative work. In a very real sense, Pasteur's sensitivity to the concerns of his audience, his ability to win them over to his side, even his skillful exploitation of the external advantages he enjoyed, show that he was in fact a "better" scientist than Toussaint.

None of this is meant to excuse the unsavory features of Pasteur's conduct in the affair of Pouilly-le-Fort, but only to render it more comprehensible. We should also keep in mind the highly competitive context of mid-nineteenth-century French academic life, one index of which is Pasteur's aggressive concern for his priority or "intellectual property." Throughout his career, Pasteur displayed a very highly developed proprietary attitude toward the concepts and techniques that were associated with his name. But if he was unusually active and successful in defending his "property" against rival claimants, he was by no means alone in his efforts to do so. Indeed, the quest for personal recognition and priority has been a powerful influence on the behavior of scientists since the Renaissance, with roots in the high value placed on originality in the scientific community.[95]

In the affair of Pouilly-le-Fort, Pasteur's concern for priority and recognition went so far as to come into conflict with another of the norms that allegedly govern the scientific enterprise—truth-telling in public discourse. The tension between these two norms has ever been a feature of modern science, and the newspapers have lately been filled with examples of the fraud that can result when the norm of truth-telling is overwhelmed by the quest for personal status and recognition.[96] In truth, however, Pasteur's deception at Pouilly-le-Fort cannot properly be compared with the more egregious (if still rare) examples of recent "scientific fraud." We do not have to do here with any outright fabrication of data. And if Pasteur did almost everything he could to convey the impression that the vaccine used at Pouilly-le-Fort had been prepared by "his" method of oxygen-attenuation, he did not quite go so far as flatly to lie: he never did say, in so many words, "the vaccine used at Pouilly-le-Fort was an oxygen-attenuated vaccine."

More than that, Pasteur was surely motivated in part by a well-founded concern that a full disclosure of the events at Pouilly-le-Fort would lead his more hostile critics to award Toussaint credit for the discovery of vaccination against anthrax, despite the very real technical differences between their procedures and results. That this concern was well founded is clear from the behavior of Pasteur's leading German rival, Robert Koch, who eventually hailed Toussaint as the worthy inventor of vaccination against anthrax, while persistently denigrating Pasteur's contributions to this and other branches of the new science of bacteriology or microbiology.[97] As this episode suggests, Pasteur knew his enemies well. In the end, it is mainly a measure of the importance attached to originality in modern science—and of the competitive environment in which Pasteur lived, moved, and had his being—that a significant and undeniable element of deception should have entered into the most celebrated public experiment by one of the greatest heroes in the history of science.

SEVEN

From Boyhood Encounter to "Private Patients": Pasteur and Rabies before the Vaccine

ON 18 OCTOBER 1831 a lone but menacing wolf left its natural habitat in the wooded foothills of the Jura mountains in eastern France and descended upon several nearby communities, attacking and biting everything in its path. The focus of its rampage was the village of Villers-Farlay, where eight of its human victims eventually died of rabies, but it also bit several people in and around the town of Arbois. Some of these terrified victims made their reluctant way to a blacksmith's shop in Arbois, there to submit to the traditional treatment for a rabid animal bite: cauterization with a red-hot iron—in effect, to have their wounds "branded." From a spot within earshot of the screaming victims, an eight-year-old neighborhood boy watched this scene in horror. That boy, the son of a local tanner, was Louis Pasteur.[1]

Half a century later, on 17 October 1885, the now famous Pasteur received a letter from the mayor of Villers-Farlay, one M. Perrot, who informed him that this village so near his home town had once again been the site of an attack by a rabid animal. This time there was only one victim, a fifteen-year-old shepherd named Jean-Baptiste Jupille, who had come forth to do battle with a rabid dog when it charged him and a half-dozen younger shepherds watching over their sheep in a meadow. Jean-Baptiste saved his comrades by killing the dog, but during the struggle was severely bitten on the hands. Like most well-informed Frenchmen, Mayor Perrot knew that Pasteur had been working on a vaccine against rabies and therefore sought his advice and help in the case of young Jupille. As we shall see more fully in the next chapter, Pasteur quickly agreed to undertake the treatment of this brave shepherd boy, who escaped the rabies to which he had once seemed doomed. For now, however, the point is that Mayor Perrot's letter reminded Pasteur of the scene he had witnessed at the blacksmith's shop in Arbois half a century ago.

In fact, Pasteur's memory of his boyhood encounter with rabies remained so vivid that he asked Mayor Perrot to conduct an inquiry into this episode from the distant past. The stated aim of the inquiry was to investigate Pasteur's suspicion that those who had died of rabies after the wolf's attack of 1831 had been bitten on their hands or face, while those whose bites were confined to clothed areas of their bodies had escaped the disease. At Pasteur's request, Mayor Perrot dutifully interviewed surviving villagers who could still recall something of that terrifying day. His interviews not only confirmed Pasteur's suspicion but also provided a riveting account of the events of 18 October 1831 that is now deposited in the Pasteur papers at the Bibliothèque Nationale in Paris.[2]

Pasteur's boyhood encounter with rabies almost surely accounts for part of his later fascination with the disease. True, he made no such explicit claim himself and sometimes offered more prosaic reasons for choosing rabies as the target of his search for the world's first laboratory vaccine against a human disease. After he had achieved that goal, and as the still famous Institut Pasteur in Paris was being built with the grateful donations that this achievement inspired, he insisted in private correspondence that he had undertaken the study of rabies "only with the thought of forcing the attention of physicians on these new doctrines"—that is to say, his still controversial germ theory of disease and the technique of vaccination through attenuated cultures.[3] At an earlier point, in August 1884, when his rabies vaccine had not yet been applied to human cases, Pasteur offered a more specific reason for his interest in the disease in an address to the International Medical Congress in Copenhagen. His prior success at producing vaccines against animal diseases—chicken cholera, anthrax, and swine fever—naturally aroused hope that vaccination could be extended to human diseases. But blocking that goal was one immense obstacle—namely, that "experimentation, while allowable on animals, is criminal on man." For this reason, vaccination could be extended to man only on the basis of a deep knowledge of animal diseases, "in particular those that afflict animals in common with man."[4] As the oldest and most striking example of a lethal disease common to man and animals, rabies held special promise in the quest to extend vaccination to human diseases.

True enough, but the rabid wolf attack of 1831 also left Pasteur with a very personal and unforgettable appreciation of the popular horror of the disease. This popular horror of rabies had no basis in its statistical or demographic significance. For rabies has always been rare in man. It probably never claimed even a hundred victims in any year in France, and French estimates for the decade just before Pasteur produced his famous vaccine indicate an annual mortality of considerably fewer than fifty.[5] Rabies was equally rare in England, where the annual mortality rate ranged from a low of one person in 1862 to a high of seventy-nine in 1877, "by far the worst year on record."[6] But despite its rarity, rabies has always held a very special place in the popular imagination. It was—until AIDS—the very model of a mysterious and horrific disease. Its usual carrier is man's best friend. Its human victims are all too often children. Its microscopic anatomical lesions and its proximate agent, a tiny filterable virus, long escaped detection and isolation, leading a few to insist that rabies could arise "spontaneously," in the absence of a rabid animal bite. One persistent theory held that the disease could result from the nervous trauma allegedly suffered by sexually frustrated dogs, and men in the throes of symptomatic rabies were sometimes said to be priapic and sexually insatiable.[7] In this and other ways, rabies became linked in the popular imagination with "animal" sexuality, bestiality, and other cultural anxieties, so much so that the appearance of rabies in a community sometimes led to panic and even to "Great Dog Massacres" that were designed to exorcise the evil disease.[8]

The terrifying spectre of "spontaneous" rabies found some, if not much, empirical sanction in the prolonged and variable interval between the bite of a rabid animal and the outbreak of symptoms in its victims. Most students of rabies had long since agreed that it was caused by a poison (or "virus") transmitted in the saliva of the attacking animal, but they had to admit that this alleged virus eluded detection and that its lethal work remained long invisible and intangible. The "incubation period" of rabies varies widely from species to species and from individual to individual. In dogs, the average is perhaps a month. In humans, the incubation period is usually a month or two, but occasionally reaches a year or more. This feature of rabies aroused profound dread in any victim of an animal bite, who could never be sure that the disease might not yet manifest itself in him or her.[9]

But it was of course the symptoms and outcome of rabies that inspired this dread. In the scarcely exaggerated popular image of the disease, rabies embodied the ultimate in agony and degradation, stripping its victims of their sanity and reducing them to quivering, convulsive shadows of their former selves. The rabies virus moves slowly but steadily from the site of the infective wound toward the organs of the central nervous system. The initial symptoms give little indication of the horror to come. Among the early signs of clinical rabies—irritability, fatigue, malaise, and other nonspecific forms of distress—perhaps the most common (though even they are by no means universal) are pain at the site of the infective wound and severe headache. Within a few days, more obvious indications of central nervous system involvement begin to manifest themselves. Difficulty in breathing, severe pain in the stomach or chest, and extreme hypersensitivity to visual stimuli (especially bright or shimmering objects) often appear as the disease continues its relentless course. Perhaps the most distinctive feature of the disease, present in the majority of rabies patients, is a pronounced aversion to liquids, which the victim often pushes aside even when desperately thirsty. This symptom gives rabies its other familiar name, "hydrophobia" (or fear of water)—though the fear is not of water per se, but rather of the pain, choking, gagging, and convulsions induced by trying to swallow the shimmering liquid.

By the time the virus reaches the brain, the effects are often such as to make its victims behave like "mad" animals themselves. An appreciable minority (perhaps 20 percent) of rabies patients exhibit a predominantly "dumb" or quiet and paralytic form of the disease. But most suffer from the "furious" form marked by episodes of extreme hyperactivity, convulsions, thrashing, hallucinations, excessive salivation, and spitting. A few even howl like forlorn dogs and try to bite anyone within reach. The quiescent periods that separate these episodes of bizarre behavior are in some ways worse yet. For the pitiful victims then often display an almost eerie lucidity, a heightened sense of affection toward relatives and others, and an exquisitely human awareness of their impending death.

Of the many horrifying features of rabies, surely the most dreaded and dreadful is its uniformly fatal outcome. Once the symptoms become manifest, once the disease has "declared" itself, the mortality rate is effectively 100 percent.[10] The only merciful feature of rabies is that its clinical course is fairly brief—the final stupor and coma ordinarily come within a few days of the outbreak of symptoms. The immediate cause of death is usually cardiac arrest or respiratory collapse. At least until the advent of mechanical respirators, which in effect only prolong the agony, all that could be done for rabies patients was to make them as comfortable as possible, usually by placing them in a darkened room and otherwise reducing external stimuli. No one who has observed a rabies patient—certainly, no physician who has stood by helplessly as the disease took its toll—is likely to forget the experience, and there can be few more poignant stories in the annals of medicine than case histories of rabies. Some sense of the full horror of the disease is captured in the remarkable and pitiful case history of John Lindsay, weaver at Fearn Gore near Bury, England, first published in 1807 and reprinted in Appendix K. Listen, finally, to the testimony of the distinguished physician and belletrist Lewis Thomas. In late 1993, as he lay on his own deathbed, Thomas spoke of his belief that, in the final momemts, death comes without agony, perhaps because the brain's last act is to release pain-killing opiates.

He could think of only one exception from his clinical experience: death from rabies, where the agony never seemed to end.[11]

By now it should be clear that rabies, however rare, was an especially dramatic disease with which to begin the effort to extend laboratory-produced vaccines to human diseases. From the outset, Pasteur knew that he would be hailed as a savior if he succeeded in this quest.[12] Here above all he displayed the theatrical flair that marked his choice of subjects to pursue and his manner of presenting the results to an audience gripped with suspense and eager to hear a happy ending. As everyone knows, Pasteur did not disappoint them. The closing act of his work on rabies was an appropriately spectacular conclusion to an already remarkable career.

PASTEUR'S WORK ON RABIES FROM 1881 THROUGH 1884: THE LABORATORY NOTEBOOKS

But behind this last great public performance lay a long and often disappointing series of rehearsals. Here again Pasteur's private laboratory notebooks will serve as a central source. In this chapter, they will be used mainly to enrich, rather than replace, the story that emerges from Pasteur's published papers on rabies between 1881 and 1884. But the notebooks from this period also reveal a dramatic and important story that left no trace whatever in Pasteur's published work. That story, told here for the first time, concerns two hitherto unknown attempts by Pasteur to cure symptomatic rabies in human cases. And it turns out that the story of these two "private patients" may be linked with the single most celebrated achievement in Pasteur's career—the application of his rabies vaccines to young Joseph Meister in July 1885, as we shall see in Chapter Nine.

Pasteur's laboratory notebooks contain at least one passing reference to rabies as early as 1876. In August of that year, on a list of "books to buy," he included Joseph Enaux and François Chaussier, *Méthode de traiter les morsures des animaux enragés* (*Method of Treating the Bites of Rabid Animals*), describing it as a "good treatise to consult."[13] In this book, published nearly a century before (in 1785), Enaux and Chaussier ascribed rabies to a poison (virus) and endorsed the classic treatment of cauterizing rabid animal bites as soon as possible. But the book also contained an extended discussion of "malignant pustule," the name by which anthrax was known when it occurred in humans.[14] At this point, in 1876, Pasteur was almost certainly more interested in this part of the book than in its main topic of rabies. Most of the other books on his list concerned anthrax, including "all the works of Davaine."[15] And, as we have seen in the preceding chapter, Pasteur was just then redeeming his long-standing pledge to begin the study of infectious diseases by focusing on anthrax.

It was not until mid-December 1880 that Pasteur began to make regular and sustained references to rabies in his laboratory notebooks. That is not to say that rabies had now become the central focus of his research. Far from it. We need only recall from the preceding chapter that just five months had passed since Toussaint had announced his discovery of a vaccine against anthrax, leading Pasteur and his collaborators to accelerate their research on that disease and to redouble their efforts to find their own vaccine against it. At a point when the famous Pouilly-le-Fort trial was still six months away, it can hardly be said that rabies had become Pasteur's dominant preoccupation. Nor did it become so for roughly three more years. In the meantime, Pasteur and his collaborators pursued research on a variety of other diseases. Besides anthrax and chicken cholera, they included septicemia, swine fever, peripneumonia, yellow fever, and "horse typhoid." Because Pasteur generally arranged his notebooks chronologically rather than topically, and because he and his collaborators were pursuing these several lines of research simultaneously, the notebook pages devoted to rabies are repeatedly interrupted by reports of work on other diseases. It is therefore often difficult to follow every twist and turn in the path of Pasteur's early research on rabies, and no systematic attempt to do so will be made here.

One thing, however, is perfectly clear. As Pasteur reported in January 1881, and as his laboratory notes confirm, his research on rabies began on 10 December 1880. On that day Dr. Lannelongue, a surgeon at the hospital of Sainte-Eugénie, informed him of the admission there of a five-year-old boy suffering from rabies. Pasteur went to see the boy at 5 o'clock that afternoon with his collaborators, Charles Chamberland and Emile Roux. They observed all the classic symptoms of declared rabies in this doomed little boy, who had been bitten on the face by a rabid dog a month before. The boy died at 10:30 the next morning, December 11. At 3 o'clock that afternoon, four hours or so after the boy's death, Pasteur used a painter's pencil to collect some mucus from his mouth. After being mixed with a small amount of ordinary water, the mucus was injected into two rabbits that were then transported to Pasteur's laboratory on the rue d'Ulm. Both rabbits succumbed to these injections within thirty-six hours.[16]

Over the next few weeks, Pasteur established that blood taken from the two rabbits could, in its turn, produce similarly rapid deaths with similar symptoms in other rabbits or dogs. He associated these deaths with a new microbe similar in form (a figure 8) to the chicken cholera microbe, but different in its physiological properties and pathological effects. He also managed to cultivate this new microbe in artificial cultural media.[17] In his laboratory notebook from this period, he sometimes referred to it as "the microbe of rabies,"[18] but that was almost surely out of convenience rather than conviction. Even in private, Pasteur never insisted upon any direct connection between this new microbe and rabies. His published accounts were even more circumspect. From the outset, in January 1881, he spoke only of "a new disease produced by the saliva of a child dead of rabies." Given the source of the saliva, he did not immediately dismiss the possibility that there might be some "hidden relation" between rabies and this new microbe, but he also stressed that the disease it produced—both in its symptoms and in the rapidity with which it killed rabbits and dogs—differed strikingly from ordinary rabies.[19] By March, Pasteur had found the new microbe in the saliva of healthy adults as well as in that taken from victims of diseases other than rabies.[20] He then firmly rejected any connection whatever between rabies and the new microbe—which he had now come to call the *organisme auréole* (the "organism with a halo") or simply the *microbe de salive*, the "saliva microbe."[21]

With this point established, Pasteur's interest in the new microbe declined sharply. He did later claim that the saliva microbe—like the chicken cholera and anthrax microbes—could be attenuated and a vaccine therefore produced by exposure to atmospheric air. But this alleged new vaccine had nothing to do with rabies and had little practical import of any sort. For by June 1881, when he announced the discovery of this vaccine, Pasteur had decided that the saliva microbe might well be entirely harmless to man, however lethal its effects when injected into rabbits or dogs.[22] Some latter-day students of Pasteur's work have identified his "saliva microbe" as a pneumococcus.[23] If so, he never recognized it as such himself.

As early as 26 January 1881 Pasteur referred in his laboratory notes to a search for "the organism of true rabies."[24] The locus of his search—in the brain tissue of rabies victims rather than in their saliva or blood, where the saliva microbe could be found—suggests that he even then very much doubted any direct link between rabies and the saliva microbe. But he did always suppose that a rabies microbe must exist and tried repeatedly to isolate it. His laboratory notes record moments of hope when he thought he had achieved that goal, but in the end he had to admit that the "true rabies microbe" continued to elude him.[25] In retrospect, we can say that Pasteur's search for this microbe was doomed to fail given the techniques at his disposal. For the rabies "microbe," like that of smallpox, is in fact a filterable virus, much too small to be detected by the microscopes then available and incapable of cultivation in any of the artificial cultural media known to Pasteur.

But Pasteur did not allow his failure on this front to block advance along other lines. With a flexibility born partly of necessity, he came increasingly to focus on a principle and technique that had played a decidedly secondary role in his work on chicken cholera and anthrax. In that work, the main goal—and success—had been to cultivate and attenuate the implicated microbe in sterile artificial media, outside the animal economy. Yet Pasteur had also long conceived of living organisms as another sort of "cultural medium," and the ultimate success of his quest for a vaccine against rabies depended crucially on his skillful exploitation of this insight. Using by turns rabbits, guinea pigs, dogs, and monkeys, Pasteur made the nervous tissue, and especially the brain, of living organisms the medium in which to cultivate and hopefully attenuate the otherwise elusive "rabies microbe."[26]

Pasteur had at least one important predecessor in his work along these lines, the now forgotten veterinarian Pierre-Victor Galtier (1846–1908), who (like the ill-fated Toussaint of Chapter Six) had studied with Auguste Chauveau and ultimately became a professor at the Veterinary School of Lyon.[27] In 1879 Galtier reported that rabies could be transmitted experimentally from dogs to rabbits with a marked reduction in the incubation period of the disease—from perhaps a month on average in dogs to an average of eighteen days in his rabbits. This result almost literally doubled the number of experiments that could be performed within a given period of time. This advantage of rabbits—along with the fact that they were relatively cheap, safe to handle, and easy to keep—quickly made them the experimental animal of choice for students of rabies, including Pasteur. Galtier also suggested that the long incubation period of rabies raised the possibility that a preventive remedy might be applied after infection with the virus but before the symptoms broke out.[28] In 1881 Galtier reported that he had transmitted rabies experimentally to guinea pigs as well as rabbits and claimed that sheep could be rendered immune to rabies by the intravenous injection of saliva from rabid dogs.[29]

In his published work, Pasteur referred only once to this claim by Galtier—the first reported example of the experimental production of immunity against rabies—and even then only to cast doubt upon it.[30] Several years later, however, Pasteur's own leading collaborator on rabies, Emile Roux, publicly confirmed Galtier's claim that sheep could be rendered immune to rabies by the intravenous injection of saliva from rabid dogs.[31] By then, however, Pasteur had produced his own vaccines against rabies in dogs and man. Small wonder that Galtier's contributions have faded from view. Yet his work surely gave Pasteur reason to hope that his own efforts to produce immunity against rabies were not entirely baseless, and in any case Galtier had established the possibility and advantages of using rabbits and guinea pigs in rabies research.

Pasteur and Roux quickly seized the opportunities opened up by Galtier's work. And they soon went well beyond anything done by Galtier, who published nothing novel on rabies after 1881.[32] From the outset, Pasteur and Roux skillfully exploited an important finding that had emerged during their earlier work on chicken cholera and anthrax: namely, that the virulence of pathogenic microbes vis-à-vis a given organism could be altered by sequential (serial) passages through the same or other appropriate living organisms. The virulence of any microbe is relative to the organism to which it is applied and the organisms through which it is successively passed. Serial passages of a microbe through one species may increase its virulence vis-à-vis a given organism, while serial passages of the same microbe through a different species may decrease its virulence vis-à-vis that same organism. For example, as we shall see more fully below, the serial passage of the rabies virus through rabbits *increases* its virulence vis-à-vis both dogs and humans, while serial passage of the virus through monkeys *decreases* its virulence vis-à-vis both dogs and humans. In general, serial passage of a microbe within a given organism increases its virulence for that organism; for example, serial passage of the rabies virus through guinea pigs increases the virulence of the virus in successive guinea pigs.[33] (See fig. 7.1.)

Actually, this effect had been known before Pasteur focused his attention on animal diseases. His great German rival, Robert Koch, for one, had drawn attention to the increasing virulence produced by serial passages in his early work on anthrax and traumatic infectious diseases. But Koch supposed that serial passages increased the virulence of microbial cultures by enhancing their "purity"; in 1878, he described the technique of serial passages as "the best and surest method of pure cultivation,"[34] and did not imagine that the intrinsic properties of the microbe had thereby been changed. Pasteur, by contrast, had come to believe that the alterations in virulence produced by serial passages resulted from real changes in the properties of the microbe itself. Through exposure to different "cultures," Pasteur gradually realized, microbes could be quite fundamentally transformed in their physiological and pathological properties.[35]

In any case, by the time Pasteur took up the study of rabies, he knew that attenuated cultures of the chicken cholera microbe could regain their original virulence in chickens by repeated passages through young or small birds. Similarly, he knew that attenuated cultures of the anthrax bacillus could be made progressively more virulent in sheep by repeated passages through young guinea pigs. He and Roux doubtless expected a similar result in the case of rabies and therefore launched a systematic program of experiments in which the rabies virus was passed sequentially from rabbit to rabbit or guinea pig to guinea pig. In the process, the virulence of the rabies virus gradually increased vis-à-vis dogs as the incubation period of the disease gradually decreased. Ultimately, they found that the incubation period,

Ordinarily, the serial passage of a given micro–organism through another organism increases the virulence of the microbe vis–à–vis that organism. But this rule is by no means universal. In fact, the virulence of any given microbe vis–à–vis other organisms is relative. Hosts vary in their response to the invasion of microbes, and they sometimes decrease rather than increase the virulence of the invading microbes vis–à–vis themselves or other organisms. The following chart provides examples of both outcomes from Pasteur's own research.

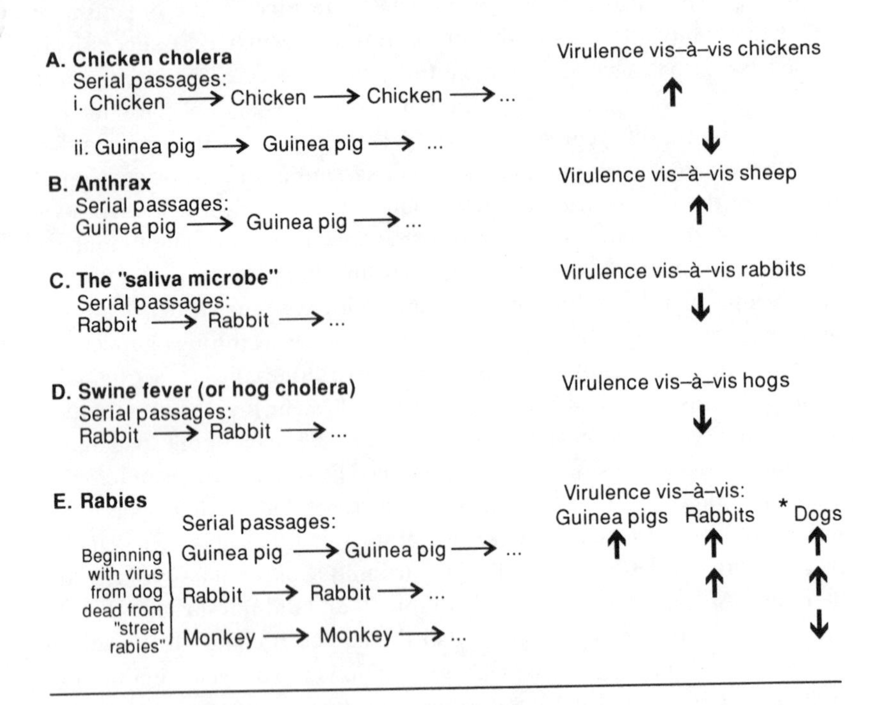

*Although Pasteur never quite said so explicitly, he assumed that the immune response in dogs would be in the same direction for humans.

↑ Indicates increase in virulence (correlated with decrease in incubation period)

↓ Indicates decrease in virulence (correlated with increase in incubation period)

Figure 7.1. On the relativity of immune responses.

perhaps a month or more in "street rabies"—as Pasteur called the rabies in dogs that acquired the disease in the ordinary way (i.e., through the bites of rabid dogs)—could be reduced to about a week in rabbits by prolonged serial passages through rabbits.[36] Pasteur and Roux thus reduced by at least half again the already abbreviated incubation period that Galtier had achieved in his earlier work on experimental rabies in rabbits and guinea pigs. In doing so, Pasteur and Roux further demonstrated the advantages of these animals in rabies research. More than that, as we shall see, the highly virulent and stable or "fixed" rabies virus that resulted from these serial passages through rabbits was eventually to become the starting point in the production of Pasteur's rabies vaccine.

But if the method of serial passages was the centerpiece of Pasteur's early work on rabies, he long kept it private. Three years passed before he disclosed any details of the technique in print. In fact, Pasteur was generally reticent about his work on rabies until mid-1884. Here again, as in his search for vaccines against chicken cholera and anthrax, Pasteur pressed onward with only occasional hints in public as to the nature and progress of his ongoing research. As late as February 1884, Pasteur had published only two brief papers on rabies. For the most part, however, the thin public record was this time in keeping with the actual state of affairs in Pasteur's laboratory. We must not forget that he and his collaborators were simultaneously at work on several other diseases, including notably anthrax. More important, their research on rabies had not yet brought much in the way of secure results. Partly because the presumed rabies microbe persistently eluded them, the Pastorians found rabies a difficult challenge.

Pasteur's notebooks throughout the first four years or more of his work on rabies provide a full and rich record of the often confusing, inconclusive, and frustrating results of this research. The notebooks also suggest that Pasteur's passion for order—in nature as in daily life—did not always serve him well in his work on rabies, where individual responses to pathogenic microbes tended to disrupt any neat pattern. Barely concealed beneath the laconic and meticulous records of his experiments is Pasteur's increasing impatience at the vagaries of disease as it manifested itself in real individual living organisms. This would have come as no surprise to a clinician—as it did not to Pasteur's medically trained collaborator, Emile Roux—but Pasteur found it frustrating. Living animals, it turned out, were rather crude and demanding "cultural media."

Pasteur's frustration becomes most evident in the notebook pages devoted specifically to efforts to find a vaccine against rabies. No attempt will be made here to give a systematic account of the early phases of this quest. But it does deserve saying even now that Pasteur's search for a rabies vaccine, like his earlier work on methods of brewing beer, was characterized by a remarkably empirical, hit-or-miss approach to the problem. Charting through the Byzantine maze of Pasteur's early laboratory notes on rabies, one eventually realizes that his basic procedure was simply to inject a variety of experimental animals—though mainly rabbits—with a wide range of cultures or substances and then watch what happened. Through late 1883, Pasteur had tried the following techniques among others in his search for a rabies vaccine: injections of saliva or blood from rabid animals, sometimes in large quantities, sometimes in small; inoculations with filtered emulsions of rabid brain tissue; and injections of emulsified rabid tissue from the medulla oblongata that had previously been treated with hydrogen peroxide (*l'eau oxygène*), perhaps a reflection of the oxygen theory of attenuation he had conceived during his work on the chicken cholera vaccine. He even tried to protect animals against rabies by infecting them with anthrax.[37] In the end, none of these approaches fulfilled Pasteur's occasional expressions of confidence in one or another of them.

But if these and other failures ever tempted Pasteur to abandon his search for a vaccine, there was always the awful specter of rabies victims to revivify his efforts. Not even the usually aloof and outwardly gruff Pasteur could ignore or forget the horror of rabies, especially in children. His otherwise impassive laboratory notebooks sometimes take on a very different tone in the face of clinical rabies in children. A notebook entry in early November of 1883, for example, records the poignant story of a seven-year-old boy who was seized with a severe headache upon leaving school and fell into convulsions upon reaching home. The night before entering the hospital where he was very soon to die of rabies, this pitiful boy had a premonition of the disaster that awaited him and beseeched his mother not to leave him alone, embracing her in "very enthusiastic and prolonged caresses." Pasteur's notebook then reverted to its usual dispassionate tone as he recorded the effects of rabies in rabbits inoculated with brain tissue taken from this boy after his death.[38]

This and other recorded encounters with doomed rabies patients repeatedly stoked Pasteur's ambition to find a way to prevent all such scenes in the future. As he and his collaborators struggled fitfully toward that goal, Pasteur occasionally disclosed their most secure results, albeit sometimes after a substantial delay and almost always briefly and vaguely. It comes as something of a surprise to discover that the entire body of Pasteur's published work on rabies barely fills one hundred pages in print.

PASTEUR'S PUBLISHED PAPERS ON RABIES FROM MAY 1881 THROUGH AUGUST 1884

In May 1881, in his first published paper on rabies per se (as distinct from the "saliva microbe"), Pasteur reported that he and his collaborators had developed a new technique for transmitting the disease with certainty. Hitherto, research on rabies had been impeded by the fact that the disease was not consistently transmitted either by the injection of rabid saliva or the bite of a rabid animal. More surprisingly, since it seemed clear that the nervous system and especially the brain was the ultimate seat of the disease, even subcutaneous injections of rabid nervous tissue did not always transmit the disease from one animal to another. Galtier himself had reported that such injections did not uniformly produce rabies in the recipient animal. Pasteur seized on Galtier's admission to emphasize that some doubt remained as to the anatomical locus of rabies and, ipso facto, the most reliable way of transmitting the disease experimentally. But, said Pasteur, he and his collaborators—actually, it was Emile Roux[39]—had at last developed a uniformly successful method of transmitting the disease from animal to animal. In this new method, cerebral matter was extracted from a rabid dog under sterile conditions and then inoculated directly onto the surface of the brain of a healthy dog through a hole drilled into its skull. Under these circumstances, Pasteur reported, the dog thus inoculated through its trephined skull invariably contracted rabies in less than three weeks, as compared to the average incubation period of a month in dogs that contracted rabies in the ordinary way through the bites of another rabid dog.[40]

More than eighteen months passed before Pasteur published a second paper on rabies. In this brief and often vague paper of December 1882, with "all details left aside for the present," Pasteur announced that rabies could also be reliably transmitted to previously healthy animals by intravenous injection. When transmitted this way, as distinguished from the previously announced method of intracranial inoculation, the virus usually produced rabies in its "paralytic" rather than "furious" form. Pasteur further claimed that the incubation period of the disease had now been reduced to somewhere between six and ten days, though he said nothing to indicate precisely how this result had been achieved. He reported that nothing had yet come of attempts to produce immunity against rabies by injecting saliva or blood from rabid animals into healthy ones. But he and his collaborators had happened upon a few dogs that were "spontaneously" or "accidentally" immune to rabies. When injected with a rabies virus that was virulent enough to kill other dogs, these innately "refractory" dogs also displayed symptoms of rabies, but then recovered from the disease and resisted subsequent injections of highly virulent rabies virus. This result established that rabies shared the distinguishing feature of the other "virus" diseases exemplified by smallpox: it did not recur in a host that had survived an initial attack of the disease. That rabies shared this feature of viral diseases had been far from certain since its victims almost always died. But Pasteur now insisted that a few "naturally" resistant dogs could indeed recover from relatively mild forms of symptomatic rabies, after which they remained forever immune to the disease. And this result encouraged hope that the search for a vaccine against rabies might eventually succeed. Pasteur concluded his paper by reporting that he and his collaborators had now carried out more than two hundred experiments on rabies in pursuit of this goal.[41]

Several hundred more experiments had been completed, with the sacrifice of several hundred more animals, by the time Pasteur published his third paper on rabies after another interval of more than a year. By this point, in February 1884, Pasteur and his collaborators had been at work on the disease for more than three years. The two papers published thus far had been brief and tantalizingly vague. Only Roux's technique of transmitting rabies by intracranial inoculation through a trephined skull had been described in any detail—though Pasteur's laboratory notebooks make it clear that even this technique did not invariably succeed.[42] But now, quite suddenly, in his third published paper on rabies, Pasteur claimed that he and his collaborators were well on their way to a solution to the problem of rabies.

In this third paper, delivered to the Académie des sciences on 25 February 1884, Pasteur did concede that he and his collaborators had still not managed to isolate and cultivate a rabies microbe in artificial media, though he continued to presume that one must exist. Insisting that a rabid brain could easily be distinguished from a normal one by the presence of numerous fine granules in the rabid medulla, he hoped that he would eventually be able to prove that these granulations were "actually the germs of rabies." But whatever the outcome of further attempts to isolate the rabies microbe, there was already much more exciting news to report. For Pasteur claimed that he and his team had now found a "method of rendering dogs resistant to rabies in numbers as large as desired." The point of departure for the new method, he reported without elaboration, was the production of rabies viruses of varying degrees of virulence. He further disclosed that he now had on hand twenty-three dogs capable of withstanding injections of the most virulent rabies virus. In principle, the problem of preventing rabies in man had also now been solved, since the dog was the ultimate source of the disease. Moreover, the lengthy incubation period of rabies gave reason to hope that a bite victim could be rendered immune before the symptoms became manifest.[43]

Three months later, on 19 May 1884, Pasteur gave a somewhat fuller account of the methods by which he and his collaborators had prepared the rabies virus in varying degrees of virulence. Pasteur now publicly disclosed, for the first time, their technique of increasing the virulence of ordinary canine rabies by serial passages through guinea pigs or rabbits. In both species, serial passages led to a gradual increase in virulence and an associated decrease in the incubation period of the disease. The shorter the incubation period, the more virulent the virus that produced it. As the virus was passed sequentially from guinea pig to guinea pig or rabbit to rabbit, the incubation period steadily declined toward a stable minimum—roughly a week in the rabbit—that corresponded with a stable or "fixed" maximum in virulence. Pasteur had reported as early as December 1882 that the incubation period in his experiments had already reached six to ten days instead of the month typical of ordinary "street rabies" in dogs. But only now, seventeen months later, did he disclose the method by which this result had been achieved.[44]

In this same paper of May 1884, Pasteur also revealed that he and his collaborators had found an organism in which serial passages produced the opposite effect on the rabies virus—decreasing rather than increasing its virulence. Actually, this attenuating effect of serial passages, like its inverse, had already been noticed and exploited during Pasteur's earlier work on other diseases. Specifically, Pasteur had found that the virulence of the saliva and swine fever microbes could be decreased as well as increased by serial passage through appropriate living organisms. Successive passages of the saliva microbe through the guinea pig, for example, made the microbe less virulent for rabbits. This result, published in September 1882, suggested the possibility that an attenuated culture—in a word, a vaccine—might be produced against any given microbial disease by successive passages of the implicated microbe through appropriate animals.[45] In November 1883, Pasteur reported that precisely this method had been used to produce a new vaccine against swine fever. The crucial step in the production of this vaccine had been the discovery that the swine fever microbe could be attenuated to the point of harmlessness for hogs by several passages through rabbits.[46]

In the case of rabies, Pasteur now reported in his paper of May 1884, the virus could be attenuated for dogs by passing it from dog to monkey and then successively from monkey to monkey. After just a few such passages through monkeys, he claimed, the rabies virus became so attenuated that its hypodermic injection into dogs never resulted in rabies. Indeed, it sometimes produced no effect even when transmitted to dogs by Roux's supposedly infallible method of intracranial inoculation. At some point in its serial passage through monkeys, the rabies virus lost its virulence for dogs and began instead to protect them from the effects of somewhat more virulent strains of the virus, which in their turn acted as vaccines against still more virulent strains until eventually dogs could be rendered immune to even the most lethal virus. If all dogs were vaccinated in this way, rabies could eventually be eliminated. But until that "distant period," wrote Pasteur, there was an obvious need for a means of preventing rabies in humans after the bite of a rabid animal. He then created great excitement by reporting that his first attempts along this line in monkeys seemed highly promising. He went so far as to say that "owing to the long incubation, I believe that we will be able to render [human] patients resistant with certainty before the disease becomes manifest." But he also emphasized that "proofs must be collected from different animal species, and almost ad infinitum, before human therapeutics can be so bold as to try this mode of prophylaxis on man himself."[47]

Pasteur closed his paper of May 1884 with a characteristic request that an official commission be appointed—in this case by the minister of public instruction—to validate the results of his research on rabies. He proposed to have this rabies commission begin its work by observing two sets of experiments that bore a striking structural resemblance to the famous Pouilly-le-Fort trial of his anthrax vaccine. First, he suggested, twenty of his vaccinated dogs should be placed with twenty unvaccinated dogs and all forty should then be subjected to the bites of rabid dogs. Second, the same experiment should be performed, except that the forty dogs should be infected with rabies through the almost infallible method of intracranial inoculation instead of through the bites of rabid dogs. Echoing almost perfectly the bold prophecy he had issued before the Pouilly-le-Fort trial, Pasteur predicted that "not one of my twenty [vaccinated] dogs will contract rabies, while the twenty control animals will."[48]

The proposed commission was duly appointed within a month. Among its members were several of Pasteur's leading colleagues and supporters from the Académie des sciences. Its chairman was his now long-standing convert, the veterinarian Henri Bouley, who had also played an important role in the Pouilly-le-Fort trial three years earlier. This French rabies commission published its initial report on 4 August 1884.[49] After two months of experiments whose results were reported to it by Pasteur, the commission found that none of his twenty-three vaccinated dogs had contracted rabies—whether from the bites of rabid dogs or from Roux's method of intracranial inoculation of the rabies virus. By contrast, two-thirds of the unvaccinated control dogs had already become rabid.[50] What the commission did not report in any detail—nor could it, since Pasteur had not supplied the pertinent information—was the method or methods by which his "refractory" dogs had been made immune. Like other readers of Pasteur's published papers, the commission members presumably supposed that his "refractory" (i.e., immune) dogs had been injected first with a rabies virus attenuated by serial passages through monkeys and then with progressively more virulent strains of the virus.

A week later, in a major address of 10 August 1884 to the International Congress of Medical Sciences at Copenhagen, Pasteur proudly cited this report of the French rabies commission in support of his claim that "rabies is no longer an insoluble riddle." Situating his work on rabies in the context of a wide-ranging and triumphant discussion of the growing evidence for his germ theory of disease, Pasteur admitted that his audience "must be feeling a great blank in my communication; I do speak of the micro-organism of rabies." The reason for the omission, he continued, is that "we have not got it. . . . long still will the art of preventing disease have to grapple with virulent maladies whose micro-organic germs escape our investigation." Nonetheless, Pasteur reported, he and his collaborators had made major strides toward solving the rabies problem. Now, at last, he described in some detail the method of intracranial inoculation and his method of preparing the rabies virus in varying degrees of virulence. He stressed that the search for an attenuating medium for the virus had been long and frustrating. Through hundreds of experiments, the animals selected as potential attenuating organisms proved instead to increase rather than attenuate the virulence of the rabies virus. Not until December 1883 had he and his team turned to the monkey and uncovered its capacity to attenuate the rabies virus. As in his paper of May 1884, Pasteur asserted that the inoculation of a rabies virus attenuated by serial passage through monkeys, followed by increasingly virulent strains of the virus, could produce a "completely refractory state" in dogs. The only obstacle to the application of this method in human cases, wrote Pasteur, was that experimentation, "if allowable on animals, is criminal in man."[51]

Once again more than a year passed before Pasteur gave another public account of his work on rabies. The next paper, delivered to the Académie des sciences on 26 October 1885, created an immediate sensation and has lived in legend ever since. It described the application of a remedy for rabies to two boys—Joseph Meister and Jean-Baptiste Jupille—who had been badly bitten by rabid dogs. This paper was filled with human drama, but even it failed to convey the full range of hope, doubt, and anxiety that Pasteur had experienced since his last public communication on rabies. Among other things, as we shall see in Chapter Nine, this famous paper gave a very misleading impression of the animal experiments that preceded the application of Pasteur's remedy to young Meister and Jupille. In this and several other respects, there are some remarkable discrepancies between the public and private versions of this celebrated story.

One such discrepancy is astonishing. It has to do not with animal experiments, but rather with two hitherto unknown cases of human experimentation that preceded Pasteur's application of his rabies vaccine to human subjects. Unlike the stories of Meister and Jupille, these two cases have left no traces in the public record. They are recorded only in Pasteur's laboratory

The charts below indicate, in chronological order, the date of publication of each of Pasteur's papers on the "saliva microbe" and on rabies per se, and a brief statement of the basic results presented in each paper.

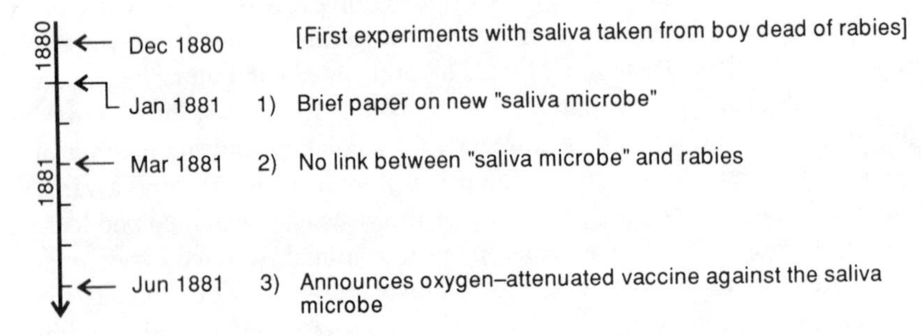
The "saliva microbe"
1880
1881
Dec 1880 [First experiments with saliva taken from boy dead of rabies]
Jan 1881 1) Brief paper on new "saliva microbe"
Mar 1881 2) No link between "saliva microbe" and rabies
Jun 1881 3) Announces oxygen–attenuated vaccine against the saliva microbe

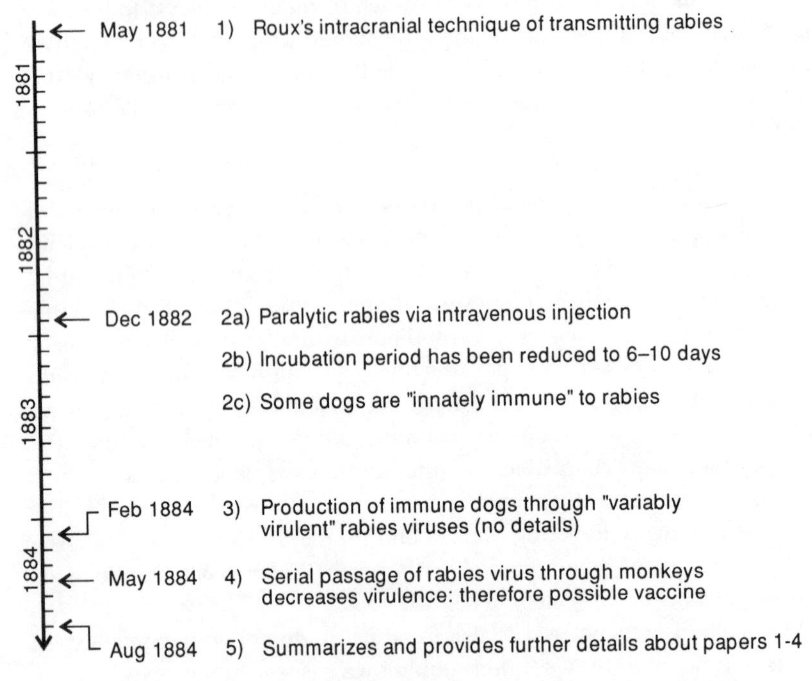
Rabies per se
1881
1882
1883
1884
May 1881 1) Roux's intracranial technique of transmitting rabies
Dec 1882 2a) Paralytic rabies via intravenous injection
2b) Incubation period has been reduced to 6–10 days
2c) Some dogs are "innately immune" to rabies
Feb 1884 3) Production of immune dogs through "variably virulent" rabies viruses (no details)
May 1884 4) Serial passage of rabies virus through monkeys decreases virulence: therefore possible vaccine
Aug 1884 5) Summarizes and provides further details about papers 1-4

Figure 7.2. Pasteur's path to his rabies vaccine: the published papers.

notes. In these two cases, it can hardly be said that Pasteur's public statements were misleading, for in fact he wrote nothing about them at all. What follows is the first published account of two attempts by Pasteur to cure rabies in patients already displaying symptoms of the disease.

PASTEUR'S "PRIVATE PATIENTS"

On the first day of May in 1885, an otherwise ordinary sixty-one-year-old Parisian named Girard presented himself at the gate of the Necker Hospital in a highly agitated state.[52] He feared he had rabies and was admitted to the service of one Dr. Rigal. Within hours, in his laboratory on the rue d'Ulm, Pasteur received a telegram informing him of Girard's admission to the hospital. The telegram came to him from Dr. Georges Dujardin-Beaumetz (1833–1896), a member of the Académie de médecine and of the Council on Hygiene and Public Health of the department of the Seine. Since 1881 Dujardin-Beaumetz had been charged with investigating and confirming all cases of rabies in the department. By order of the prefect of police, the director of each hospital in Paris was required to notify Dujardin-Beaumetz of every admission for rabies. He then conducted an inquiry into each case of suspected rabies, reporting his results to the Council on Hygiene and Public Health.[53] When he informed Pasteur of Girard's admission to the Necker Hospital, Dujardin-Beaumetz knew that the celebrated scientist from the Ecole Normale was in eager pursuit of a rabies vaccine. He presumably sent the telegram in hopes of somehow advancing that cause.

At 10 o'clock on the morning of 2 May 1885, Pasteur went to see Girard with the attending physician, Dr. Rigal. Girard told them that he had been bitten on the knee by a wandering dog sometime in March. His wound had been thoroughly cauterized and had healed without difficulty. He had been well until now. He spoke very lucidly but complained of a severe headache and stomach pain. He drank a large cup of milk but could not bear the sight of water or wine. His legs shook and he could not eat. That same afternoon, having secured authorization to do so from Dr. Rigal, Pasteur returned to the hospital with two of his associates, Adrien Loir and Dr. Emile Roux. When they reached Girard, only an orderly (*interne de garde*) remained on duty with him. Pasteur and his assistants then exposed the right side of Girard's body and injected him with one full Pravaz syringe (one cubic centimeter) of a preparation they had brought with them from the laboratory on the rue d'Ulm. Since Pasteur was not medically qualified, the actual injection was presumably performed by Roux.

Rage

Homme rabique - Necker. Service du Dr Rigal - 62

66

2 mai - Je reçois de M. Dujardin-Beaumetz le télégramme suivant:
« Directeur de Necker m'informe qu'un sieur Girard, atteint d'hydrophobie, vient d'entrer dans le service de M. Rigal » - Le [illegible] m'a dit qu'il [illegible] la veille [illegible], disant qu'il était enragé - très agité -

à 10h du matin je vais voir M. Rigal. Nous constatons que cet homme âgé de 61 ans a été mordu dans le [illegible] de mars, par un chien errant au genou (intérieur) que la plaie est très bien cicatrisée. Il ne peut dire le jour de mars où il a été mordu - Il est très lucide. Il se plaint de grand mal de tête et au ventre - Il n'a jamais été malade - Il avale grande tasse de lait, mais ne peut voir ni le vin, ni l'eau. [illegible] pas boire! on lui prend [illegible] et quand le [illegible] touche ses lèvres il le repousse, souffre évidemment dans voies de déglutition - Il a [illegible] tremblement des membres supérieurs [illegible].

L'après-midi, après avoir eu l'autorisation du Dr Rigal je vais à l'hôpital avec Roux et [illegible]. Il n'y a pas [illegible] de service. et [illegible] l'interne de garde. On inocule sous la dure-mère [illegible] une seringue pleine Pravaz du bulbe délayé de lapin 52 de la série p. 52.

[illegible] pour recommencer le soir à 10h avec cette [illegible] pleine de la même bulbe mis depuis 3h dans la glace.

Le 3 mai à 5h du matin j'inocule [illegible]: de cette moelle conservée à sec en flacon sec - j'inocule [illegible] à 3h et à 10h soir

Le 4 mai j'inocule [illegible] des inoculations de [illegible]: de bulbe lapin [illegible] p. 88 de 75e passage à [illegible] et [illegible].

mais le 2 mai à 10h du soir le [illegible] qui avait été [illegible] [illegible] [illegible] s'est opposé à toute seconde inoculation [illegible] Rigal n'était pas prévenu. Donc je n'ai plus [illegible] que le 2 au soir, [illegible] 3, 4, le 4 mai.

- Le 3 mai, le malade a été plus malade
- le 4 mai, [illegible]; il a eu 6 [illegible] des tremblements des membres supérieurs. Il a eu [illegible] - accès de tremblements [illegible]. Je demande s'il avait la rage - a parlé de la rage de son [illegible] -

- Le 5 mai, cette nuit il a eu [illegible] de tremblements dans les membres supérieurs. Il se plaint alors au cou [illegible] à 9h. mais il est très calme, très lucide, [illegible] : quand il quittera l'hôpital, il ne [illegible] comment [illegible]!
Il a ce matin demandé de lui-même du bouillon qu'il a bien pris - Il prend aussi [illegible] du lait. Il ne veut entendre parler d'eau et de vin.

- Le 6 mai, même état.
- Le 7 mai, même état. Toujours dégoût du vin et de l'eau. Une crise très prolongée de tremblements des membres et de grattage pendant la nuit. Il prend toujours volontiers du lait et du bouillon. Il [illegible] [illegible] bien parfois - [illegible] mal [illegible] qu'il [illegible] et de ce qu'il a éprouvé. Il a causé ce matin avec Roux très raisonnablement, et sa figure est bien naturelle.

- Le 8 mai, cet homme est visité par M. Dujardin-Beaumetz - Il va [illegible], n'a-t-il ... [illegible] de tremblement. [illegible]. Il sortira bientôt de l'hôpital.
- Le 9 mai [illegible] bouillon, cependant il ne prend toujours que du lait, ou du bouillon et du lait, [illegible] toujours vin et eau [illegible]

- Le 22 mai, au Conseil de Salubrité, j'apprends de Duj. Beaumetz que l'homme est sorti guéri de Necker, que M. Rigal [illegible] qu'il avait eu des symptômes rabiques! L'observation de la maladie qui est dans le service de l'hôpital et que je viens de lire [illegible] pour M. Duj. B. [illegible] diagnostic de rage. Il [illegible] ce que j'ai fait et il dort [illegible] [illegible] une série: seringue pleine à 1 [illegible] de virus rabique atténué.
Je lui écris le 23 mai qu'il [illegible] pour rapporter le cas au Conseil - Il me répond qu'il [illegible] le rapport [illegible]!

25 mai Il faut attendre, en effet, de savoir si Girard va continuer de se bien porter. S'il devenait rabique, il pourrait être soit de sa morsure, soit de l'inoculation. On le saurait par l'étude de la virulence après la mort.
Je crois, au contraire, que l'inoculation a pu le guérir de la rage déjà déclarée. C'est l'opinion de M. Duj. B. et également, [illegible] du Dr Rigal que je n'ai pas entendu cependant ces jours derniers, depuis qu'il est venu à l'idée que Girard est entré rabique à Necker.

Figure 7.3. Pasteur's laboratory notes on the presumably rabid M. Girard, his first "private patient." The record begins on 2 May [1885]. Pasteur, *Cahier 94*, fols. 62–62v (using Pasteur's handwritten pagination). (Papiers Pasteur, Bibliothèque Nationale, Paris)

Pasteur and his collaborators then made the necessary preparations to repeat this injection into Girard at 10 o'clock that night. In fact, they planned to give him a series of six additional injections over the next two days, of which the first was to be identical with the one he had already received. The subsequent five injections were to be made with preparations that differed from the first in degree of virulence. In the event, however, these plans were thwarted by the public authorities. At 10 o'clock on the night of that same day, 2 May, Pasteur returned to the hospital to oversee Girard's second injection. But during the several hours he had been absent, the hospital authorities had evidently become concerned about the propriety of the afternoon's events. For some reason, at any rate, they had consulted with the Ministry of Public Assistance, and Pasteur was now told that Girard could undergo no further injections in the absence of his attending physician, Dr. Rigal. Girard was then abandoned to his fate without further treatment.

The outcome was remarkable. On 3 May, the day following his injection, Girard's condition deteriorated, and it was still worse the day after that. His arms trembled, he was in pain, and he asked if he had rabies. During the night of 4 May he was seized with fits of trembling in his upper limbs, at which time he also rubbed his neck. Yet another trembling fit struck him the next morning at 9 o'clock. When it ended, though, he was very calm and lucid, expressing appreciation for the care he had received. He asked for bouillon and consumed it without difficulty. He also drank some milk, but still wanted to hear nothing of wine or water.

Girard remained in the same general condition until 9 o'clock on the night of 6 May, when he suffered a prolonged attack of trembling in his limbs, during which time he also scratched at his body. This attack lasted until 4 o'clock the next morning, but he took milk and bouillon that day too. He also slept well enough at times, though he was disturbed by a nightmare about all that he had suffered. On the morning of 7 May he conversed rationally with Roux, and his countenance seemed normal. On 8 May Dujardin-Beaumetz visited Girard and found him doing well. He had experienced no further attacks of trembling and spoke lucidly. A week had passed since Girard's admission to the hospital, and his condition was now such as to give hope that he might soon be released. He seemed equally well on 9 May, though even then he continued to consume only milk and bouillon, rejecting wine and water as well as all solid food.

Two weeks later, on 22 May 1885, at a meeting of the Paris Council on Hygiene and Public Health, Pasteur learned from Dujardin-Beaumetz that Girard had been discharged from the Necker Hospital, presumably having been cured. Dujardin-Beaumetz also reported that Dr. Rigal, who had become skeptical of the original diagnosis of rabies, was once again prepared to endorse it. In Dujardin-Beaumetz's own judgment, the diagnosis of rabies was fully justified by the hospital dossier and other evidence. Finally, Dujardin-Beaumetz asked Pasteur to specify the nature of the injection given Girard on 2 May. In his response, presumably solicited for inclusion in one of Dujardin-Beaumetz's reports to the Council on Hygiene and Public Health, Pasteur disclosed only that the injection had consisted of "one full Pravaz syringe [i.e., one cubic centimeter] of attenuated rabies virus." He did not specify the method by which this attenuation had been achieved.

The next day, 23 May 1885, Pasteur sent Dujardin-Beaumetz a letter asking him to postpone his report on Girard to the Council. Dujardin-Beaumetz immediately agreed to honor Pasteur's wishes. Before any such report was sent, Pasteur wanted enough time to have passed so that no doubt could remain that Girard might yet succumb to rabies. If Girard did eventually die of rabies, Pasteur continued, it would be important to determine whether his death had resulted from the dog bite or rather from the injection he had received at the Necker Hospital. That issue could be settled by transmitting tissue from Girard's brain to susceptible animals and then observing whether their clinical response was typical of ordinary rabies or rather of the altered virus that had been injected into him by Pasteur and his collaborators. But Pasteur did not think it would come to that. On 25 May 1885, in his laboratory notebook, he recorded his belief that the injection of 2 May had cured Girard of symptomatic rabies.[54]

Within a month, Pasteur was treating a second case of "declared" rabies, this time in an eleven-year-old girl named Julie-Antoinette Poughon. She had been bitten on the upper lip by her own puppy sometime in May and had been admitted to the Hospital of St. Denis on the morning of 22 June 1885 after suffering for two days from severe headache. Pasteur and the doctor in charge of her case agreed that she was clearly suffering from rabies. At Pasteur's suggestion, her doctor injected Julie-Antoinette with one full Pravaz syringe of a substance previously prepared in Pasteur's laboratory. At midnight she was given a second injection, which differed from the first in degree of attenuation. The next morning, 23 June, Pasteur returned to the Hospital of St. Denis at 10 o'clock with his nephew Adrien Loir. They barely reached Julie-Antoinette before she died at 10:30. The symptoms of rabies had quickly overtaken her despite the two injections of the previous day.[55]

The dramatic stories of Girard and Julie-Antoinette find no trace whatever in the published record. Except in his laboratory notebook and in his correspondence with Dujardin-Beaumetz, Pasteur never wrote about them

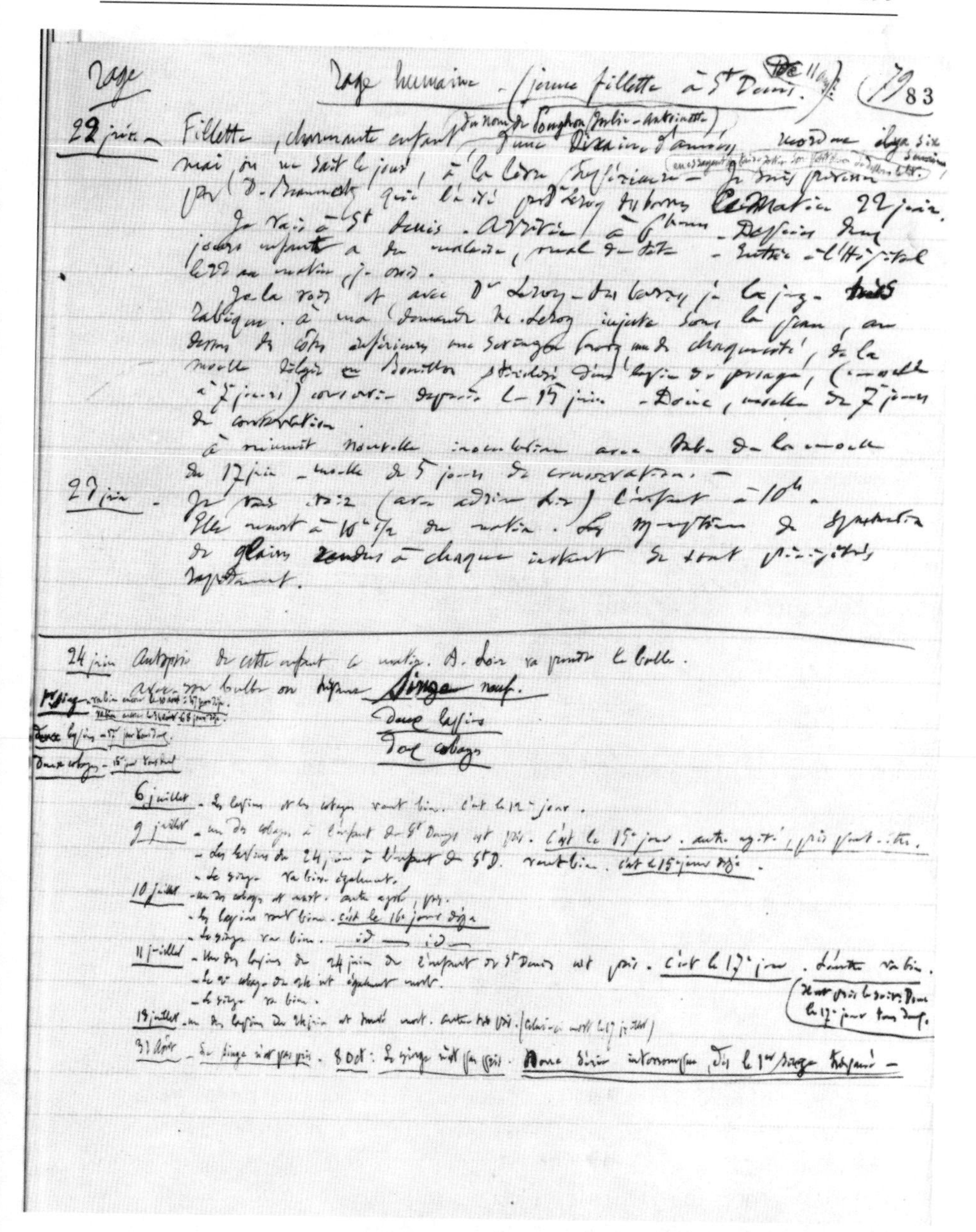

Figure 7.4. Pasteur's laboratory notes on Julie-Antoinette Poughon, his second "private patient." The record begins on 22 June [1885] and ends just a day later with an entry reporting the girl's death from rabies. Pasteur, *Cahier 94*, fol. 79 (using Pasteur's handwritten pagination). (Papiers Pasteur, Bibliothèque Nationale, Paris)

at all. Of Julie-Antoinette, there was perhaps little to be said. She had died of rabies despite Pasteur's desperate efforts to save her. But the public silence about Girard is impossible to understand unless Pasteur somehow came to doubt the reality of his alleged cure. Had he continued to believe that he had cured Girard, it is unthinkable that Pasteur would have kept such a monumental achievement out of the public eye.

Unfortunately, Pasteur's laboratory notebook tells us nothing about Girard's subsequent fate. The page devoted to the case ends abruptly on 25 May 1885 with the assertion that Drs. Rigal and Dujardin-Beaumetz now shared Pasteur's belief that his injection had cured Girard of symptomatic rabies. Something obviously happened later to destroy that belief, but Pasteur's notebook does not reveal what it was.[56] So a puzzle remains. It is unclear from the documents cited here whether or not the Girard treated by Pasteur died of rabies, and if so, exactly when. But that may not be an especially crucial point. For if this Girard did eventually die of rabies, his death must have come at a point when Pasteur had already ceased to believe in his cure, thus helping to explain his apparent lack of interest in the news. And even if this Girard was still alive on 23 June 1885, Pasteur would have had another powerful reason for abandoning his belief in Girard's cure: the death of young Julie-Antoinette Poughon despite Pasteur's attempt to cure her by a method very similar to that used in the case of Girard.

Pasteur's real, if temporary, belief that he had cured Girard of symptomatic rabies is testimony to the power of wishful thinking in the face of dread disease. There was astonishingly little basis for that belief in any of Pasteur's animal experiments to that point, as we shall see. But quite apart from the results of his animal experiments, Pasteur's belief that he had cured Girard of clinical rabies would have been met with profound skepticism for another reason—namely, the uncertainty surrounding the diagnosis of rabies. Had Pasteur published an account of his alleged cure of Girard, the vast majority of physicians would have scoffed at the diagnosis. For death within days of the outbreak of symptoms had always been part and parcel of the very definition of rabies. In effect, "recovery" from clinical rabies was a contradiction in terms. Indeed, Girard's own attending physician, Dr. Rigal, at least briefly disavowed his initial diagnosis of rabies when Girard showed signs of recovering.[57] Other physicians would have been more insistent still that Girard must have suffered from another disease.

Pasteur himself later pointed out some of the uncertainties surrounding the diagnosis of rabies. Two years after *l'affair Girard*, for example, he spoke to the Académie des sciences about several cases of "false rabies." Relying on the authority of one Dr. Trousseau, Pasteur cited two cases in which symptoms of the disease had been induced solely by fear. In one case, a man suddenly displayed several of the classic features of rabies—including throat spasms, chest pain, extreme anxiety, and other nervous symptoms—merely because the disease had become the subject of a lunchtime conversation. And this man had never even confronted a rabid animal. Presumably more common was the second case, that of a magistrate whose hand had long before been licked by a dog later suspected of rabies. Upon learning that several animals bitten by this dog had died of rabies, the magistrate became extremely agitated, even delirious, and displayed a horror of water. His symptoms disappeared ten days later, when his physician persuaded him that he would already be dead had he been afflicted with true rabies.[58]

In this same address, Pasteur commented upon a recently published case history of "false rabies." Partly because it includes an arresting account of the classic symptoms of rabies, his commentary deserves quoting at length. As recorded in the *Comptes rendus* of the Académie des sciences for 17 October 1887, Pasteur spoke as follows:

> The patient to whom Mesnet refers in his brochure was an alcoholic who, having seen some sort of deposit in his glass during lunch, was seized by a feeling of horror toward the liquid and by a constriction of the throat, followed by headache and by lameness and fatigue in all his limbs. He spent Sunday in this state.
>
> During that night and during the day on Monday and Tuesday, no sleep, a fit of suffocation, throat spasms, and a horror of liquids, which he pushed aside in his glass. His countenance expressed disquiet. His eyes were fixed, glazed, the pupils greatly dilated. His speech was brief, jerky, rapid. He had difficulty breathing. When he was offered a glass of water, he pushed it aside with terror, and suffered fits of suffocation and of constriction of the throat. Bright objects and light were particularly disagreeable to him. He was painfully affected when the air was agitated in front of his face. He died Wednesday night after having suffered from a violent delirium, with extreme agitation, howls and cries, extremely abundant salivation, spitting, biting his bedsheets, and trying also to bite the person taking care of him. In short, this man displayed all the features of furious rabies [*l'hydrophobie furieuse*]. But he did not die of rabies. He had never been bitten and on several occasions, at long intervals, had already displayed symptoms analogous to false rabies. This man was an alcoholic and belonged, moreover, to a family in which one member had died of insanity [*aliénation mentale*].[59]

By October 1887, when he gave this address, Pasteur had a vested interest in emphasizing the difficulty of diagnosing rabies. For he was then defending himself against allegations that his rabies vaccine not only sometimes failed to protect those who submitted to it, but in some cases was itself the cause of rabies and therefore death. A few hostile critics were insisting that some people died of rabies not only *despite* Pasteur's vaccine but *because* of it, and they tried to make Pasteur and his treatment responsible for the death of anyone who displayed any symptoms of nervous disease. In defense of his vaccine, Pasteur now emphasized the extent to which symptoms like those of rabies could appear in patients who did not have the disease. He therefore insisted that a diagnosis of rabies could only be established with confidence by experiments in which tissue from the victim's brain was transmitted to animals susceptible to the disease.[60]

But the uncertainty surrounding the diagnosis of rabies, which here served Pasteur's interests, could equally well have been turned against him had he publicized his alleged cure of Girard. In Girard's case, or in any other case of apparent recovery from rabies-like symptoms, no reliable diagnostic test existed. The animal experiments that Pasteur and others considered the most reliable diagnostic tool could be performed only after the death of a presumably rabid patient. Only then could nervous tissue be extracted from the patient's brain and injected into susceptible animals to see if they succumbed to rabies. Such postmortem tests obviously did nothing to reduce the uncertainty of diagnosing rabies in living patients.

Even today, rabies can be difficult to diagnose, as is made dramatically and tragically clear by four recent cases of human-to-human transmission of rabies through corneal transplants. In all four of these cases—two in Thailand, one in the United States, and one ironically in Pasteur's native region of France—the existence of rabies in the deceased corneal donors had gone unsuspected until the unfortunate recipients died of the disease. In two of the recipients, moreover, a firm diagnosis of rabies was not established until well after their deaths.[61] Rabies may seem to be a very distinctive clinical entity, but it can also be present in the absence of its usual dramatic symptoms, and at least some of those symptoms can appear in the absence of the disease. And if rabies can be missed even today despite the full panoply of current histiological and immunological diagnostic techniques, it was obviously much more difficult to diagnose in Pasteur's day.

All of this suggests that Pasteur would have faced a no-win situation had he tried to persuade others that he had cured Girard of rabies. Only if Girard died of rabies would most physicians have accepted the diagnosis in the first place. But in that case, of course, Pasteur's "cure" would have failed. Small wonder, perhaps, that he never publicly disclosed his belief that he had cured Girard. And when young Julie-Antoinette Poughon died of rabies after submitting to a remedy very similar to Girard's "cure," even Pasteur himself may have decided that his first "private patient," M. Girard, had been just another example of mistaken diagnosis—of "false rabies," in short.

CONCLUSION

Pasteur's belief that he had cured Girard had no detectable basis in prior animal experiments. If Drs. Rigal and Dujardin-Beaumetz briefly joined Pasteur in that belief, it was surely because of his general scientific eminence. They doubtless presumed that he had good grounds for his claim, but they would have felt otherwise if they had had access to Pasteur's laboratory notebooks. Without those notebooks and in the face of Pasteur's reticence, they had no way of knowing exactly what substance had been injected into Girard or to what extent, and with what success, it had been tested on animals. Pasteur's letter to Dujardin-Beaumetz concerning Girard disclosed only that he had been injected with one cubic centimeter of "an attenuated rabies virus."[62] Had Pasteur told Drs. Dujardin-Beaumetz and Rigal the full story of his animal experiments up to that point, they would have been surprised to learn of the precise preparation he had applied to Girard and curious to hear exactly how he proposed to justify his confidence in his alleged cure.

Unlike Drs. Dujardin-Beaumetz and Rigal, we now enjoy the privilege of direct access to Pasteur's once-private laboratory notes on rabies. They reveal, first of all, that Girard had been injected with a preparation that Pasteur had not yet described in print—namely, an emulsified spinal cord that had been extracted from a rabbit dead of experimental rabies and left to dry in a sealed flask for roughly two weeks.[63] This desiccated spinal cord was, then, the source of that "attenuated rabies virus" to which Pasteur referred in his correspondence with Dr. Dujardin-Beaumetz. The laboratory notes further reveal that six weeks later, on 22 June 1885, young Julie-Antoinette Poughon was treated by the same method, that is, with an injection prepared from a dried rabid cord—in her case, to no avail.[64] In his published papers up to this point, Pasteur nowhere mentioned experiments with dried spinal cords. He had written only of attenuated rabies viruses produced by serial passage through monkeys. The preparations actually used on Girard and Julie-Anoinette would thus have come as a surprise to Drs. Dujardin-Beaumetz and Rigal—or anyone outside Pasteur's tiny inner circle.

Outsiders would have been still more surprised to learn that Pasteur had never tried, not even once, to cure symptomatic rabies *in animals* by *any* method before he decided to treat Girard—or so it seems to me from an analysis of his laboratory notebooks. A few days *after* Girard's treatment had begun, Pasteur did try to cure a rabbit of symptomatic rabies, but the animal died three days after the first series of its injections.[65] In the six weeks that passed between this unsuccessful animal experiment and Pasteur's equally unsuccessful attempt to cure Julie-Antoinette Poughon, his laboratory notes record no other attempts to treat animals suffering from symptomatic rabies. At the least, there is no evidence that Pasteur undertook any sustained program of experiments to treat "declared" rabies in animals before he undertook his treatment of Girard and Julie-Antoinette Poughon.

What, then, are we to make of Pasteur's attempts to cure rabies in these two "private patients"? To begin with the obvious, they represent examples of human experimentation. More than that, M. Girard and Julie-Antoinette Poughon were treated by a method that had apparently never been successfully tested on animals with symptomatic rabies. Even so, it should be emphasized, there was nothing unethical about Pasteur's interventions in the case of these two apparently doomed rabies patients. Even in his day, the distinction between therapeutic experiments and unethical human experimentation was perfectly clear. Everyone agreed that "therapeutic experiments"—those undertaken in the hope of benefiting the person submitting to them—were fully justified.[66] Pasteur's desperate attempts to save Girard and Julie-Antoinette Poughon from "declared" rabies did not violate any accepted ethical standards.

Nor did Pasteur violate any ethical precept by declining to publish accounts of these two "clinical trials." Indeed, in the case of Girard, it might even be said that Pasteur properly resisted the temptation to issue a "premature" announcement of his presumed cure. The subsequent death of Julie-Antoinette, despite Pasteur's attempt to cure her by a similar method, probably led him to believe that the diagnosis of rabies had been mistaken in the case of Girard. By his reticence in the meantime, he had prevented false hopes of cure in other victims of symptomatic rabies.

Yet it is hard to resist the judgment that Pasteur—whatever his formal ethical obligations—would have performed a valuable public service had he ultimately revealed the full stories of his two "private patients." Had he done so, clinicians would have become aware of another and especially arresting example of the uncertainty surrounding the diagnosis and clinical features of rabies. Such a public disclosure would also have served ever after as a striking illustration of the power of wishful thinking in the face of dread disease. Instead, these episodes shared the fate of most unsuccessful clinical trials: they were buried along with Julie-Antoinette Poughon.

In the end, the stories of Girard and Julie-Antoinette may have a much greater and distinctly ironic significance. For they were closely linked, both in time and technique, with a radical shift in the approach by which Pasteur sought to develop a vaccine against rabies, as we shall see more fully in Chapter Nine. For now, let us merely highlight the suggestive chronological sequence. Pasteur undertook his treatment of Girard on 1 May 1885. Up to that point, he had used several different methods in his attempts to produce a safe and effective rabies vaccine for animals, with variable and confusing results. By the time he undertook his treatment of Girard, Pasteur and his collaborator Emile Roux were already beginning to focus on the injection into dogs of emulsified rabid spinal cords. But until May 1885 Pasteur usually injected the rabid spinal cords in a very different—indeed precisely opposite—sequence from the one he would eventually use.

On 28 May 1885, just three days after recording his belief that he had cured Girard, Pasteur launched a systematic program of animal experiments to try to produce a vaccine by what he called in his laboratory notebook "the other method" in comparison to what had gone before—as we will see in Chapter Nine. By 23 June 1885, when Julie-Antoinette Poughon died of rabies despite Pasteur's effort to cure her, he was growing increasingly confident about the results of this "other method" in animal experiments. And just two weeks later, on 6 July 1885, it was precisely this "other method" that Pasteur used for the first time in a human case when he undertook the treatment of the badly bitten but thus far asymptomatic boy named Joseph Meister.

In other words, there is circumstantial evidence to suggest that Pasteur's radical shift in approach—his sudden turn to the eventually successful "other method"—in the search for a safe and effective vaccine was inspired by his presumed cure of Girard. If so, Joseph Meister became just the first of thousands to benefit from what was almost surely a case of mistaken diagnosis. Even here, in the case of Pasteur's greatest triumph, it might thus seem that a "lucky mistake" had once again put him on the path to success. We shall see, however, that the story is vastly more complicated than this sketch might suggest. And when that story is read out full and clear, we will have new grounds for appreciating the very real wisdom in Pasteur's own famous maxim that "chance favors only the prepared mind."

EIGHT

Public Triumphs and Forgotten Critics: The Debate over Pasteur's Early Use of Rabies Vaccines in Human Cases

ON MONDAY, 6 July 1885, three frightened and unexpected visitors made their way to Pasteur's laboratory at 45 rue d'Ulm in Paris. They had come to Paris by train from a village in Alsace, where two days before, on 4 July, two of them had been attacked by a dog displaying all the classic signs of rabies. One of the victims was the dog's owner, a grocer named Théodore Vone. His dog had bruised his arms, but without penetrating his shirt or skin. Pasteur sent him home with the assurance that he had nothing to fear. The other two visitors were a nine-year-old peasant boy named Joseph Meister and his fretful mother, who had not been attacked by the dog but was there to be with her badly bitten son. The boy had been bitten a dozen times or more, with severe wounds on the middle finger of his right hand and on his thighs and calves, some of them so deep that he could hardly walk. His trousers had been ripped to shreds. His condition might have been worse yet—indeed, the still rampaging dog might have killed him—had he not been rescued by two men who cornered and captured the dog, which was then destroyed by its master, M. Vone. An autopsy of the attacking dog revealed that its stomach contained hay, straw, and chips of wood, as was typical of rabid dogs. The worst of young Meister's bites had been cauterized with carbolic acid by a local doctor, but not until twelve hours after the attack.[1]

In the afternoon of that same day, 6 July 1885, Pasteur went as usual to the weekly meeting of the Académie des sciences. There he spoke of young Meister to his Académie colleague E.F.A. Vulpian, who had often lent support to Pasteur's causes and was now a member of the French rabies commission that had been appointed the year before at Pasteur's request. Pasteur asked Dr. Vulpian to examine Meister in consultation with Dr. Joseph

Grancher, clinical professor of children's diseases at the Paris Faculté de médecine and a recent recruit to the Pastorian team. Upon examining the boy's wounds, Drs. Vulpian and Grancher concluded that he almost surely faced death from rabies.[2] Pasteur then decided to treat young Meister by a method that he had thus far tried only on dogs. Since the boy was reluctant to go to a hospital, Pasteur arranged for him and his mother to be installed at an annex of the laboratory two blocks away, on the rue Vauquelin. At eight o'clock that same night, 6 July, young Meister submitted to the first of thirteen injections he would undergo over the next eleven days. He survived the injections and escaped the death from rabies to which he had once seemed doomed.[3]

Three months later, on 16 October 1885, the mayor of the village of Villers-Farlay near Pasteur's home town of Arbois sent him the letter referred to at the beginning of the previous chapter. Mayor Perrot's letter told Pasteur of the brave fifteen-year-old shepherd, Jean-Baptiste Jupille, who had been attacked and badly bitten two days earlier by a rabid dog while protecting several younger boys. The selfless courage of young Jupille gained in drama from the sorry circumstances of his family. The Jupille family, which included four or five other children, had fallen on hard times after the father lost his arm in a railway accident. Upon losing his arm, the father also lost his job with the railroad company for which he was then working. And since he was declared personally responsible for the accident, he had received no compensation for the injury. To enable the family to survive, Mayor Perrot had named the father village policeman for Villers-Farlay, but the salary barely sufficed to sustain the family. Jean-Baptiste, the eldest Jupille child, had therefore been sent to work as soon as possible as a shepherd for a local farmer. And now the poor family faced the prospect of losing him to rabies.[4]

Pasteur received Mayor Perrot's first letter about Jupille on 17 October 1885, just one day after it had been sent. He responded immediately, telling the mayor the happy story of Joseph Meister and offering to treat young Jupille by the same method. Pasteur's letter continued as follows:

> I should tell you, however, that the conditions are less favorable in this case. According to your letter, Jupille was bitten on the 14th of this month. This letter will reach you on the 18th. The boy will get here the morning of the 20th or the night of the 19th. The bites will already be six days old [by then]; those of little Meister had been only sixty hours old, and I do not yet know from my experiments at what point following the moment of [rabid] bites I can begin the treatment. Nonetheless I ought to tell you that I have succeeded in rendering some dogs immune to rabies six and eight days after their bites. . . .

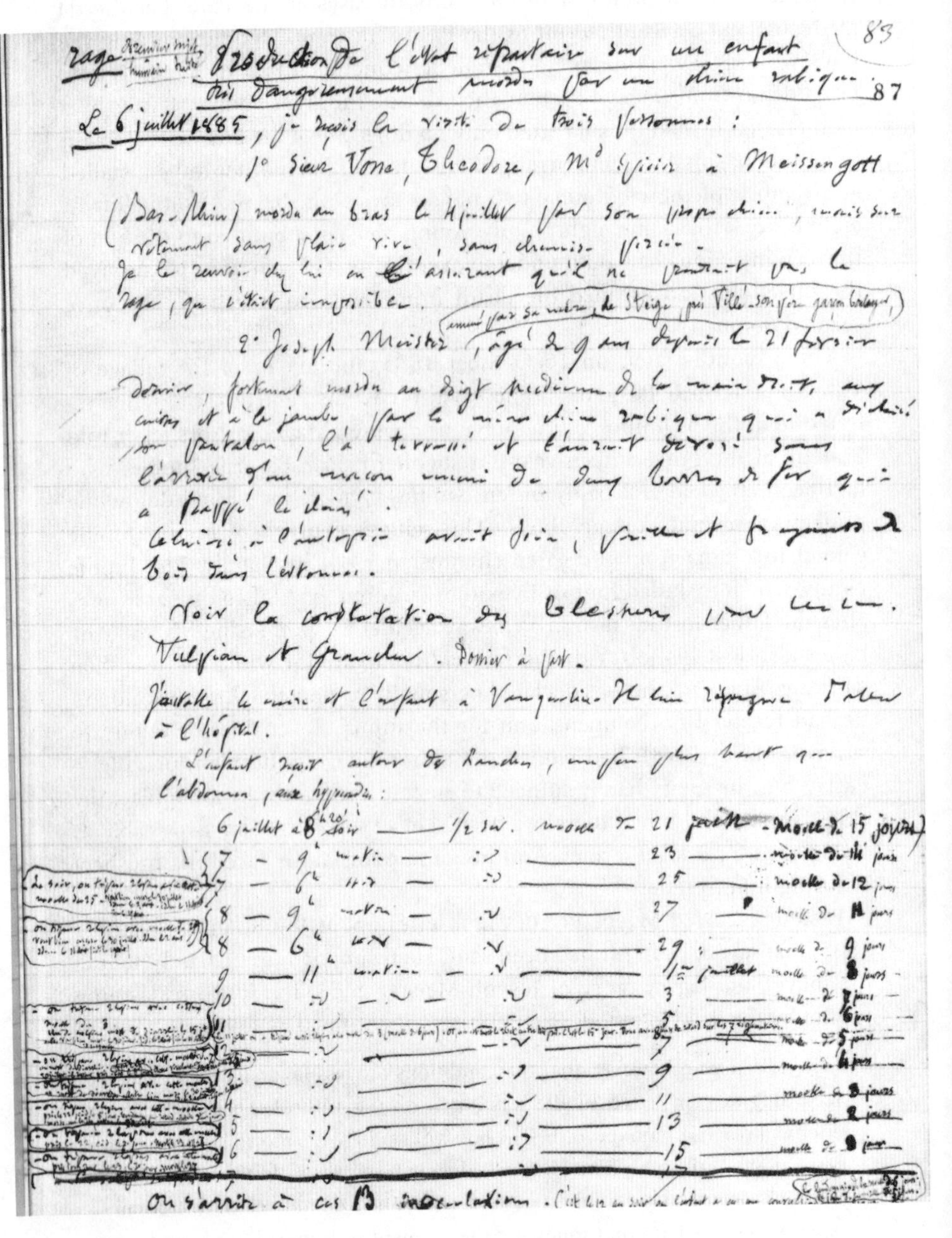

rage — Prévention de l'état rabique sur un enfant très dangereusement mordu par un chien rabique — 87

Le 6 juillet 1885, je reçois la visite de trois personnes:

1° Sieur Vone, Théodore, Md épicier à Meissengott (Bas-Rhin) mordu au bras le 4 juillet par son propre chien, mais sur vêtement dans plaie vive, sans chemise percée.
Je le renvoie chez lui en l'assurant qu'il ne prendrait pas la rage, que c'était impossible.

2° Joseph Meister, âgé de 9 ans depuis le 21 février dernier, fortement mordu au doigt médium de la main droite, aux cuisses et à la jambe par le même chien rabique qui a déchiré son pantalon, l'a terrassé et l'aurait dévoré sans l'arrivée d'un maçon armé de deux barres de fer qui a frappé le chien.
Celui-ci à l'autopsie avait du foin, paille et fragments de bois dans l'estomac.

Voir la constatation des blessures ... Vulpian et Grancher. Dossier à part.
J'installe la mère et l'enfant à Vauquelin. Je leur épargne d'aller à l'hôpital.

L'enfant reçoit autour des hanches, un peu plus haut que l'abdomen, aux hypocondres:

6 juillet à 8h20 soir — 1/2 ser. moelle du 21 juin — moelle de 15 jours
7 — 9h matin — " — 23 — moelle de 14 jours
7 — 6h soir — " — 25 — moelle de 12 jours
8 — 9h matin — " — 27 — moelle de 11 jours
8 — 6h soir — " — 29 — moelle de 9 jours
9 — 11h matin — " — 1er juillet — moelle de 8 jours
10 — " — " — 3 — moelle de 7 jours
11 — " — " — 5 — moelle de 6 jours
12 — " — " — 7 — moelle de 5 jours
13 — " — " — 9 — moelle de 4 jours
14 — " — " — 11 — moelle de 3 jours
15 — " — " — 13 — moelle de 2 jours
16 — " — " — 15 — moelle de 1 jour

On s'arrête à ces 13 inoculations. ...

Figure 8.1a.

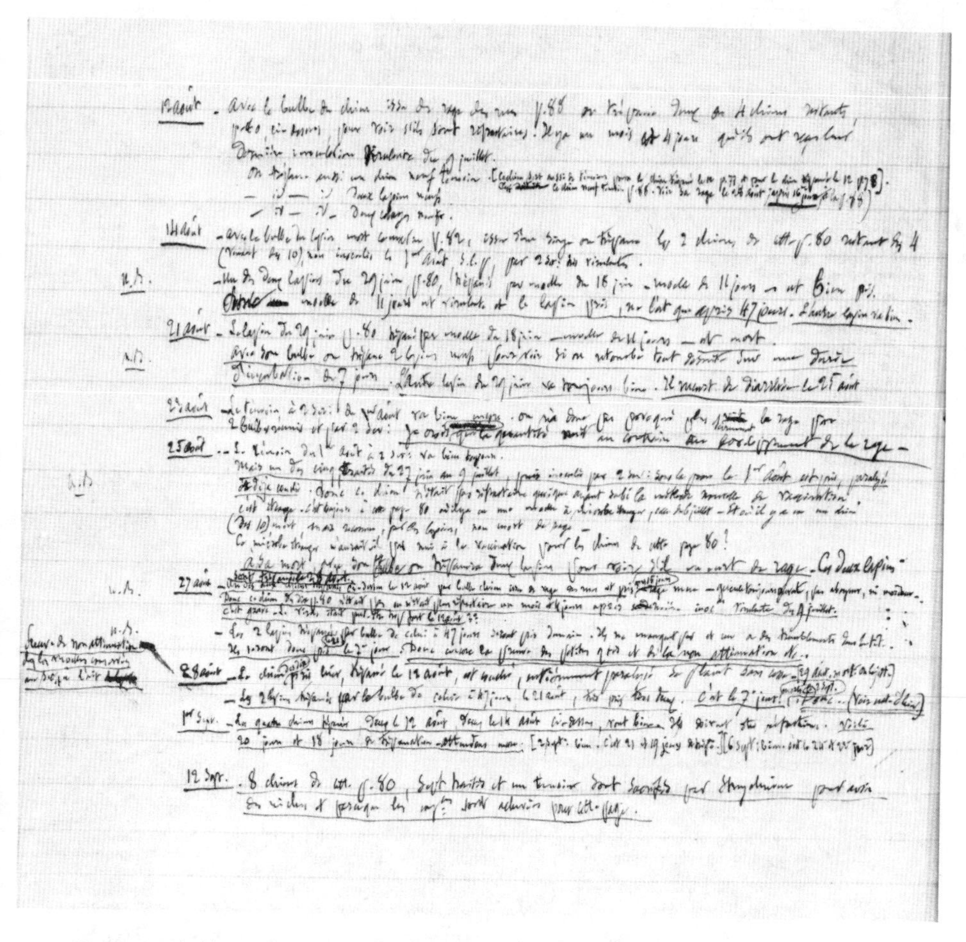

Figure 8.1 (a,b). Pasteur's laboratory notes on the treatment of Joseph Meister. The record begins on 6 July 1885. Not surprisingly, Pasteur gave special attention to these two pages in his notebook, beginning with the heading "Production of the refractory state in a child very dangerously bitten by a rabid dog." Pasteur, *Cahier 94*, fols. 83–83v (using Pasteur's handwritten pagination). (Papiers Pasteur, Bibliothèque Nationale, Paris)

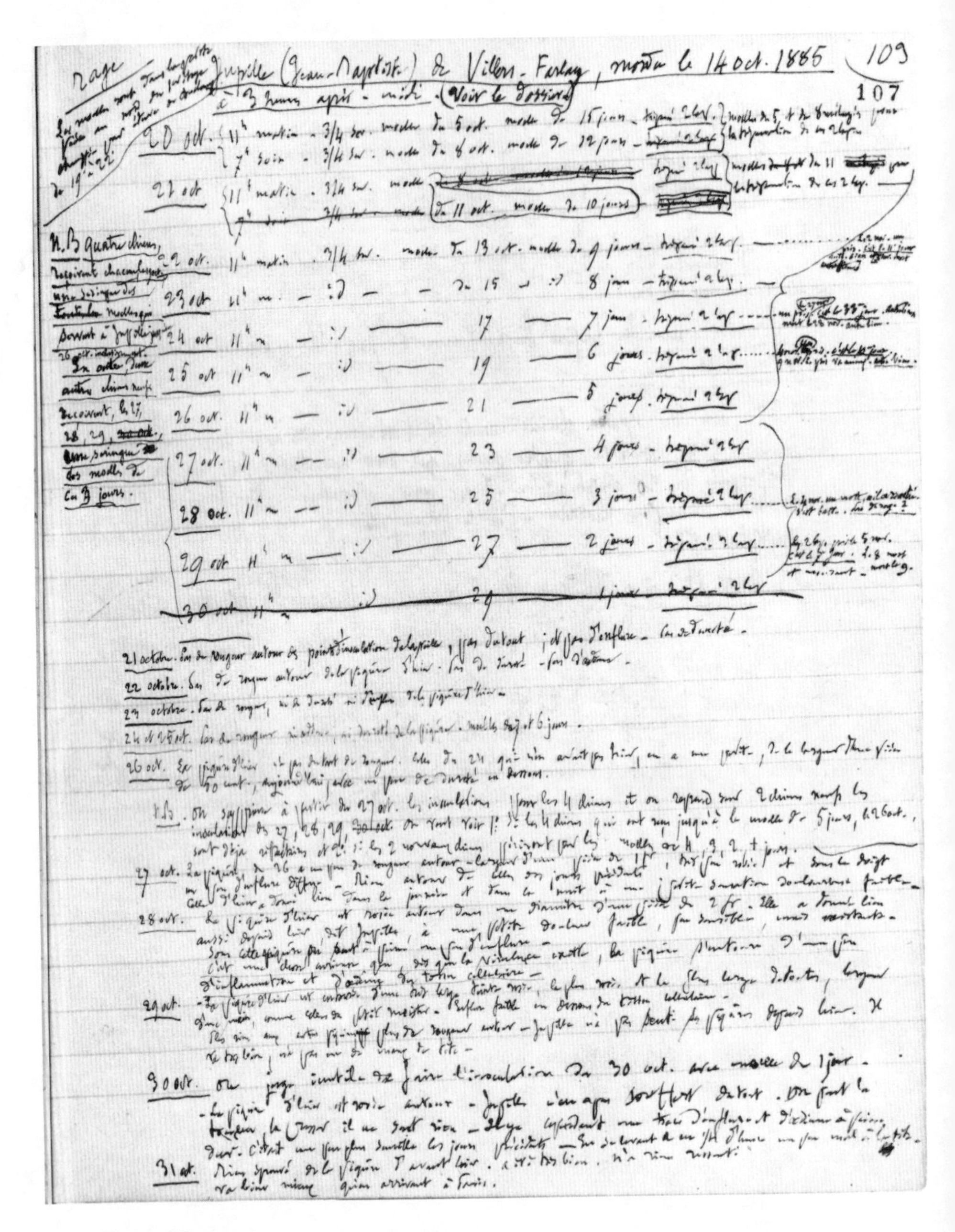

Figure 8.2a.

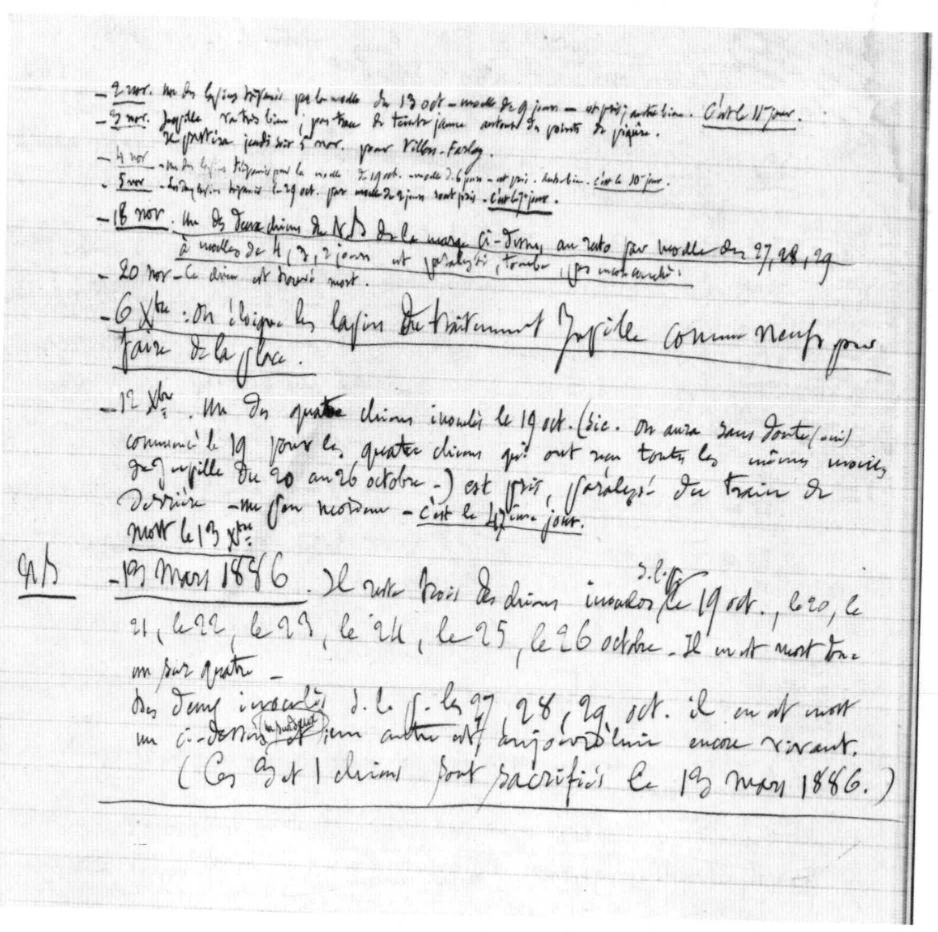

Figure 8.2 (a,b). Pasteur's laboratory notes on the treatment of Jean-Baptiste Jupille, beginning on 20 October 1885. Pasteur, *Cahier 94*, fols. 103–103v (using Pasteur's handwritten pagination). (Papiers Pasteur, Bibliothèque Nationale, Paris)

> As [Jupille] is doubtless not rich, I will keep him with me, in a room in my laboratory. . . . The village will only pay the expenses of his round-trip voyage. I covered all the expenses to which I allude for young Meister [as well]. It is true that I do so in view of these being my very first trials. Later, I think, the municipalities or general councils will be asked to intervene [fiscally].[5]

In his response to this letter from Pasteur, Mayor Perrot reported that he had communicated its contents to Jupille's parents. They had been reluctant at first to send their son off to Paris, having heard conflicting advice from all sides. But the mayor told them that the veterinarians' report on the attacking dog left no doubt that it had been rabid and that "their son was lost unless they accepted the generous offer of M. Pasteur, who alone knew how to save him from the horrible death that threatened him." In the end, Mayor Perrot's counsel prevailed. Having secured parental consent to do so, he put young Jean-Baptiste on the next train to Paris with enough money to cover the expenses of his journey.[6]

The treatment of young Jupille began at 11 A.M. on 20 October 1885. Like Joseph Meister before him, he was to undergo a long series of daily injections of Pasteur's new rabies vaccine.[7] On 26 October, the day on which young Jupille submitted to the eighth injection in the series, Pasteur went to the Académie des sciences to deliver the famous paper in which he announced the application of his rabies vaccine to human cases.

PASTEUR'S FAMOUS PAPER OF 26 OCTOBER 1885

Pasteur began his celebrated paper of 26 October 1885 by reporting that his experiments on rabies had taken an important new turn since his last public communication of 10 August 1884. While insisting that his previously announced method of preventing rabies in dogs—namely, by injecting them with a rabies virus attenuated by serial passage through monkeys—had marked a real advance, Pasteur did now concede that the advance had been "more scientific than practical." This method, he now revealed, led to "various accidents," with the result that "not more than fifteen or sixteen dogs in twenty could be made resistant to rabies with certainty." The method had still other practical defects. It took three or four months to be sure that the injected animals had been rendered immune, and the monkey-attenuated virus could not easily be produced and applied at a moment's notice, thereby limiting its utility in the emergencies faced by the victims of the "casual and unforeseen" bites of rabid animals.[8]

Pasteur and his collaborators had therefore pressed onward in search of another method of prevention that was both more rapid and "capable of giving . . . a state of perfect security in the dog." Until that goal was achieved, wrote Pasteur, "it was impossible to think of making any trial of the method on man." But he had great news to report: "After, I may say, innumerable experiments, I have at last found a method of prophylaxis both practical and rapid, and one that has already proved successful in the dog so constantly in so many cases that I feel confident of its general applicability to all animals and to man himself."[9]

The point of departure for the new method was the technique of serial passages through rabbits via Roux's method of intracranial inoculation, which produced a stable or "fixed" rabies virus of maximum virulence and minimum incubation period for both rabbits and dogs. A series of rabbit-to-rabbit passages begun in November 1882 and continued without interruption in the three intervening years had now reached its ninetieth rabbit. The first rabbit in the series had been infected with the spinal marrow of an ordinary rabid dog by Roux's method of direct inoculation onto the exposed brain. The virus was then passed directly from rabbit to rabbit, always by the method of intracranial inoculation. Over the course of the first forty or fifty passages, the incubation period had declined from about fifteen to seven days, where it still remained in the ninetieth passage, though a slight tendency toward a six-day incubation was beginning to emerge. "Nothing is easier, therefore," wrote Pasteur, "than to have constantly at one's disposal, for considerable lengths of time, a virus of perfect purity and always identical with itself." The technique of serial passages was "virtually the whole secret of the method."[10]

Well, not quite. For Pasteur now revealed an even more crucial feature of his new method of preventing rabies. During the past year, he and his collaborators had developed a new technique for altering the virulence of the rabies virus. Instead of passing the virus through monkeys, they now attenuated it by extracting spinal cords from rabbits dead of the "fixed" rabies virus, cutting them into strips several centimeters long, and then suspending these spinal strips from a thread inside a flask with two cotton-stoppered holes at the top and near the bottom. To extract moisture from the filtered air that circulated through the flasks, Pasteur placed caustic potash inside them (see fig. 8.3). Infected rabbit spinal strips that were suspended in this filtered, desiccated air gradually lost their virulence vis-à-vis dogs, becoming harmless after a period of time that varied somewhat with the thickness of the strips but especially with the ambient temperature. Ordinarily, two weeks sufficed to render the suspended strips harmless to

Figure 8.3. The Roux-Pasteur technique for preserving spinal marrow from a rabid rabbit. (Musée Pasteur, Paris)

dogs. "These points," wrote Pasteur, "constitute the scientific part of the method."[11]

Pasteur then described in some detail the precise manner in which he and his collaborators applied these principles in order to render dogs immune to rabies quickly and surely.

> In a series of flasks, the air inside which is kept dry by dropping pieces of caustic potash into them, suspend every day a portion of fresh spinal marrow taken from a rabbit that has died of rabies of seven-days incubation. Every day also inject under the skin of the dog to be rendered immune a full Pravaz hypodermic syringe of sterilized broth in which a small piece of one of the drying marrows has previously been ground up. Begin with a marrow old enough to make sure that it is not at all virulent. . . . On the succeeding days proceed in the same manner with fresher marrows, and use those of every second day, until finally we inoculate a last and very virulent one that has been drying only one or two days.[12]

By this method, Pasteur reported, he had rendered "fifty dogs of all ages and all races immune to rabies without a single failure" when young Joseph Meister unexpectedly appeared at his laboratory door. He decided, "not without profound anxiety," to apply the method to the apparently doomed boy. At 8 P.M. on 6 July 1885, with Drs. Vulpian and Grancher in attendance, Meister was injected "under a fold made in the skin of the upper right abdomen with one-half Pravaz syringe [i.e., one-half cubic centimeter] of the marrow of a rabbit that died of rabies on June 21st." During the fifteen days from 21 June to 6 July, the rabid spinal marrow had been drying out in one of Pasteur's special flasks. Young Meister submitted to twelve additional injections over the next ten days, always in the abdomen, alternately on the right and left sides, and always one-half Pravaz syringe in amount. On the second and third days, 7 and 8 July, he was injected four times, twice each day, with a broth containing infected spinal cords that had been drying out for fourteen, twelve, eleven, and nine days, respectively. Every succeeding day through 16 July, Meister received one injection each day with a cord that had been drying out for one day less than its predecessor. The last injection was made with the most virulent rabies virus available—that contained in a fresh spinal cord from a rabbit dead of a rabies virus that had been repeatedly passed through rabbits.[13]

If this method seemed dangerous, especially toward the end, Pasteur justified it by pointing to the results of his experiments on "the fifty dogs already mentioned." Once a state of immunity had been achieved, he wrote, increasingly virulent injections were without risk and indeed seemed only to enhance the level of immunity. And the highly virulent injections that came toward the end of Meister's treatment had another important advantage: they drastically reduced the period of time during which one might fear the eventual appearance of rabies. Given the virulence of the rabid spinal cords deployed toward the end, rabies would doubtless appear quickly, presumably within weeks, or else it would not appear at all. And so, as early as mid-August of 1885, five or six weeks after Meister's treatment had begun, Pasteur felt confident that the young Alsatian lad would escape the effects both of the bites of the rabid dog and of the virulent injections to which he had been exposed during the last several days of his treatment. By 26 October, when Pasteur delivered this famous paper to the Académie des sciences, Meister had been in perfect health for three months and three weeks. It therefore seemed almost certain that the threat of rabies had long since passed in his case.

The human interest of Pasteur's paper then gave way to a theoretical digression about how the results of his new method of preventing rabies might be explained—a crucial issue to which we shall return in the next chapter. Pasteur's paper regained its dramatic tone at the end, as he turned to the story of Jean-Baptiste Jupille. The transition came in the form of the statement that "probably the most anxious question at the moment is that of how much time may be allowed to elapse between the bite and the application of the treatment." In the case of Joseph Meister, this interval had been only two days or so, but "it will certainly be considerably longer in a large number of cases." And as a matter of fact, Pasteur now reported, he had already begun treating just such a case a week ago. In this second attempt to apply his new remedy to human cases, once again "obligingly assisted by MM. Vulpian and Grancher," Pasteur was trying to save a fifteen-year-old boy who had been bitten "in circumstances of peculiar gravity" a full six days before his treatment began. Pasteur continued—and concluded—as follows:

> The Académie [des sciences] will hear, not without some emotion, the story of the deed of bravery and cool-headedness accomplished by the boy whose treatment I took in hand last Tuesday. Jean-Baptiste Jupille is a shepherd boy hailing from Villers-Farlay in the department of the Jura. Seeing a powerful dog with suspicious gait attacking a group of six of his comrades, all younger than himself, he seized his whip and rushed forward to meet the animal. The dog at once caught hold of Jupille by the left hand. There followed a hand-to-hand battle, so to speak, the boy finally throwing the animal down and pinning him to the ground under his knee. Next, with his right hand he forced open the jaws of the beast—all the while sustaining new bites—and, taking the thong of

> his whip, he tied the muzzle of his enemy and beat him to death with one of his wooden shoes.[14]

As Pasteur sat down, his colleagues at the Académie des sciences rose in applause. Three of them asked for the floor. The first speaker, predictably enough, was Dr. Vulpian, who was not only a member of the French rabies commission but had also participated in the decisions to treat young Meister and Jupille. Vulpian expressed his immense admiration for Pasteur's paper, an admiration that he was convinced would soon be shared "by the entire medical world":

> Rabies, that terrible disease against which all therapeutic efforts have hitherto failed, has finally found its remedy. M. Pasteur, who has had no precursor in this line except himself, has been led by a series of researches pursued without interruption for several years to create a method of treatment by which one may prevent, with certainty, the development of rabies in humans who have recently been bitten by a rabid dog. I say "with certainty" [*à coup sûr*] because, after what I have seen in M. Pasteur's laboratory, I have no doubt as to the constant success of this treatment, when it is put into practice in all its particulars [*dans toute sa teneur*] a few days after a rabid bite.
>
> Given all this, we must now preoccupy ourselves with the organization of a treatment service for rabies by Pasteur's method. It is essential that everyone bitten by a rabid dog should benefit from this great discovery, which puts the finishing touch on the glory of our illustrious colleague and adds the most distinguished luster to our country.[15]

The second speaker, one M. Larrey, was so taken by Pasteur's account of young Jupille's courage that he successfully urged the Académie des sciences to award him a national prize for virtue (*prix de vertu*).[16] Last to speak was the veterinarian Henri Bouley, an erstwhile critic of Pasteur's doctrines who had become a convert a decade ago and who now served both as chairman of the French rabies commission and as president of the Académie des sciences. Bouley predicted that Pasteur's report of his latest achievement, "one of the greatest advances ever accomplished in the domain of medicine," would make the date 26 October 1885 "forever memorable in the history of medicine and forever glorious for French science."[17]

Pasteur's exciting news spread throughout the world with astonishing speed. Already famous for his vaccine against anthrax in sheep, Pasteur now became a full-fledged international hero. Victims of animal bites soon flocked to Paris from near and far, from as far away as Russia to the East and America to the West, to benefit from the new treatment. It won lavish praise from nearly all who submitted to it, and centers for the treatment quickly spread to other countries. By November 1886, little more than a year after young Joseph Meister first went through the series of injections, Pasteur's treatment had been applied to nearly twenty-five hundred people in Paris alone.[18] From virtually everywhere in the world there also came a flood of monetary contributions, large and small, from emperors and schoolchildren, to support Pasteur's center for rabies treatment. The celebrated Institut Pasteur, built and initially sustained by these private donations, was officially inaugurated in November 1888, just three years after Pasteur announced the application of his rabies vaccine to human cases. A statue in front of the building depicts a pitched battle between a rabid dog and the brave young shepherd Jean-Baptiste Jupille. By the time Pasteur died in 1895, some twenty thousand people had submitted to his rabies treatment at centers throughout the world.[19] And the Institut Pasteur in Paris, designed from the first to be a center for basic research as well as for the treatment of rabies, has loomed large ever since in the history of science and medicine.

This is the familiar and triumphal version of the story. But a fuller, more complicated, and somewhat less heroic version deserves to be told. In fact, this revised account is so different from the usual story that it will occupy the rest of this chapter and all of the next one. Here, too, the most compelling material is drawn from Pasteur's private papers and laboratory notebooks, as will become especially clear in the next chapter. But we can anticipate some of the issues to be addressed there if we first amplify the now faint voices of a neglected set of historical actors: those who dared to criticize Pasteur and his treatment for rabies.

FORGOTTEN CRITICS

From the outset, a few scientists and more than a few physicians insisted that Pasteur's new treatment for rabies was ill-founded in principle and downright dangerous in practice. A separate stream of criticism came from anti-vivisectionists and anti-vaccinationists, who were to be found almost exclusively in England.[20] When noticed at all, such critics have been dismissed as benighted obstacles on the path to scientific and medical progress—precisely the reputation that Pasteur and his allies worked hard to pin on them. In 1889, as English critics became increasingly shrill in opposition to a proposed anti-rabies institution in London, their exasperated compatriot T. H. Huxley sallied forth to excoriate them in his inimitable style:

> But the opposition which, as I see from the English papers, is threatened has really for the most part nothing to do either with M. Pasteur's merits or with the efficacy of his method of treating hydrophobia. It proceeds partly from the fanatics of *laissez faire*, who think it better to rot and die than to be kept whole and lively by State interference, partly from the blind opponents of properly-conducted physiological experimentation, who prefer that men should suffer rather than rabbits or dogs, and partly from those who for other but not less powerful motives hate everything which contributes to prove the value of strictly scientific methods of inquiry in all those questions which affect the welfare of society.[21]

This lovely bit of invective allows us to see just how much was at stake here. For Huxley and other scientistic spokesmen, Pasteur's treatment for rabies offered powerful new evidence of the therapeutic utility of an ascendant "scientific" medicine. More than that, it was a symbolic rallying point in a wider struggle for cultural authority and power—between scientific knowledge and clinical experience in medicine; between "strictly scientific methods of inquiry" and traditional sources of authority "in all those questions which affect the welfare of society"; and even—in the English context—between "state interference" and "laissez faire" in politics writ large.

As "Darwin's bulldog," Huxley had already served on the front lines in one major skirmish in this wider cultural battle. Now, three decades later, he was ready to deploy his polemical talents on behalf of the Pastorian enterprise. Nuance and concession played no part in Huxley's rhetorical strategy. Critics of Pasteur's treatment for rabies were to be lumped together with all the other forces of darkness, and all were to be pushed aside in pursuit of a larger project: to secure the cultural dominion of modern "professional" science. There was no need to pay close attention to the actual content of the critiques directed against Pasteur and his rabies vaccine; it was enough to focus on the dubious motives that allegedly inspired them. The success of this strategy is evident in the story line of all standard histories of bacteriology: Pasteur was right and a master of "scientific method"; his critics were not only wrong but also incompetent and desperate defenders of a fading cultural regime.

In France, the Pastorian juggernaut was fueled partly by nationalism. We have already heard from Pasteur's colleagues at the Académie des sciences, where it was said that his rabies vaccine "adds the most distinguished luster to our country" and that the date of its announcement, 26 October 1885, would be "forever glorious for French science." Not for the first time, nor for the last, Pasteur and his allies appealed to French national pride in support of his research. Pasteur himself had already called the germ theory of fermentation a "French" discovery, and he insisted that any beers that might be manufactured under his patents should be called "bières françaises" for domestic consumption and "bières de la revanche nationale" abroad. He also wrote that he would have been "inconsolable" had experimental vaccination been anything but "a French discovery."[22]

Anti-Pastorian sentiment did exist, even in France, but it was confined mainly to the "popular," leftist or anti-establishment press.[23] In "official" French circles—the Académie des sciences and the Académie de médecine, in both of which Pasteur was an honored member—the Pastorian treatment for rabies went almost unchallenged. The only notable or, rather, notorious exception was the clinician Dr. Michel Peter, a member of the Académie de médecine, about whom we shall soon hear a good deal more. By 1887, another quixotic French critic of the new rabies vaccine, Dr. Auguste Lutaud, was thoroughly frustrated by the Pastorian success at playing the nationalist card. In France, wrote Lutaud, "One can be an anarchist, a communist, or a nihilist, but not an anti-Pastorian; a simple question of science has been made into a question of patriotism."[24]

For the most part, Pasteur was lucky in his critics, both at home and abroad. Few in number and ineffectual in strategy, they were quickly overwhelmed by the Pastorian forces. They did their cause no favor by adopting a strident, hectoring tone that betrayed their personal hostility toward Pasteur. The clinicians among them were too obviously self-serving when they complained about the intrusion of this "mere chemist" into their traditional domain. By the late 1880s it was no longer enough simply to assert, as these doctors did, that the proper foundation of medicine was clinical experience, not animal experiments—that medicine was an "art," not a science.

And yet, for all of that, these few and forgotten critics did sometimes hit a raw Pastorian nerve, and some of their objections were more telling than the public record suggests. Even as they went down to defeat, these critics caused Pasteur some real concern and embarrassment, almost all of it hidden from public view. The critiques can be divided into three broad categories, as indeed they were at the time:[25] (1) *experimental or "strictly scientific" issues*—most obviously when experiments elsewhere did not fully confirm Pasteur's results, but also when the participants could not agree about what counted as a properly "scientific" approach to the issues in dispute; (2) *clinical concerns*—including the challenge of diagnosing rabies, especially in patients who displayed nervous symptoms after having been bitten by rabid animals and then submitting to Pasteur's "prophylactic" treatment; and (3) *statistical arguments*, perhaps the most disputatious arena of all. Actually, we can add a fourth, still broader category of dispute, which implicitly links the other three: *ethical concerns* about Pasteur's conduct and about the safety and efficacy of his treatment.

THE 1887 DEBATES IN THE ACADÉMIE DE MÉDECINE OVER PASTEUR'S WORK ON RABIES

We need not cast our net very widely to capture the specific issues in dispute under these broad categories. They were conveniently brought together in a series of heated debates at the Académie de médecine between January and July 1887.[26] Perhaps the word "debates" is a bit misleading here, for the deck was heavily stacked in Pasteur's favor. Indeed, one of the most striking features of the controversy, as recorded in the *Bulletin* of the Académie de médecine, is how completely it overturns the widespread notion that Pasteur had to battle fiercely against a conservative medical establishment. For in fact, almost all members of the Académie de médecine were openly enthusiastic about Pasteur's work on rabies and his vaccine against it.

Pasteur's critics, in striking contrast, were represented by one lonely voice in the Académie de médecine: Dr. Michel Peter, who was in fact Pasteur's cousin-by-marriage, a relationship that may have given him access to "insider information" from members of the tight-knit Pastorian inner circle.[27] An elegant man of the world and a clinician of the old school, Dr. Peter was responsible for instigating and prolonging the debates. He was in some ways the worst possible spokesman for Pasteur's critics. He wasted too much time on anecdotal "case histories" of patients who had allegedly died of rabies after submitting to Pasteur's treatment. His arguments quickly became repetitive and tiresome. And his relentlessly hostile and accusatory tone toward Pasteur won him no friends in the Académie de médecine or anywhere else in the French medical and scientific establishment.

During the debates at the Académie de médecine, Peter was not merely outnumbered and outwitted, he was also outmaneuvered and even hissed and booed. The president and the perpetual secretary of the Académie clearly arranged things in Pasteur's favor,[28] and we can be virtual witnesses of the general audience response thanks to the Académie's charming practice of including crowd noises as part of the published account of its meetings. From the account published in the *Bulletin* of the Académie de médecine it is clear that its members greeted Pasteur and his many defenders with respect and applause, while Peter's disquisitions were punctuated with muttering and hissing, with nary a single recorded indication of applause.[29] At the last debate, on 12 July 1887, when Charcot as president gave a ringing and concluding defense of Pasteur and his work, the audience responded with "prolonged applause."[30] By then, Dr. Peter must have felt disheartened and even beleaguered. He was certainly defeated, indeed overwhelmed, like so many other would-be critics of Pasteur.

Insofar as Dr. Peter's campaign against Pasteur did attract any attention outside the Académie de médecine, it was mainly because several people had recently died of rabies after undergoing Pasteur's treatment. Much of the debate, especially at first, concerned the particular circumstances surrounding these deaths and their implications for the safety and efficacy of Pasteur's vaccine. At one point, Dr. Peter went so far as to accuse Pasteur of direct responsibility for at least one such death. Dr. Vulpian expressed his sense of the gravity, if not the accuracy, of Peter's assertions by saying that they amounted to a charge of "involuntary homicide" against Pasteur.[31]

Dr. Peter's reckless accusations infuriated most members of the Académie de médecine and doubtless most of France and the international scientific community as well. Such rhetorical disasters undermined Peter's more general and sometimes more telling case against Pasteur's work on rabies, which attracted little attention at the time and has been almost entirely ignored in the century since. Yet Peter's critique, however overwrought and ill-advised in tone, included some intriguing challenges to Pasteur's work on rabies. It covered the full range of experimental, clinical, statistical, and ethical issues. And even some of Peter's most outrageous accusations were not entirely unfounded, as will become clear in the next chapter.

When it came to experimental issues, Dr. Peter had to rely on the work of others, for he had no experience or credentials of his own in experimental research, as Pasteur was quick to point out in his disdainful replies to Peter's attacks. Yet Dr. Peter displayed no great concern about his experimental "incompetence," saying that it put him in the good company of 99 percent of the members of the Académie de médecine.[32] More important, Peter's own experimental expertise was not at issue here. How, he asked in effect, did Pasteur propose to refute the serious experimental critiques published by more competent scientists? In a surprisingly clever move, Peter made the question more pointed by drawing special attention to the independent critiques produced by two scientists—one from Italy, the other from Austria—who had come to Pasteur's own laboratory in Paris to learn his techniques at first hand, only to find that they were unable to replicate the Pastorian results upon their return to their native laboratories. In particular, they were often unable to prevent rabies in experimental animals outside of France, even when they began with emigré French-born laboratory animals given them by Pasteur himself and tried to follow Pasteur-perfect techniques.

Here Dr. Peter unwittingly displayed his prescience as a sociologist of knowledge. A century ago, he taunted Pasteur as if already aware of the contingencies of experimental knowledge—as if he already knew how hard it was to translate "local knowledge" from one experimental setting to another, no matter how carefully scientists at other sites tried to imitate the technical gestures of the original laboratory, including here even the "same" laboratory animals. But Dr. Peter was, of course, "ahead of his time," and Pasteur knew how to deal with such cultural precocity—if only because he was himself the best sociologist of knowledge around, as Bruno Latour has been trying to persuade us for years.[33]

Even so, Pasteur had to recover from a sort of philosophical or sociological *faux pas*, though he was pushed before he slipped. Under pressure, he had laid himself open to "the replicability problem." The pressure came in the form of renewed complaints about his secrecy—complaints that echoed those voiced five years earlier during his research on the anthrax vaccine, as we have seen in the story of the "secret of Pouilly-le-Fort." This time Pasteur decided to meet the complaints head on. He allowed outsiders, strangers, even foreigners, to observe the Pastorian techniques for preventing rabies on the spot, behind the usually closed doors of his laboratory. He did so, it seems, with complete confidence that all of these outsiders would go away convinced of the merits of his treatment for rabies. He was mistaken, and oddly so for such a sophisticated sociologist of knowledge.

In a letter of 22 July 1886, Pasteur told the vice-president of the Municipal Council of Paris that he could easily refute "the odious falsehood" that "*Pasteur keeps his method secret*." For not only were there Frenchmen "who know all the details of [my] method" of preventing rabies, there were also a number of "foreign doctors" who had studied the method in his laboratory, some of whom had "already founded institutes to apply it in their respective countries." And far from keeping his method secret from these foreign doctors, he had even given some of them the "initial material for [their] inoculations." He listed eight such foreign visitors, who had come to his laboratory from as far away as Odessa to the east and New York City to the west.[34]

But at least two foreign doctors were soon to give Pasteur cause to regret his hospitality to them: Dr. Amoroso of the First Medical Clinic at the University of Naples, and Professor Anton von Frisch of Vienna. Dr. Peter cited both of them as part of his attack on Pasteur at the Académie de médecine in 1887. Amoroso published at least two brief critiques of Pasteur's method of preventing rabies. He reported that he had been unable to replicate all of Pasteur's results despite having studied the Pastorian techniques for three weeks in Paris, and despite having begun with two rabid animals that Pasteur himself had given him to take back to his laboratory in Naples. Amoroso's experiments on rabbits, guinea pigs, and dogs led him to two conclusions, of which the first was entirely in keeping with Pastorian results: rabies was invariably transmitted from one animal to another by Roux's method of intracranial inoculation. But it was Amoroso's second conclusion that Dr. Peter was eager to announce: that Pasteur's method of treating rabies was totally ineffectual in animals that had been inoculated with the virus by Roux's method. In response, Pasteur and his allies resorted to the petty complaint that Dr. Amoroso had inflated his credentials by calling himself "professor," and otherwise discredited his two brief critiques.[35]

Professor von Frisch of Vienna was less easily dismissed. In 1887 von Frisch published an extensive and impressively detailed critique of the Pastorian treatment for rabies in a book that covered more pages than the entire corpus of Pasteur's published papers on the subject: *Die Behandlung der Wuthkrankheit: Eine experimentelle Kritic des Pasteur'schen Verfahrens* (*The Treatment of Rabies: An Experimental Critique of the Pastorian Techniques*).[36] The book appeared just in time for Dr. Peter to enlist von Frisch as a witness for the prosecution, so to speak. Because von Frisch's scientific credentials were solid, and because his critique was so extensive, Pasteur privately expressed considerable concern and irritation in the face of this challenge.[37] For current purposes, we need not give von Frisch's critique the full analysis it deserves. We will confine attention here to its fate in the debates at the Académie de médecine.

Von Frisch's critique ranged widely. It included discussions on the clinical and statistical issues in dispute. In these domains, von Frisch did not have anything strikingly original to say, although he was an insightful critic of the Pastorian statistics, which he called "totally worthless," mainly because there was reason to doubt that the Pastorians had kept their vow to treat only people who had been seriously bitten by a certifiably rabid animal.[38] But the centerpiece of von Frisch's critique was his experimental case against the Pastorian vaccine. Von Frisch, it deserves repeating, was one of the few scientists outside Pasteur's inner circle who had been allowed to observe at first hand the way in which the Pastorian team actually went about its day-to-day work on rabies. Yet despite the crucial "craft knowledge" he had thus obtained, von Frisch reported that he had been unable to replicate the apparently decisive results claimed by the Pastorians. Perhaps most disconcertingly to the Pastorians, he insisted that he—like Dr. Amoroso—had been unable to prevent rabies in dogs that had been inoculated with the rabies virus by Roux's intracranial method.[39]

In responding to von Frisch's critique, Pasteur resorted to a familiar ploy: he expressed doubts about the Austrian's technical competence, suggesting that von Frisch lacked the skill to achieve sterile conditions in his experiments and insisting that he had made a fundamental mistake by using dogs instead of rabbits in his attempts to produce a rabies vaccine. Von Frisch responded with understandable outrage at Pasteur's ad hominem attack:

Who says I'm incompetent? Only Pasteur. And Pasteur says he is opposing his positive results to my negative findings. But who is defining "positive" and "negative" here? Once again, only Pasteur, who defines as "positive" those experiments that support his vaccine and "negative" those that do not.[40] Like Pouchet before him in the spontaneous generation debate, von Frisch was objecting to Pasteur's high-handed way of *defining* a priori which experimental results were to count and which were not.[41]

And like Pouchet before him, von Frisch lost the debate, at least in France. Nor was von Frisch, any more than Pouchet, unjustly robbed of a victory that clearly should have belonged to him. Pasteur and the Pastorians were very effective. In public, they kept their composure. And in defending himself against von Frisch and other critics, Pasteur made exceptionally clever use of another wide-ranging analysis of his work on rabies: the Report of the English Commission on Rabies, also published in 1887 and therefore available for use in the debates at the Académie de médecine.[42] In fact, no more judicious assessment of Pasteur's method of treatment appeared during his lifetime. The English commission report was solomonic in its judgments. It expressed admiration for Pasteur's experiments, some of which the commission had repeated successfully—conducted, it deserves emphasizing, under Pasteur's close supervision.[43] The report also expressed the belief that Pasteur's new treatment for rabies had probably saved many lives.

But the English commission also drew attention to the uncertainty of all statistics on rabies, citing the difficulty of establishing that the attacking animal had in fact been rabid as well as the variable effects of the location and depth of bites, of differences in the lethality of rabid animal bites in different species and races, and of the possible prophylactic effects of cauterization or other treatments applied to bitten victims before they submitted to Pasteur's treatment. The commission also suspected that at least one man may have died as a direct result of the Pastorian injections, and in the end it favored strict regulations on potentially rabid animals (muzzling and quarantine) over Pasteur's more drastic remedy.[44] Indeed, such "police" measures were already operating with striking success in Australia and Germany. And though many English pet lovers objected to such state interference, especially to laws that required them to muzzle their dogs, the eventual adoption of such measures in England virtually eliminated rabies there by the turn of the century.[45]

Despite these reservations, Pasteur seized on the Report of the English Commission on Rabies as a weapon in his battle with Dr. Peter (and his outside foreign experts). He managed to make it sound like a ringing endorsement of all his work on rabies,[46] and Peter proved unable to take advantage of the less positive parts of the Report. So the English rabies commission, presumably unbiased, helped the Pastorians to push aside yet another set of critics. In the end, as we have already seen, Pasteur carried the day in the 1887 debates at the Académie de médecine, overwhelmingly so. But his victory was due not so much to any decisive experimental evidence nor even to the endorsement of the English commission. It was rather a testament to his hard-earned scientific authority and rhetorical skills, and—not least—to a rapidly mounting body of statistical evidence that seemed clearly to show the safety and efficacy of the Pastorian vaccine in human cases.

In a way, it is a shame that Pasteur's victory in the 1887 debates was so overwhelming. For that outcome obscured some interesting and important issues that were lurking just beneath the surface of the debates—and once or twice surfaced in Peter's otherwise hapless attacks. In particular, Peter struggled unsuccessfully to draw Pasteur and his allies into a quasi-philosophical discussion as to whether or not Pasteur's work on rabies was truly "scientific" and, more important, ethical.[47]

Peter, of course, insisted that Pasteur's rabies vaccine was not truly "scientific" and that it was unethical to boot. He argued, first of all, that Pasteur's work on rabies was not properly scientific because he kept the details of his experiments secret. Like all purveyers of "secret remedies," Peter charged, Pasteur said both too much and too little about his treatment for rabies—enough to attract fame and funds, but too little to allow independent evaluation or replication of his claims.[48] Here Dr. Peter was echoing complaints about Pasteur's secrecy that had already surfaced in the Académie de médecine seven years before, during the debate over his work on anthrax and "the secret of Pouilly-le-Fort," as discussed in detail in Chapter Six. At that time, Roux had warned Pasteur that many physicians, in particular, considered his laboratory an improperly "secret sanctuary." And now, in 1887, Pasteur's early accounts of his work on rabies, through their reticence about the details of his experiments and his techniques for producing the vaccine, invited similar complaints.

To all such charges, Pasteur responded by insisting on the need for careful quality control and by denying that his motives were in any way mercenary. He pointed to the risk of a fatal disaster if the details of his method became known to those less experienced than he and his collaborators, and he reminded his critics that he dispensed his rabies treatment for free. He also emphasized, as we have already seen, that he had revealed the details of his techniques to dozens of scientists, including several foreigners who came to his laboratory to learn those practices on the spot.[49]

Even so, Dr. Peter complained, Pasteur's vaccine against rabies could not be considered properly scientific because it lacked any theoretical foundation. Pasteur was famous for his germ theory of disease, but he had failed to isolate or cultivate the microbe allegedly responsible for rabies. In developing his other vaccines, all for animal diseases, Pasteur had cultivated an attenuated strain of the specific microbe to which he ascribed each disease. But his rabies vaccine had been obtained by mere "empirical" manipulations of rabid spinal cords. For Dr. Peter, Pasteur's research on rabies represented nothing but "empiricism embellished by contradiction." He accused Pasteur of yielding to a "deceiving induction" in extending his experiments from rabbits to dogs and then from dogs to humans. And if Pasteur's original method of vaccination was "scientific," he hinted darkly, why had he modified it more than once since treating Joseph Meister in July 1885?[50]

In their responses to such criticism, Pasteur and his allies took advantage of a confusion that is still very much with us—the confusion between that which works, that which is true, and that which is scientific. One of Pasteur's supporters, the distinguished Dr. Brouardel, offered this blatantly utilitarian defense of the treatment:

> As to the reproach directed against the method as being anti-scientific, I avow that I do not understand it. . . . In my opinion, that alone is anti-scientific which is not true. If someone demonstrated to me that rabies could be cured by the use of a fantastic omelette or oyster-shells, I would still find the thing scientific. In the end, those who seek this quarrel with Pasteur simply ask him the how and the why of the method. M. Pasteur will tell us that when we have found the answer to the question posed by our great comic [Molière]: why does opium produce sleep?[51]

Early on in the debate, Pasteur himself blandly asserted that the "scientific basis" of his treatment lay in "the possibility of conferring immunity against the virus of street rabies in animals by the sub-cutaneous injection of increasingly virulent rabbit spinal cords."[52] But this was merely to transform a raw empirical result into a "scientific" foundation for his treatment. Peter was asking for more. But Pasteur, whose work on rabies was indeed more "empirical" than usual for him—though far less so than his unsuccessful efforts to improve the quality of French beer—displayed no interest in Peter's quasi-philosophical concerns. He and his allies deflected attention from this and other concerns by pointing to the overwhelmingly favorable statistics that Pasteur could marshall on behalf of his vaccine. In particular, the Pastorians emphasized that the mortality rate after his treatment was less than one percent, compared to 15 to 20 percent for untreated victims of rabid dog bites. What more could one ask?

Although Pasteur displayed no public concern about the matter, he did find it somewhat harder to shrug off Dr. Peter's charge that the application of his new vaccine to human cases was not only unscientific, but also, and more importantly, unethical. One ethical concern was based on a simple but fundamental feature of Pasteur's vaccines that he had himself done much to emphasize. Unlike the vaccine against smallpox that Jenner had taken from nature almost a century ago (in the form of cowpox), Pasteur's vaccines against rabies and other diseases were products of the laboratory. Pasteur was proud of this difference between Jenner's vaccine and those he and his collaborators had produced against chicken cholera, anthrax, swine fever, and now rabies. As products of the laboratory, Pasteur insisted, his vaccines were more susceptible to human manipulation and control than Jenner's smallpox vaccine.[53]

But a few critics, notably Drs. Peter and Lutaud, were quick to point out that Pasteur's "artificial" viruses might be the source of a novel disease of Pasteur's own making—*artificial or laboratory rabies* or even *la rage Pastorian*. Like current critics of recombinant DNA research or other forms of "genetic engineering," Drs. Peter and Lutaud expressed concern that Pasteur's treatment made use of altered rabies viruses of uncertain and potentially lethal properties. As evidence that this fear had some basis in reality, Dr. Peter and others pointed to the frequency with which the paralytic form of rabies seemed to appear in those who submitted to Pasteur's vaccine, as opposed to its rarity under natural conditions.[54]

Dr. Peter was especially alarmed by the dangers of the modified "intensive" method of treatment that Pasteur had introduced for certain cases, especially for victims of severe wolf bites or those who came for treatment only a long time after they had been bitten by a rabid animal. Pasteur's "intensive" method involved earlier and more frequent injections of virulent spinal cords, and Peter considered it reckless and wholly unjustified. He and other clinicians also complained that Pasteur's use of "live" (if inactivated) rabies viruses complicated the uncertainties of diagnosing rabies in any vaccinated person who later developed a nervous disorder of uncertain origin and character. And despite the glowing statistics that Pasteur cited in support of his vaccine, a small but steady trickle of people did become paralytic or even died after submitting to the treatment.[55]

At first sight, all such clinical objections to Pasteur's rabies vaccine seem ludicrous in view of its overwhelming statistical success. Surely a few difficulties, complications, and even deaths were inevitable upon the introduction of any novel treatment for a lethal disease. Surely Dr. Peter and his allies were demanding that Pasteur's rabies vaccine meet unreasonably high standards of uniformity, reliability, safety, and efficacy.[56] And surely Dr. Peter would have resisted the application of such exalted standards to the ordinary therapeutic measures that he and his fellow clinicians deployed every day.

But Peter could respond by insisting that rabies posed very special ethical problems. This is not to say that Peter's concerns found explicit expression in the formal language of current "bioethics." But he and others—including indeed Pasteur himself—recognized that rabies and Pasteur's vaccine against it posed a unique ethical dilemma. In fact, rabies is unlike any other disease on earth, and ethical postures appropriate to less peculiar maladies are not always equally appropriate to it. In this chapter, we will focus on those timeless features of rabies that raise a special ethical dilemma for *any* attempt to treat it. And we will end by noting Pasteur's own stated position about these ethical issues up to the point at which he decided to apply his vaccine to young Joseph Meister.

THE TIMELESS ETHICAL DILEMMAS RAISED BY RABIES

Rabies is, of course, a horrible and invariably fatal disease. As of 1977, the U.S. Center for Disease Control had recorded only three cases of presumed recovery from symptomatic rabies in all of human history.[57] But it is also very rare in humans, at least in the industrialized world. Moreover, rabies is not a "communicable" disease in the usual sense; it is not transmitted from person to person. As a rare and noncommunicable disease, rabies has never seemed to justify the risks or the intrusion on individual rights that compulsory vaccination against any disease entails. Vaccination against rabies makes sense only in the case of very small and well-defined populations—namely, those who are exposed to an exceptionally high risk of contracting the disease either because they work in a rabies-saturated environment (for example, in laboratories conducting research on the disease) or, much more usually, those who have already been bitten by a certifiably rabid animal. Ordinary preventive measures, including vaccines against smallpox, polio, and other infectious diseases, can be encouraged and justified on the grounds of their potential benefit to society at large as well as the individual submitting to them. Rabies vaccination, by contrast, can be justified solely on the basis of its potential benefit to the vaccinated person, who poses no threat to others.

But if Pasteur's vaccine was therefore unlike ordinary preventive measures, so too was it unlike ordinary therapeutic measures. For even in the case of a person already bitten, the situation is far from straightforward. In the first place, it is sometimes impossible to capture the attacking animal and establish that it was indeed rabid. More important, there is no way to be sure that even the bites of a certifiably rabid animal will lead to rabies in the victim. In fact, the level of uncertainty is high. The mortality rate of "declared" or symptomatic rabies is effectively 100 percent, but the threat of death from the bites of a rabid animal is vastly less. Depending on such factors as the species of attacking animal (wolf and cat bites, for example, seem to pose a much higher risk than dog bites), the depth and location of the bites (bites on the face are much more lethal than those on the hands or limbs), and the application and timing of cauterization or other treatments for the bites, estimates of the risk of contracting rabies from the bites of a certifiably rabid animal range from as high as 60 percent to as low as 5 percent. It is perhaps futile to try to settle on a meaningful "average" figure within this wide range, but it is worth emphasizing that Pasteur himself estimated that only 15 to 20 percent of people bitten by rabid dogs would eventually die of rabies if they would not or could not submit to his treatment.[58]

In short, the great majority of the victims of rabid animal bites could forgo Pasteur's treatment without experiencing any untoward consequences in the future. And they had to decide whether or not to submit to the treatment at a point when they had no symptoms of the disease. For the efficacy and very possibility of Pasteur's vaccine depended on the peculiarly long incubation period that separates the infective bites of a rabid animal from the outbreak of symptoms. At some point during this incubation period, perhaps as soon as a week or two, the vaccine loses its capacity to prevent the disease from taking its natural and invariably fatal course. Once the symptoms become manifest, neither Pasteur's vaccine nor any other is of any use. But those who choose to undergo the series of vaccinal injections have no way of knowing whether or not they would ever have fallen victim to rabies had they made the opposite decision. There is simply no way to be sure that the rabies vaccine is even potentially beneficial to the vaccinated individual. In this crucial respect, Pasteur's "treatment" was unlike ordinary therapeutic measures, undertaken for the immediate sake of a person already suffering from a disease. When he vaccinated asymptomatic victims of animal bites, Pasteur was subjecting them to a painful and inherently risky series of injections even as he knew that many and probably most of them would escape the disease anyhow.

Not even AIDS, the only other human disease with a mortality rate of 100 percent (or so we believe thus far), poses such a unique ethical dilemma.

True, the symptoms of full-blown AIDS, like those of rabies, appear only after a long incubation period of the implicated virus (HIV)—a similarity that explains why several teams of investigators are now in frantic pursuit of a safe and effective "postexposure" vaccine against AIDS. Pasteur's treatment for rabies, like all its successors, was also based on "postexposure" vaccines. But in the case of AIDS, one can predict the eventual emergence of the disease with a high, if imperfect, degree of confidence because the HIV virus can be detected in the blood. In striking contrast, Pasteur had no secure way of knowing that asymptomatic victims of rabid animal bites had been infected with the rabies virus. In effect, every decision to treat an asymptomatic victim of rabid animal bites entails an exquisite "moral calculus," in which a low probability of infection must be balanced against a 100 percent fatality rate once the symptoms appear.

If these distinctions now seem overly precious, that is only because the accumulated statistical evidence of a century suggests that the risk of death or serious harm from rabies vaccination is much less than the risk of death from the bite of rabid animal. True, there have always been thoughtful critics of this superficially convincing statistical evidence for Pasteur's vaccine,[59] and even the Pastorians themselves became concerned about the persistent if rare "accidents" (paralysis or death) that followed the treatment. Indeed, by the time Pasteur died in 1895, the Institut Pasteur itself had switched from his original "live" vaccine to an inactivated carbolic acid vaccine, and the rabies vaccines developed since differ even more radically from Pasteur's initial version.[60] Even so, as the statistical evidence available to Pasteur seemed increasingly to justify his vaccine, he was able to claim a sort of retrospective ethical sanction for it.

But the situation was entirely different when Pasteur first applied his treatment to young Joseph Meister. At that point, obviously, there was no "statistical" evidence of the safety and efficacy of his vaccine. Three months earlier, as described in the previous chapter, Pasteur undertook his secret attempts to cure M. Girard and Julie-Antoinette Poughon of symptomatic rabies. In their case, however, he faced a much less difficult ethical dilemma. For he was then undertaking a clearly *therapeutic* trial on two patients whom he had every reason to believe would otherwise face certain death. But when Pasteur decided to treat the asymptomatic Joseph Meister, he was conducting an experimental trial of a "live" rabies vaccine on a human "subject" who had some real if indeterminate chance of surviving without it—an unusually risky form of human experimentation in which there was no fully secure way of knowing whether the trial was even potentially of benefit to the individual submitting to it.

PASTEUR ON THE ETHICAL ISSUES RAISED BY RABIES

From the outset of his quest for a rabies vaccine, Pasteur clearly appreciated the problems posed by the ethical strictures against nontherapeutic human experimentation. More than once, he addressed the issue explicitly. On 15 May 1884, for example, he told an audience from the Friendly Association of Former Students of the Ecole Centrale des Arts et Manufactures that he and his collaborators had managed to produce an attenuated strain of the rabies virus—in a word, a vaccine—that was yielding very promising results in tests on dogs, and he held out the prospect that such a vaccine might soon be applied to humans. He also reported that increasing public awareness of his quest for a rabies vaccine had already brought him numerous requests for treatment from anxious victims of animal bites; he would doubtless receive many more such appeals in the future. But, he insisted, any clinical trial of a rabies vaccine would perforce pose ethical concerns about human experimentation. First, he would need to secure the aid of a physician, since he did not possess an M.D. degree. He would ask a doctor to join him in any human trials "so as not to engage in illegal medical practice." More important, Pasteur emphasized that he would undertake such trials only after extensive and decisive experiments on animals. Not only would he need first to "acquire the certainty of being able to prevent the disease in dogs"; he would also forgo any human trials until "after having multiplied the same proofs in animals, on dogs, monkeys, and particularly on the bovine species, which seems to contract rabies as a result of bites much more easily than man or the dog."[61]

In August 1884 he delivered a similar message to the International Medical Congress in Copenhagen. He told his audience that he had undertaken the study of rabies precisely because it offered the possibility of a way around the accepted precept that "experimentation, while allowable in animals, is criminal in man." Rabies was the most striking example of an invariably lethal disease common to man and animals, and prior experiments on animals could therefore be used to establish the safety and efficacy of a rabies vaccine before it was applied to human cases. But even then, "proofs must be multiplied *ad infinitum* on diverse animal species before human therapeutics should dare to try this mode of prophylaxis on man himself."[62] And as late as December 1884, Pasteur resisted a written plea to treat a bitten child because his method had not yet been securely established in the case of dogs already bitten. Even should he be able to achieve that goal, wrote Pasteur in his response to this plea, his hand would "tremble" before applying the treatment to humans, "for what is possible in the dog may not be so in man."[63] As late as 12 June 1885 he declined to treat a bitten father and his child on the grounds that his researches had not yet reached the point that would allow him to apply it to man.[64]

Yet a mere three weeks later, on 6 July 1885, Pasteur made the opposite decision in the case of young Joseph Meister, with the happy outcome now known to all. What had happened in the meantime to change Pasteur's mind? Had he achieved his goal of "multiple proofs from diverse animal species" as to the safety and efficacy of his rabies vaccine? Put another way, had he met his own criteria for an ethical human trial of his treatment?

NINE

Private Doubts and Ethical Dilemmas: Pasteur, Roux, and the Early Human Trials of Pasteur's Rabies Vaccine

ONE DAY in the mid-1880s, the "independent" research of Pasteur and his leading collaborator on rabies, Emile Roux, came too close for comfort. On that day, or so we are told by Pasteur's nephew and research assistant Adrien Loir, he prepared some cultures of the swine fever microbe, working as always under Pasteur's watchful eye, and carried them to a laboratory stove. Since Loir's hands were filled with flasks, Pasteur opened the door of the stove for him. As Loir went about his usual tasks, Pasteur noticed an unusual flask in the stove: a flask of 150 cubic centimeters supplied with two tubules open to the ambient atmosphere, one above the other and so arranged as to produce a continuous stream of ordinary air inside the flask (see fig. 8.3). Loir's account continues as follows:

> In this flask a strip of rabbit spinal cord was suspended by a thread. The sight of this flask, which [Pasteur] held aloft, seemed to absorb [him] so much that I did not want to disturb him. . . . After a long silence, he asked me, "Who put this flask here?" I answered that "it could only be M. Roux," for "this is his rack." [Pasteur] took the flask and went down the hall. He raised it above his head, and set himself to look at it in the full light of day for a long, long time. Then he returned to put the flask back in its place [on Roux's rack in the stove] without saying a word.[1]

But if Pasteur said little to Loir about Roux's unusual flask, he did immediately order the construction of a dozen similar flasks—stipulating, however, that they should differ from Roux's flask in two ways: they should be much larger in volume, and they should contain caustic potash in order to dry the air flowing through them. By adding caustic potash, which Roux had not done, Pasteur hoped to prevent the spinal strip from putrefying in ordinary air. Under those conditions, any attenuation of the rabies virus in the spinal strip could be ascribed to the effect of "allowing [atmospheric] oxygen time to attenuate the virus"—in keeping with Pasteur's preference for oxygen-attenuated vaccines.[2]

The very next day Pasteur began suspending strips of rabbit spinal cord in his new desiccating flasks, which he let stand at ordinary room temperature instead of depositing them in the stove, as Roux had done. That afternoon, Roux noticed three of these new flasks sitting on a table in the laboratory. He sent for Loir:

> "Who put those three flasks there," he asked me while pointing to the table. "M. Pasteur," I answered. "He went to the stove?" [asked Roux]. "Yes" [I replied]. Without saying another word, Roux put on his hat, went down the stairs, and left by the door on the rue d'Ulm, slamming it shut as he [always] did when angry.[3]

According to Loir, Roux never said a word to Pasteur about this incident. But thereafter, he claimed, Roux came to the laboratory only at night, when he knew he would not cross paths with Pasteur. And from that moment, Loir continued, rabies became a "dead letter" for Roux.[4]

Here, as often elsewhere in his reminiscences, Loir provides no exact date—not even a year—for this anecdote. But Loir surely did not intend his last sentence to be taken literally. For Roux did not become permanently estranged from the Pastorian rabies project. Elsewhere, Loir himself describes Roux's return to Pasteur's laboratory and his crucial contributions to its work on rabies. Even so, Loir's anecdote is a striking illustration of a more general theme: the tension between Pasteur and Roux. The exact nature of the relationship between them has long been an object of discussion and speculation. To judge from the most credible accounts, this was not a simple case of an affectionate disciple working happily under the master's yoke.[5]

From time to time in the rest of this chapter, I will suggest that at least some of the discord between Pasteur and Roux over rabies can be traced to differences in their professional formation and orientation. Here Pasteur as life-long experimental scientist is contrasted with Roux as a former medical man who never forgot the lessons of his brief career in clinical medicine and who carried part of that professional ethos with him when he joined the Pastorian team, especially when it came to the application of rabies vaccines to human cases. Admittedly, Pasteur and Roux somehow managed to put aside, or paper over, their differences when push came to shove. Even during periods when they were apparently most at odds, their correspondence is stiffly affectionate or at least formally correct in tone. Nor is it always easy to disentangle the scientific vs. clinical split between Pasteur and Roux from other sources of conflict between them. But the task is worth pursuing, not least because it may provide yet another example of the persistent divide between scientific and clinical approaches to the problems of disease, animal experiments, and the ethics of human experimentation.[6]

THE TENSION BETWEEN PASTEUR AND ROUX

No small part of the tension between Pasteur and Roux was "merely" personal. In their physical appearance, political views, and everyday mode of life, they were an odd couple indeed. Pasteur, a sturdily built, financially secure family man with conservative political leanings, was the quintessential "bourgeois"; Roux, a tubercular, ascetic but mercurial "confirmed" bachelor of vaguely leftist or transcendental political views, was the quintessential "bohemian" by contrast. Roux, it might even be said, was a sort of Don Quixote to Pasteur's Napoleon.[7]

Given the personal differences between them, Pasteur and Roux were perhaps bound to clash. Even the personal traits they did have in common pointed toward that outcome: both were stubborn, aloof, severe, demanding of others, quick to take offense, and given to outbursts of temper. And once Roux joined the Pastorian team, their personal differences were exacerbated by a sense of rivalry between master and employee as they worked toward vaccines against anthrax and rabies. Behind the scenes, they were sometimes competing with each other as much as they were collaborating, and there are signs that Roux resented his subordinate role and Pasteur's high-handed treatment of him.

Actually, it is in some ways surprising that Roux ever became part of the Pastorian enterprise in the first place. When he joined Pasteur's laboratory in 1878 at the age of twenty-five, Roux had not yet received the M.D. degree toward which he was struggling despite his straitened financial circumstances. He had been a student of Pasteur's own disciple, Emile Duclaux, at the medical college at Clermont-Ferrand, after which he pursued clinical training in Paris. The French army covered the costs of his medical studies and paid him a modest stipend on the understanding that he would serve as a military physician for ten years after completing his training. In 1877, however, Roux was dismissed from the army for "disciplinary reasons," presumably some form of insubordination.[8]

After his discharge from the army, Roux was making his way, if just barely, by treating poor people for varicose veins, when Duclaux recommended him to Pasteur. Up to that point, Pasteur had selected his research assistants from the pool of postgraduate "agrégés-préparateurs" in the physical sciences at the Ecole Normale Supérieure, in which capacity he had himself served in his youth. Quite deliberately, Pasteur had not yet allowed a medical man to join his team.[9] It is too often forgotten that Pasteur had no M.D. and was not legally qualified to practice medicine. Perhaps partly for that reason, he was openly disdainful of doctors, saying that they were too interested in making money and in high society to meet the rigorous demands of experimental scientific research. Yet now, in 1878, Pasteur decided to expand his tight research circle to include this feisty doctor-in-training who had just been dismissed from the army for insubordination. Why?

The decisive factor, surely, was that Roux had been recommended by Duclaux, Pasteur's favorite disciple and collaborator. But Pasteur had also come to see the need for a veterinarian or medical man as he began to direct the resources of his laboratory toward a frontal assault on the infectious diseases, beginning with anthrax, a lethal and economically significant disease of sheep. A host of experiments on living animals was now in prospect, and Pasteur wanted a research assistant who was at least skilled in the techniques of injection. Thus Roux began his career with Pasteur in 1878 as an animal "inoculator."[10] From the beginning, he performed superbly at his technical tasks, and he was soon participating in the search for attenuated anthrax cultures as well as injecting them into experimental animals.

As we have seen in Chapter Six, visible signs of discord between Pasteur and Roux surfaced during the famous trial of an anthrax vaccine at Pouilly-le-Fort in 1881. The master's conduct in that affair could not have soothed any prior tension between them, and it also gave Roux a clear appreciation of just how boldly, even recklessly, Pasteur was willing to apply vaccines in the face of ambiguous experimental evidence about their safety or efficacy. In this quest for vaccines, as in his earlier research, Pasteur displayed the scientist's attraction to "signals" amid the "noise," and he exuded the bold self-confidence that is often found in scientists who have revealed such patterns to outside acclaim.

Roux, in sharp contrast, proceeded with what I choose to call a clinician's caution in the face of inconvenient or anomalous evidence. In his own research on vaccines, Roux tended to draw carefully limited conclusions from the experimental evidence at hand. When it came to the results of injecting vaccines into living animals, he (unlike Pasteur) expected and even appreciated all the vagaries of their individual responses. As we shall see, Roux was especially circumspect in the case of the application of rabies vaccines to human beings, much to Pasteur's exasperation. As they worked toward a vaccine against rabies, Pasteur and Roux were also headed toward a series of conflicts that once or twice brought them to the verge of complete and permanent rupture. The issues that divided them most deeply had to do with the ethics of human experimentation: specifically, how much evidence of what sort and what degree of reliability should be required from animal experiments before one could justify the application of vaccines to human victims of rabid animal bites?

The most visible sign of an open split between Pasteur and Roux over these issues came at the single most dramatic moment in Pasteur's career: his decision, in early July 1885, to treat Joseph Meister with a vaccine that had thus far been tested only on dogs. For current purposes, the most striking point to notice is Roux's conspicuous absence from the Meister story, which is odd, to say the least. Not only was he Pasteur's leading collaborator on rabies; by then, he had also attained his M.D. degree and was (unlike Pasteur) qualified to practice medicine. He could have treated Meister, had he been asked and willing to do so. In fact, it seems very likely that Roux simply refused to participate in Meister's treatment in any way. And it is equally likely that he did so because he considered Pasteur's treatment of Meister to be a form of unjustified human experimentation.[11] Roux's clinical caution or scruples thus kept him from taking part in what would become the most glorious episode in the Pastorian saga.

Since Pasteur could not himself legally perform the injections on Meister, and since Roux presumably refused to do so, Pasteur had to find more obliging medical men to play that role. As we have seen in Chapter Eight, Pasteur found them in Drs. Vulpian and Grancher. In fact, it was Dr. Grancher, not so incidentally Pasteur's employee, who actually performed the injections on Meister.[12] The participation of Vulpian and Grancher in the treatment of Meister might seem to pose a problem for my suggestion that Roux's clinical background helps to explain his disagreements with Pasteur. After all, Vulpian and Grancher were doctors, too. Like Roux, they had been exposed to the clinical mentality or ethos, and yet they seemed to have few qualms about the proposed treatment of Meister.

But neither Vulpian nor Grancher had Roux's deep experience with rabies. More important, they also lacked Roux's intimate knowledge of the contents of Pasteur's laboratory notebooks. Except for Pasteur himself, no one knew better than Roux just how much and what sort of experimental evidence then existed as to the safety and efficacy of the vaccine used to treat young Meister. In Roux's eyes, quite clearly, the evidence did not justify Pasteur's decision to treat young Joseph Meister with the vaccine in question.

In his famous paper of 26 October 1885, Pasteur tried to meet in advance any ethical concerns about his decision to treat Meister by insisting that he had already made fifty dogs immune to rabies, without a single failure, by the same method he then used to treat Meister beginning on 6 July 1995. Pasteur continued with the following crucial passage: "*My set of 50 dogs, to be sure, had not been bitten before they were made refractory [i.e., immune] to rabies; but that objection had no share in my preoccupations, for I had already, in the course of other experiments, rendered a large number of dogs refractory after they had been bitten.*"[13]

This claim leads us toward a close, if not exhaustively detailed, analysis of Pasteur's laboratory notebooks in order to address three compelling questions about the results of his animal experiments at the time he decided to treat Joseph Meister: (1) Exactly how many dogs had been rendered immune to rabies *after* they had already been bitten by rabid animals? (2) By what method or methods had these dogs been rendered immune and with what rate of success? And (3) exactly what meaning can be attached to Pasteur's claim that he had already rendered fifty dogs immune to rabies "without a single failure" by the same method used on young Joseph Meister? The attentive reader will recall that very similar questions were raised, explicitly or implicitly, by Dr. Michel Peter during the famous 1887 debates at the Académie de médecine.

PASTEUR'S LABORATORY NOTES ON RABIES VACCINES

In Chapter Seven, we were introduced to Pasteur's remarkably empirical, "hit-or-miss" efforts to find a reliable rabies vaccine. Before rabid spinal cords became the focus of his attention, he tested a wide variety of other techniques as well, including the injection into dogs of various quantities of blood and nervous tissue taken from animals dead of rabies. Throughout these early and almost haphazard trials, Pasteur did sometimes produce immune dogs, even when other dogs injected simultaneously by the same method died of rabies. In one fairly typical example from late June 1884—unusual only by virtue of its relatively grand scale—Pasteur injected fourteen dogs subcutaneously with a broth prepared from the brain of a rabbit just dead of a highly virulent rabies virus that had been passed sequentially through fifty-six earlier rabbits. Of the fourteen dogs so inoculated, nine died of rabies but the other five survived and proved resistant to subsequent injections of virulent rabies.[14]

Whenever and however an immune dog emerged from such experiments, Pasteur considered it "vaccinated." By August 1884, he had about twenty-five such dogs, whose immunity he then demonstrated in experiments before the French Rabies Commission, which was appointed that same year at his request. But none of these dogs had sustained rabid animal bites *before* their inoculations, and the methods used on them often resulted in rabies when applied to other dogs. No one outside the Pastorian circle had any way of knowing this fact, including presumably the members of the official French Rabies Commission. By keeping what he called the "details" of his experiments out of public view, Pasteur repeatedly conveyed a misleadingly optimistic impression of the actual results recorded in his laboratory notebooks.

That judgment applies with full force to the results of Pasteur's *post-bite* trials on dogs.[15] Among Dr. Peter's explicit complaints was that Pasteur failed to specify what he meant when he claimed that "a large number of dogs" had been rendered immune to rabies after sustaining rabid animal bites. The first remarkable conclusion to emerge from a close study of Pasteur's laboratory books is that this "large number" was in fact less than twenty. More important, in the course of producing immunity in these bitten dogs—no more than sixteen, by my count—Pasteur failed to save ten dogs treated at the same time and by the same methods. In the case of three or four of the dogs that died despite their treatments, Pasteur believed their deaths resulted from some cause other than rabies and therefore imagined that they could be counted as "successes." This is but one striking example of the wishful thinking, or self-deception, found scattered throughout his laboratory notebooks on rabies. There was obviously no basis for including these dogs among the successfully vaccinated, for they never had a chance to demonstrate their alleged immunity to rabies. At best, a case could be made for excluding them from any list of failures, but only if they were discounted entirely.

More than that, the success rate in these dogs treated after sustaining rabid bites was essentially no different from the survival rate of otherwise similar dogs that were simply left alone after their bites. Actually, in these experimental trials of rabies vaccines, Pasteur hardly lived up to his reputation as a rigorous practitioner of the "controlled experiment." In most cases, he did not employ control dogs at all. While conducting his trials on twenty-six bitten dogs, he used only seven controls. Of these seven dogs left to suffer their fate without treatment, five were still alive at the time Pasteur treated Joseph Meister.[16] One of the surviving five control dogs did eventually die of rabies in September 1885, but by then one of Pasteur's sixteen

Table 9.1 Results of Pasteur's "post-exposure" experimental trials on dogs after they had been bitten by a rabid dog, August 1884 through May 1885

Date	*No. of Dogs Treated after Bitten by Rabid Dog*	*No. of Dogs Succumbing to Rabies*
August 1884	3	0
October 1884	3	2
November 1884	1	1
January 1885	2	1
February 1885	1	0
March 1885	5	2
April 1885	5	3
May 1885	6	1
Total	26	10

"Success" rate: 16/26 = 62%

Controls: Dogs Left Untreated after Bitten

Date	*No. of Untreated Controls*	*No. Succumbing to Rabies*
October 1884	2	0
November 1884	1	1
March 1885	4	2
Total	7	3

"Survival" rate: 4/7 = 57%

allegedly "vaccinated" dogs had also died of the disease after an unusually long incubation period. At any rate, four of the control dogs apparently never did develop rabies. Choosing the most favorable and least favorable interpretations of Pasteur's results, and depending on the precise moment of calculation, it turns out that the survival rates for the two sets of dogs fall into the following ranges: for the dogs treated by Pasteur, 50 to 78 percent; for the untreated control dogs, 57 to 71 percent. (See table 9.1.)

Given the small number of dogs in question (especially in the case of the controls) and the uncertainties of diagnosis and incubation period, the apparent precision of these survival rates is more than a bit specious. But there can be no doubt that the results of these post-bite trials on twenty-six dogs were ambiguous at best. Had Dr. Peter or other critics been aware of these "details," they surely would have asked Pasteur to explain exactly how his post-bite trials provided any justification for the decision to treat Joseph Meister. And the question would have been hard for Pasteur to ignore. For in his famous paper of 26 October 1885 on Meister and Jupille, it deserves repeating here, he openly admitted that of the last fifty dogs he had vaccinated "without a single failure" before treating Meister, *none* had been previously exposed to rabid dog bites. It was, he said, precisely because of the "large number" of other dogs he had already rendered immune after rabid bites that he felt able to put this concern out of his mind.

If this claim already seems odd in view of the actual results of Pasteur's post-bite trials, it becomes more suspect still when close attention is paid to the methods applied to these twenty-six bitten dogs. As we have almost come to expect, Pasteur evaded the issue in public. *When speaking of the dogs he had rendered immune after rabid bites, he said not a word about the method or methods by which this feat had been accomplished.* But the implication, surely, was that they had been treated with injections of desiccated spinal cords. For otherwise, his post-bite trials would seem devoid of any pertinence to Meister's case. Unless the immune dogs had been treated by desiccated cords, why would they have given him any reassurance as he prepared to treat Joseph Meister by that method? True, Pasteur did imply that some sort of distinction could be drawn between the treatment applied to his bitten dogs and the treatment applied to Meister after invariably successful results in the last fifty (unbitten) dogs.[17] But he left the nature of that distinction entirely unclear. In the face of such reticence, it was natural to assume that Pasteur had applied the same method in both cases, but had perfected it in the (unspecified) interval between his post-bite trials and his experiments on the last fifty dogs.

In fact, however, Pasteur had switched to a radically new method in his experiments on this last group of fifty (or perhaps forty) unbitten dogs. It was essentially the technique applied to Joseph Meister beginning on 6 July 1885. But it differed drastically from the methods previously used to treat the twenty-six bitten dogs. As only Pasteur's laboratory notebooks reveal, not a single one of those twenty-six dogs, including of course the sixteen that did develop immunity to rabies, was treated by the method later applied to young Meister.[18] *Actually, the bitten dogs were treated by three different methods, none of which was ever described in print.*

Until 26 October 1885, when Pasteur reported that he had treated Meister and Jupille by injecting them first with dried rabid cords and then with

10. Pasteur observing rabbits injected with the rabies virus. From *La Science illustré*, 15 September 1888. (Musée Pasteur, Paris)

11. Joseph Meister in 1885. (Burndy Library, Dibner Center, Cambridge, Mass.)

12. Jean-Baptiste Jupille in 1885. (Burndy Library, Dibner Center, Cambridge, Mass.)

13. From the famous painting by Rixens of Pasteur's jubilee at the Sorbonne. (Musée Pasteur, Paris)

14. Pasteur, in 1892, with his grandson. (Musée Pasteur, Paris)

15. The original building of the Institut Pasteur, inaugurated in November 1888. (Musée Pasteur, Paris)

16. Pasteur in 1895, the last photograph taken of him in the gardens of the Institut Pasteur. (Musée Pasteur, Paris)

17. Pasteur's funeral procession through the streets of Paris, 5 October 1895. (Musée Pasteur, Paris)

18. Pasteur's mausoleum at the Institut Pasteur. (Musée Pasteur, Paris)

19. "The Death of Pasteur. Exhibition of the Body at the Institut Pasteur."
(Musée Pasteur, Paris)

20. "La mort du Pasteur," *Le Journal illustré*, 6 October 1895.
(Musée Pasteur, Paris)

(Collection Musée Pasteur - Paris)

21. "Pasteur est eternal." (Musée Pasteur, Paris)

22. Pasteur as "Benefactor of Humanity." Frontispiece from Fr. Bournard, *Un bienfaiteur de l'humanité: Pasteur, sa vie, son oeuvre.* (Collection of the Library, Wellcome Institute for the History of Medicine, London)

23. “National Homage: From France to Louis Pasteur.”

24. "Pasteur Destroys the Theory of Spontaneous Generation." Advertising card for La Chocolaterie d'Aiguebelle. (The William H. Helfand Collection)

25. "Pasteur Discovers the Rabies Vaccine." Advertising card for La Chocolaterie d'Aiguebelle. (The William H. Helfand Collection)

26. Pasteur seated in his laboratory. Advertising card for the Urodonal Company in honor of the centenary of Pasteur's Birth. (The William H. Helfand Collection)

27. "Wine Is the Healthiest and Most Hygienic of Beverages." Advertisement on the official map of the Métro subway system. (The William H. Helfand Collection)

progressively fresher cords, the only announced method was the injection of rabid nervous tissue after it had been attenuated by serial passage through monkeys. When he disclosed this technique in May 1884, Pasteur claimed that the monkey-attenuated vaccine was yielding highly promising results in experiments on dogs.[19] But none of those promising results, it turns out, came from experiments on dogs already exposed to rabid bites. The three methods that Pasteur in fact applied to his bitten dogs are worth revealing here, especially since the third *did* involve the injection of dried spinal cords, but in a manner that differed strikingly from the one used later on Meister. And the special features of this third method will soon lead us into a discussion of Pasteur's theoretical views on immunity, which underwent a dramatic shift as a result of his work on rabies.

PASTEUR AND HIS FIRST METHOD WITH RABID SPINAL CORDS: FROM MOST VIRULENT TO LEAST VIRULENT

Pasteur's post-bite trials, recorded in widely scattered entries in two of his laboratory notebooks, ranged in date of origin from August 1884 to mid-May 1885. His first two methods need not detain us for long. First, in the case of the first seven of the twenty-six treated dogs, the initial inoculation was prepared from the brain of rabbits just dead of a rabies virus that had been augmented in virulence by serial passage through other rabbits. Four of these seven dogs were dead by January 1885, though Pasteur had reason to believe that at least two and perhaps three had died of some cause other than rabies. The three surviving dogs proved immune to subsequent inoculations of virulent rabies.[20] Second, in the next eight treated dogs, the first injection was prepared from the brain of a guinea pig just dead of rabies of more or less ordinary virulence. Of these eight dogs, three soon died of rabies. Once again, the survivors had been rendered immune to rabies.[21]

On 13 April 1885, when the sixteenth bitten dog sustained its first injection, Pasteur began a systematic program of taking spinal cords from rabbits dead of "fixed" or highly virulent rabies and suspending them in desiccated air. From that point through the next five weeks, up until 22 May 1885, when a last group of six dogs received their final injections, Pasteur used these suspended spinal cords as part of a regular series of injections that he hoped would prevent rabies in these last eleven bitten dogs. Seven of the dogs, including five of the last six, were still alive on 16 June 1885. On that day, roughly three weeks after the last six dogs had received their final injections and three weeks before Joseph Meister appeared at his laboratory door, Pasteur "sacrificed" the five survivors so that he could use their cages for

The chart below indicates, in chronological order, some of Pasteur's most significant animal experiments and human trials on potential rabies vaccines using desiccated rabid spinal cords.

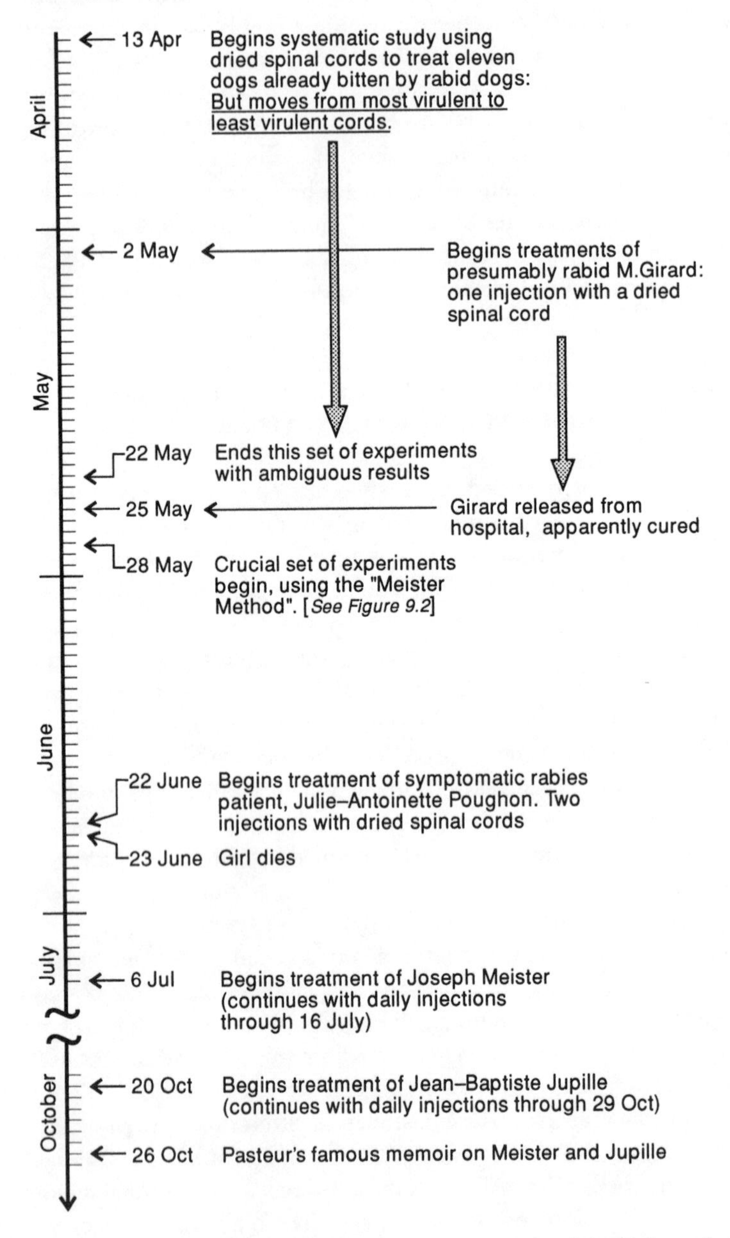

Figure 9.1. Pasteur's path to his rabies vaccine, 13 April 1885 through 6 July 1885: Animal experiments and human trials with dried spinal cords.

other experimental animals.[22] As his experiments multiplied, this practice became increasingly common, the "sacrificed" dogs being dispatched by lethal injections of strychnine. If necessary because space was lacking, this practice nonetheless came at a cost, for these dogs might have developed rabies after an unusually prolonged period of incubation—as some other animals certainly did.

But Pasteur's laboratory notes reveal a much more remarkable and more significant feature of his experimental trials on these last eleven bitten dogs. In all eleven, as noted, injections were prepared from suspended rabid spinal cords. *But here the cords were deployed in a sequence precisely the reverse of the one soon to be adopted in the case of young Joseph Meister.* In Meister's case, Pasteur began with cords that had been drying out for roughly two weeks and then moved to cords that were progressively less dry until, finally, he reached a fresh and highly virulent cord. In the case of the eleven bitten dogs, he began with a fresh cord and then moved to drier and drier cords until, finally, he reached a fully dried-out cord. To anyone familiar with Pasteur's earlier work on other vaccines, this latter modus operandi is astonishing. In developing his vaccines against chicken cholera, anthrax, and swine fever, he had first injected attenuated strains of the implicated microbes and then moved to progressively more virulent strains. Yet here, in these trials with suspended spinal cords on already bitten dogs, he began with fresh, highly virulent cords and only then moved to drier, more attenuated cords. His attempts to prevent rabies in these bitten dogs had now taken a direction precisely the opposite of that followed in all his earlier work on vaccines.

But this *volte-face* is not quite so mysterious as it seems at first sight. For it was associated with a dramatic shift in Pasteur's conception of immunity. In the course of his work on rabies, Pasteur switched from a biological theory of immunity to a modified chemical theory of a sort he had often disparaged when it had been advanced by his critics and competitors. He did so in an attempt to make sense of the variable and sometimes confusing effects that his experimental animals displayed after infection with the rabies virus. The conclusions that Pasteur drew from these confusing effects were themselves more than a bit confusing and susceptible to widely divergent interpretations. But they also bespeak a remarkable flexibility of mind in the now aging Pasteur.

Actually, Pasteur never did invest as much time and energy in efforts to establish a theoretical basis for attenuation and immunity as he did in his more pragmatic, even "empirical," search for effective vaccines. But throughout his work on chicken cholera, anthrax, and swine fever, he linked immunity with the biological, and particularly the nutritional, requirements of the pathogenic organism. In the case of animals inherently immune to a given disease, he suggested that they presented the invading microbe with an internal "economy," "culture," or "environment" that was inimical to its development, either because their temperature was too high or because they lacked some substance essential to the microbe's life and nutrition. In animals rendered immune by recovery from a prior attack by preventive inoculations (Pasteur's "vaccines"), he supposed that each invasion by a given microbe (even in an attenuated state) removed a portion or all of some essential nutrient, thereby rendering subsequent cultivation of the same microbe difficult or impossible.[23]

But at some point during his work on rabies, Pasteur began to doubt the validity of this biological "exhaustion" theory, at first in the case of rabies and then more generally. According to his own retrospective account, he began to adopt a chemical "toxin" theory for rabies as early as January 1884.[24] A year later, his conversion was largely complete and no longer confined to rabies alone, as is clear from a long and unusually explicit theoretical entry of 29 January 1885 in his laboratory notebook.[25] By then, he was growing increasingly confident that he had made an "immense discovery" of potentially "great generality"—namely, that the living rabies virus produced a dead, soluble, chemical "vaccinal substance" inimical to the further cultivation of the virus and therefore capable of producing immunity to rabies. Thus far, however, Pasteur chose to reveal this new theory only to "those who work alongside me"—that is, Charles Chamberland and Emile Roux, saying that he did not know how to "hide my ideas" from them. Sensibly enough, he planned to expose his theory to others only after it had been thoroughly tested by experiments "already underway."[26]

For present purposes, there is no need to explore the precise extent to which Pasteur's new position was justified by the evidence at hand. Nor is there any need to follow every twist and turn in his experimental and conceptual path to this conclusion. For now, it will suffice to draw attention to the sorts of considerations that lay behind his theoretical conversion and that can help us to understand why he ever tried to treat bitten dogs by moving from virulent (or fresh) to attenuated (or dried) spinal cords instead of the other way around.

The first step in solving the puzzle is to notice Pasteur's increasing focus on the effects of injecting different *quantities* of the same virus into his experimental animals. In trying to make sense of the variable response of individual living organisms to infection with the rabies virus, he began to suspect that the variations depended more on the *amount* of virus injected than on its *intrinsic virulence*. As Pasteur reported in his unusually reflective (i.e., "theoretical") notebook entry of 29 January 1885, he had been led to this belief by two interrelated generalizations that seemed to be emerging from his experimental evidence: (1) injecting large quantities of a virus of given virulence produced a higher proportion of immune dogs than smaller quantities of the same virus—at least twice as high, by his reckoning; and (2) even when large quantities of a given virus did produce rabies in the inoculated animal, the disease often appeared much later than was usual with smaller quantities of the same virus. This second generalization upset Pasteur's prior assumption that length of incubation depended only on the inherent virulence of the injected virus. Both pieces of evidence thus pointed in the same direction: for a rabies virus of given virulence, the injection of large quantities seemed to produce a higher level of immunity than did the injection of small quantities. Pasteur also suggested that this generalization could explain why rabid dog bites so rarely produced immunity in the bitten dogs, whereas subcutaneous injections of this same "street rabies" into healthy dogs quite often did. The significant difference was that smaller quantities of the rabies virus were transmitted through bites than through subcutaneous injections.

To Pasteur, such results seemed explicable only on the assumption that the rabies virus "manufactured" a nonliving vaccinal substance inimical to its own development. If immunity depended only on the intrinsic and inherited virulence of a living, reproducing rabies virus, then small quantities should produce the same effects as large. Pasteur had not yet managed—nor, indeed, did he ever manage—to separate this hypothetical chemical "vaccinal substance" from the rabies virus that presumably produced it. But as early as January 1885, this was his ultimate hope and goal. At the same time he pondered the possibility that a similar vaccinal substance was produced by the developing anthrax bacillus. In the case of rabies, Pasteur hoped to capture this chemical substance separately from the living virus by filtration. In the case of anthrax, he hoped that the hypothetical chemical vaccine could be found in vitro after the anthrax bacillus had been killed by heating at appropriate temperatures for appropriate periods of time. In both cases, Pasteur had quite suddenly become a convert to the modified chemical theory of immunity that he had so effectively criticized when it was advanced by Auguste Chauveau, Casimir Davaine, and Henri Toussaint, among others. Indeed, the techniques by which Pasteur now sought to isolate a nonliving vaccine against anthrax bear a striking resemblance to the techniques once deployed by his already deceased competitor, Toussaint—though Pasteur declined to say so out loud.[27]

At any rate, Pasteur's inability to separate the hypothetical vaccinal substance from the living rabies virus left him with a delicate task. The goal, of course, was to inject a maximum amount of the alleged vaccinal substance and a minimum amount of living rabies virus. But since no way could be found to separate the two, the results of any given injection would depend on the relative amounts of living virus and hypothetical vaccinal substance. And since the virus was the presumed source of the vaccinal substance, the quantity of this vaccinal substance perforce depended partly on the amount of virus injected along with it. If the amount of injected virus was too small—as in the case of rabid dog bites—so too would the quantity of vaccinal substance be too small to produce immunity. In such a case, the supply of vaccinal substance would be inadequate to prevent the further development of the virus, and rabies would thus eventually appear in the inoculated animal.

Although Pasteur was understandably reluctant to say so himself, this interpretation of his results had the advantage for him of being almost infinitely flexible. Almost any result could be explained by adopting appropriate—and unverifiable—assumptions about the relative amounts of living virus and associated vaccinal substance. By the time Pasteur presented his modified chemical theory of rabies immunity in print—briefly in the famous memoir of 26 October 1885 on Meister and Jupille, and more extensively in a paper of January 1887[28]—he had adopted the technique of beginning with dry rabid spinal cords and moving to progressively fresher ones. As Pasteur pointed out, most commentators assumed that this technique was equivalent to beginning with a highly attenuated virus and only then moving to more virulent strains. But he argued instead that the vaccinal properties of his cords depended not on the inherent virulence of the virus they contained—indeed, the virulence might be the same in all of the cords, dry or fresh—but rather on the relative amounts of living virus and vaccinal substance in them. Specifically, Pasteur suggested that the drying process might somehow reduce the *amount* of living virus—without changing its virulence—more rapidly than it reduced the amount of nonliving vaccinal substance. And so, after a period of roughly two weeks, there might remain enough vaccinal substance to prevent the reduced amount of living rabies virus from developing further and thus giving rise to rabies. Ideally, of course, one would prefer to use spinal strips in which all of the living virus had been destroyed while some vaccinal substance still remained. And Pasteur predicted that such a "dead" vaccine against rabies would one day be found, though he had not yet been able to perfect one himself.

But in January 1885, when Pasteur also expressed the hope that he might someday isolate a "dead" rabies vaccine, his interpretation of rabies immunity was very different from the one he had settled on two years later. So, too, were the techniques by which he then sought to produce immunity in his experimental animals. His laboratory notes from early 1885 make it abundantly clear that a reliable rabies vaccine continued to elude him. Well into the spring of 1885, he had still not settled on any one approach to the problem. He continued to inject dogs, bitten and unbitten, with several very different sorts of potential vaccines—and the results were inconclusive and confusing.[29] True, Pasteur had for some time displayed a special and growing interest in the possibilities of a vaccine prepared from desiccated spinal cords. In his notebook entry of 29 January 1885, Pasteur even referred to experiments with desiccated spinal cords of low virulence as perhaps the most important test for his new chemical theory of rabies immunity. But he had not yet begun systematic trials of such potential vaccines. And if his laboratory notebook thereafter devotes increasing attention to desiccated spinal cords, it also reveals that he long remained uncertain about the precise point at which desiccated cords might become at once nonlethal and capable of producing immunity when injected into dogs.

In fact, the experiments actually recorded in Pasteur's laboratory notebook through mid-May 1885, including especially his trials on bitten dogs, suggest that even then he remained uncertain about the basic issues raised in his notebook entry of 29 January 1885. From that point on, he made several more or less systematic attempts to compare the effects of injecting large and small quantities of rabid nervous tissue of presumably constant virulence—the very issue that had pointed him toward his new chemical theory of rabies immunity in the first place. Another related issue—more salient for the moment—concerned the speed with which immunity had to be achieved if there was to be any chance of success in the life-and-death struggle against the rabies virus.

In his notebook entry of 29 January 1885, Pasteur endorsed the position that immunity had to be established quickly—perhaps as soon as the eighth day, certainly no later than the fifteenth—if a dog was to escape the lethal effects of exposure to the rabies virus.[30] To judge from the experiments recorded in his laboratory notes from that point through mid-May 1885, Pasteur seemed then to assume that virulent strains of the rabies virus—or, more precisely, fresh rabid spinal cords—might produce immunity more quickly than drier cords. At this point, unlike two years later, Pasteur presumably thought that fresh rabid spinal cords might contain a greater quantity of his hypothetical vaccinal substance than drier cords. In any case, he often chose to begin his series of preventive inoculations with a very fresh cord (what he would, at other times, call "a highly virulent" virus), presumably in the hope that it would produce immunity quickly. A striking example of this practice is found in his last eleven post-bite trials on dogs. In all of them he began the series of injections with a highly virulent (fresh) rabid spinal cord and only then moved to less and less virulent (i.e., or drier and drier) cords.[31]

Within a few months, however—certainly by April 1885—Pasteur began to notice that the incubation period of rabies in at least some of his experimental animals was more prolonged when they were injected with dry instead of fresh cords, which presumably meant that dry cords conferred some degree of immunity in the case of some animals.[32] For quite some time, Roux had noticed the same trend, although a range of experimental contingencies, including especially the ambient temperature, could easily obscure any clear pattern.[33]

But could Pasteur have had this vaguely emerging pattern in mind when, on 2 May 1885, he decided to treat M. Girard, his first rabid "private patient"? The evidence is circumstantial, to be sure, and Pasteur's laboratory notebooks do not explicitly indicate that the results of such animal experiments lay behind his decision to treat Girard with a highly desiccated spinal cord. What we do know for sure is that within three days of Girard's release from the hospital—presumably "cured" of rabies by just one such injection—Pasteur suddenly undertook a systematic series of experiments in which dogs were "treated" by a sequence of injections that began with very dry spinal cords and ended with very fresh cords.

If Girard's presumed "cure" did inspire or encourage this new experimental program (to repeat a suggestion made in Chapter Seven), it would seem that Pasteur was once again exceptionally lucky, especially given that the diagnosis of rabies in M. Girard was almost surely mistaken. But I suspect that Pasteur, were he here to defend his work, would insist yet again not only that chance favors the prepared mind, but also that "luck comes to the bold."[34]

PASTEUR'S EXPERIMENTS ON DOGS BY THE "MEISTER METHOD": LEAST VIRULENT TO MOST VIRULENT SPINAL CORDS

In any case, Pasteur's laboratory notebooks amply confirm that, at the time he undertook to treat Meister, he had not yet produced anything remotely approaching "multiple proofs" of the efficacy of his method on "diverse animal species." But that is the least of it. For the notebooks also reveal that Pasteur had not yet met the much less demanding criteria to which he referred in his famous paper on the Meister case, three months after the boy's treatment had been completed.

In fact, the notebooks provide no evidence that Pasteur had actually completed the animal experiments to which he appealed in justification of his

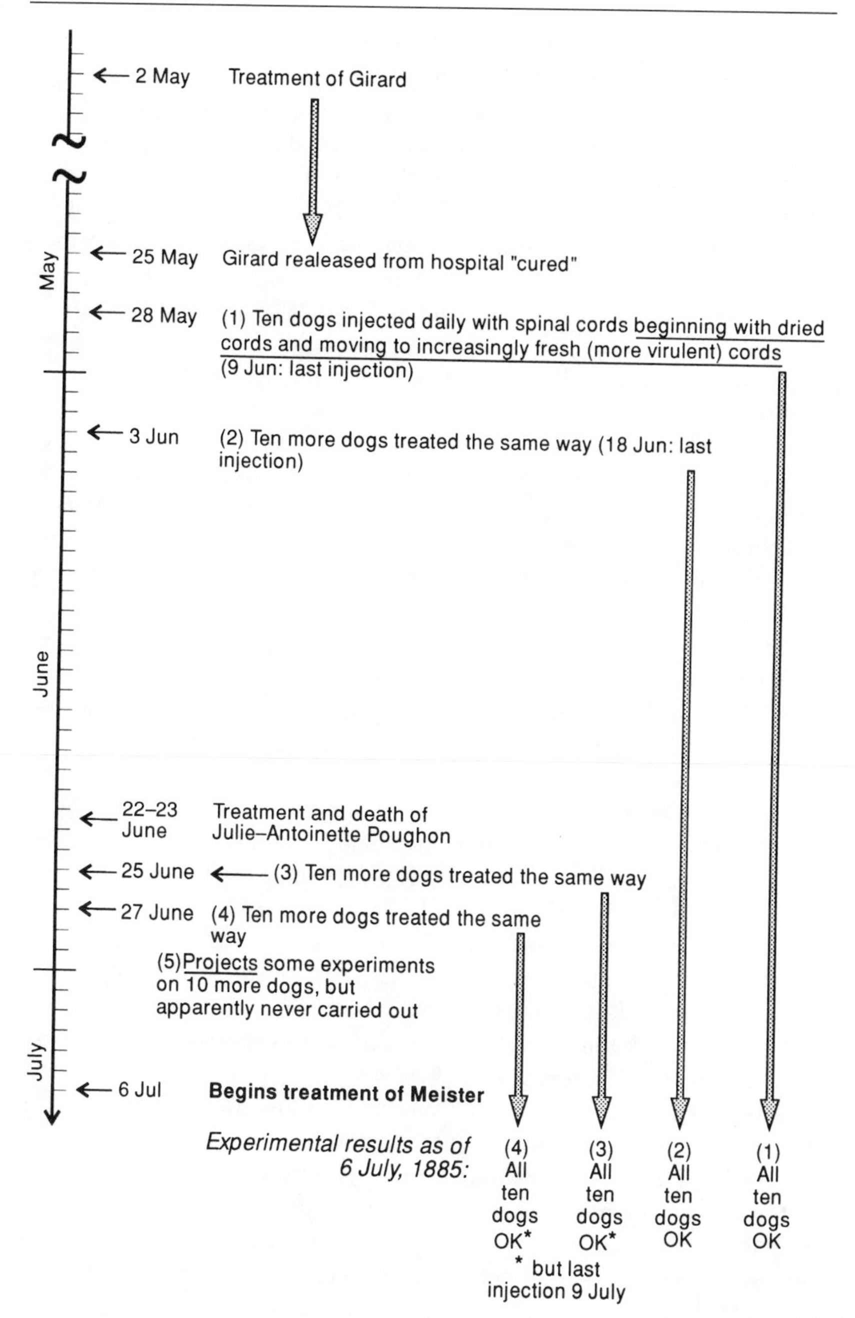

Figure 9.2. The results of Pasteur's experiments on dogs treated by the "Meister Method," 28 May 1885 through 6 July 1885.

decision to treat Meister. Rather, they show that as of 6 July 1885, when Meister's treatment began, Pasteur had just *begun* a series of vaguely comparable experiments on forty dogs (and conceivably on fifty, though I have not yet been able to identify these last ten dogs). As of that date, according to the laboratory notebooks, only twenty of the forty to fifty experimental dogs had even completed the full series of "vaccinal" injections. And none of the dogs had survived as long as thirty days since their last (and highly lethal) injection. (See fig. 9.2.) From a few earlier experiments, Pasteur might reasonably have surmised that rabies symptoms typically appeared between the seventeenth and twenty-sixth day in dogs inoculated with highly virulent rabies virus. That these twenty dogs had not yet displayed fatal symptoms of rabies, three to four weeks (twenty-three to thirty days) after they had been injected with a highly virulent rabies virus, was the best evidence Pasteur had of the safety and efficacy of his antirabies vaccine at the time he decided to treat young Joseph Meister.[35] Furthermore, as Pasteur himself conceded, not a single one of these experimental dogs had first been bitten or otherwise inoculated with rabies before being "treated" by the method used on Meister.

Against this background, it should come as no great surprise that Pasteur never did publicly disclose the state of his animal experiments on the "Meister method" as they stood at the point at which he decided to treat the boy. Nor, indeed, have they been revealed in print until now. They are recorded only in Pasteur's private notebook of that period, which, like the other one hundred laboratory notebooks he left behind at his death in 1895, remained in the hands or control of his immediate family until the mid-1970s. Even now, the notebooks have only begun to be subjected to the close scrutiny and analysis they deserve.

But it is already clear, and should not surprise us, that the most acute critics of Pasteur's treatment for rabies were medical men. Even Dr. Grancher, who performed the injections on Meister and other early subjects of the Pastorian treatment, later admitted that "the great majority of doctors did not believe in [Pasteur's] antirabies vaccine."[36] If some of these critical doctors were motivated in part by personal hostility toward Pasteur and by their concern over the intrusion of the new experimental science into their traditional domain, they also directed sometimes telling attention to the pertinent ethical issues, and their cautious skepticism clearly owed something to the clinical ethos or mentality they shared with Roux. In fact, as Dr. Peter suspected and as Dr. Roux knew full well, the decision to treat Meister was ethically dubious by then prevailing standards, as was some of the rest of Pasteur's conduct in his headlong and headstrong quest for vaccines.

ROUX AND PASTEUR AFTER MEISTER: PARADOXES AND PUZZLES

The story just told leaves one or two puzzles unresolved. For if Roux had such deep and long-standing misgivings about Pasteur's conduct, including notably the decision to treat young Joseph Meister, why did he return to the master's laboratory a few months later to participate in its subsequent work on rabies? And why did he keep his misgivings private, even after Pasteur's death? Despite Roux's alleged concern with ethical issues, did he not himself take part in a lifelong "cover-up" of the real Pasteur and the real story of his work on vaccines?

Let us begin with the first of these questions, which is perhaps the easiest to answer. Why *did* Roux return to Pasteur's laboratory and its work on rabies? To ethical absolutists or conspiratorial muckrakers, the answer may come as something of a disappointment. For Roux's return is probably best explained by the simple fact that he came to believe in the overall safety and efficacy of the original Pastorian vaccine. To be sure, Roux continued to have serious differences with Pasteur over matters of detail and about particular cases. Even when he did rejoin the Pastorian rabies team, he retained much of his clinical skepticism. On balance, however, he had become a convert to Pasteur's cause.

One powerful factor, of course, was the increasingly evident success of Pasteur's vaccine in almost all human cases. But Roux may have been even more impressed by the rapidly expanding body of favorable evidence from animal experiments. For Pasteur had by no means abandoned or curtailed his animal research on rabies in the wake of his celebrated success with Meister. And the evidence from those later animal experiments seemed to vindicate Pasteur's original intuition. Once again, or so Roux had now come to believe, Pasteur had been "on the right track" even before his experimental evidence was fully convincing to others. Luckily for Pasteur, Roux's "conversion" came just in time to offset a swelling tide of criticism from Dr. Peter and other clinicians.

In a very revealing letter of 4 January 1887, on the eve of the debates with Dr. Peter at the Académie de Médicine, Roux advised Pasteur that he could spare himself much "trouble and fatigue simply by extracting from your notebooks the details of the experiments on the vaccination of dogs already bitten [i.e., healthy dogs that survived rabies after having been inoculated with the virus through the bites of rabid dogs]." Those experiments, Roux continued, "are capital and justify the application of the method to man."[37] Inexplicably, Pasteur never did follow this sage piece of advice.

In any case, Roux's letter suggests that by January 1887 he had become convinced that the accumulated evidence from animal experiments was now sufficient to establish the basic safety and efficacy of Pasteur's treatment for rabies. By then, somewhat paradoxically, Pasteur had already benefited from Roux's prior skepticism about the treatment, which was well known to those within and close to the Pastorian circle. The most spectacular example of this paradoxical benefit came in the case of one of Pasteur's most blatant "failures," a boy who had died of rabies in October 1886 in spite of, or even because of, the Pastorian treatment. Here again Pasteur's conduct seems ethically dubious, and here again the episode remained private until disclosed a half century later by his nephew Adrien Loir.

According to Loir, whose basic credibility we now have good reasons to accept, Roux discovered, through animal experiments carried out with material taken from the boy's brain upon autopsy, that the boy had died of rabies. Without knowing of this evidence, the boy's aggrieved and angry father had already accused Pasteur and his collaborators of killing his son and threatened to sue. Loir reported that Pasteur, then resting at a villa in Italy for the sake of his fading health, listened calmly to the circumstances of this case, with "serene" confidence in his method of treatment. Given his usual caution and clinical mentality, Roux was almost surely less serene, but he nonetheless placed himself on Pasteur's side at this crucial juncture. With the collusion of other authorities, Pasteur and Roux managed to keep the full circumstances of the boy's death out of the public eye, and no legal action was taken. Toward this end, Roux's participation was crucial.[38]

Even so, Roux continued to display his clinical caution. He and Pasteur still disagreed, especially because Pasteur had introduced a modified version of his original treatment in cases where subjects had been severely bitten (especially by wolves) or had presented themselves for treatment only after a long delay. Roux was clearly skeptical about this new "intensive method" of treatment, as Pasteur called it. It seems likely that Roux's skepticism was based partly on his usual concern for convincing evidence from prior animal experiments. He was especially concerned about Pasteur's cavalier resort to highly virulent cords in such cases. In a letter of 10 April 1887 to Dr. Grancher, having perhaps heard once too often of Roux's reservations about the "intensive method," Pasteur wrote that "Roux is decidedly too timid." "I understand his scruples," Pasteur continued, "without accepting them [*sans les approuver*]."[39] For me, no single piece of documentary evidence better captures the difference between Pasteur's scientific as opposed to Roux's clinical mentality. It is powerfully reinforced by the testimony of Dr. Grancher, who several years after treating Joseph Meister had this to say about Pasteur's approach to rabies vaccines: "*Pasteur lacked prudence in medical matters. He had made no reservations as to the possibility of partial failures [of his rabies vaccine]. Had he been a doctor, he would have instinctively taken some precautions by foreseeing the possibility of [occasional] failures."*[40]

ROUX'S PUBLIC RETICENCE ABOUT PASTEUR'S CONDUCT: ANOTHER SIGN OF HIS CLINICAL MENTALITY?

This brings us, finally, to the other puzzles posed at the outset of the preceding section. Those questions can be collapsed into one: Why did Roux remain forever in the Pastorian fold and forever silent about Pasteur's ethical indiscretions, some of which came at his own expense? This question, which has no easy answer, gains in force when we recall that Roux did not merely choose to conceal what he knew about the less savory features of Pasteur's conduct in the quest for vaccines. Quite the opposite. Roux played an active part in the construction of the heroic legend of Louis Pasteur. Whatever he may have said to his own disciples in private conversation, Roux was a staunch public defender of the Pastorian faith.

Surely part of the explanation lies in the fact that Roux's own career and reputation were so closely linked with Pasteur's. While it seems unlikely that the bohemian Roux was concerned about "job security" in any usual sense, he clearly did become increasingly protective of the reputation of the enterprise with which he had been associated throughout his career and which was, after all, the main source of his claim to fame.

In the end, however, I would like to suggest that another part of Roux's protective public stance toward Pasteur can be ascribed to the very clinical sensibility that brought him into conflict with the master in the first place. To the extent that Roux retained vestiges of that mentality, he would have been sensitive to the sometimes irrational forces that drove the ill and aging Pasteur. To the same extent, he would have been reluctant to disclose the master's ethical indiscretions after Pasteur's death. Most important, perhaps, Roux's "clinical" tolerance for ambiguity may have allowed him to appreciate the virtues of the Pastorian enterprise as a whole even if he sometimes objected to the means by which its founder had achieved his ends. Perhaps he appreciated, more than Pasteur himself, the exquisite ethical dilemmas the master had faced.

For the sake of history and his own place in it, Roux's clinical mentality, if that's the right word for it, came at a cost. Like his students Charles Nicolle and Emile Lagrange, historians may wish that Roux had been less "scrupulous," or more forthcoming, about his long-standing disagreements with Pasteur. Had he chosen to do so, Roux could easily have produced a revealing, even scandalous, public exposé of Pasteur's conduct. By choosing to do otherwise, indeed the opposite, Roux may well have confirmed Pasteur's judgment that he was "decidedly too timid." But we can appreciate, in a way that Pasteur could not, just how much the Pastorian enterprise would benefit from Roux's clinical sensibilities. And we would not expect Roux to display that mentality vis-à-vis Meister only to abandon it in the case of Pasteur himself.

PART IV

The Pastorian Myth

For Louis Pasteur

"Who is Apollo?"—College student

How shall a generation know its story
If it will know no others? When, among
The scoffers at the Institute, Pasteur
Heard one deny the cause of child-birth fever,
Indignantly he drew upon the blackboard,
For all to see, the Streptococcus chain.
His mind was like Odysseus and Plato
Exploring a new cosmos in the old
As if he wrote a poem—his enemy
Suffering, disease and death, the battleground
His introspection. "Science and peace," he said,
"Will win out over ignorance and war,"
But then, the virus mutant in his vein,
"Death to the Prussian!" and "revenge, revenge!"
. . . . .

Two wars later, the Prussians, once again
The son of Mars, in Paris, Joseph Meister—
The first boy cured of rabies, now the keeper
Of Pasteur's mausoleum—when commanded
To open it for them, though over seventy,
Lest he betray the master, took his life.

I like to think of Pasteur in Elysium
Beneath the sunny palm of ripe Provence
Tenderly raising black sheep, butterflies,
Silkworms and a new culture, for delight,
Teaching his daughter to use a microscope
And musing through a wonder—sacred passion,
Practice and metaphysics all the same.
. . . .

—Edgar Bowers, *For Louis Pasteur* (1989)

TEN

The Myth of Pasteur

DEATH CAME to Pasteur in the late afternoon of Saturday, 28 September 1895, at the age of 72, in a simple bedroom at Villeneuve l'Etang, near Garches, an annex of the Institut Pasteur roughly a dozen kilometers northeast of Paris. Pasteur had presumably received the last rites of the Catholic Church from a priest of the Dominican order. Even so, he probably died as he lived, a Christian "believer" without any deep attachment to the specific doctrinal content or rituals of the Roman Catholic Church.[1] Pasteur's body was embalmed and transported from Garches to a makeshift chapel at the Institut Pasteur on the rue Dutot in Paris, where his family and disciples gathered for a private ceremony and then opened the Institut doors to the public, which filed by the casket in a massive wave of devotion.[2]

By formal decree, Pasteur's funeral was designated a national event at state expense.[3] On 5 October 1895, a large and distinguished crowd filled the Cathedral at Nôtre Dame for High Mass. Among the mourners were François Félix Faure, the new president of the Third Republic; Grand Duke Constantine of Russia; and Prince Nicolas of Greece. The ceremony, at once solemn and grand, reminded observers of the funeral the year before for Faure's assassinated predecessor, President Sadi Carnot. In his funereal *éloge*, Raymond Poincaré, minister of public instruction and future president of France, reportedly moved his listeners to tears with these words:

> Adieu, dear and illustrious master! Science, which you have so grandly served—sovereign and immortal science, become more sovereign still through you—will transmit to the most distant ages the indelible imprint of your genius. France, which you loved so much, will proudly preserve your venerated memory as a national good, as a consolation, as a hope. Humanity, which you have helped, will surround your glory in a unanimous and imperishable cult wherever national rivalries dissolve, and wherever the common faith in unlimited progress is kept alive and strong.[4]

As the rest of the world learned of Pasteur's death, telegrams of condolence flooded into Paris from near and far, and every faction in France was briefly united in a national outpouring of grief and praise for its latest fallen hero. The Parisian newspapers, even the cheap and sensationalistic "scandal sheets," were filled with glowing obituaries and tributes. "Pasteur is eternal," blared one leading tabloid, which, like its rivals, reproduced photographs and other heroic visual images of Pasteur. The iconography of Pasteur has yet to find its scholar, but it is easy enough to decode the meaning of pictures of Pasteur with muses gathered at his feet or as a savior with a halo above his head, sometimes bedecked with wings, suffering the little children to come unto him. (See plates 19–23.)

In another, more exalted form of official national recognition, Pasteur had been offered one of the precious places reserved for the remains of French heroes in the Panthéon, near his old laboratory at the rue d'Ulm. But the family had already decided that he would be buried beneath the new Institut Pasteur in what was then a remote part of Paris. Following a long cortège through the jammed streets of Paris and a ceremony with full military honors, Pasteur's body was temporarily placed in one of the chapels at Nôtre Dame. Four months later, in January 1896, his casket was transferred to a resplendent new crypt at the Institut Pasteur, where his wife was interred beside him upon her death in 1910.[5] (See plates 17, 18.)

The national outpouring of grief upon Pasteur's death came as no surprise. In a sense, it had been rehearsed for a decade or more. Long the recipient of major scientific honors and prizes, Pasteur had been a full-fledged national hero at least since the mid-1870s, by which time his efforts to deploy scientific knowledge and techniques in the solution of practical problems had gained wide publicity. In 1874, when Pasteur was barely past his fiftieth birthday, the National Assembly had awarded him an annual state pension of 12,000 francs. His discovery of a vaccine against anthrax in 1881 had brought him widespread fame, and the application of his rabies vaccine to human cases in 1885 transformed him into an international living legend.

From the early 1880s on, Pasteur was invited to one celebration after another in his honor. In 1881 he was awarded the Grand Cross of the Legion of Honor and in 1882 he was elected to the Académie française, that body of forty "immortals" (or life-tenure members) which has carried official responsibility for the purity of the French language since its foundation by Richelieu in 1635. In 1882 Pasteur was awarded a second national recompense, increasing his annual state pension to 25,000 francs and making it transferable upon his death first to his wife and then to his children.[6] A year later he was honored with an official state celebration at Dole, where a commemorative plaque was placed on the house of his birth and on the occasion of which he gave a moving speech in memory of his parents.[7] Thereafter, on triumphal tours abroad, Pasteur and his expanding entourage basked in applause—notably at meetings of the International Congress of Medical Sciences at London in 1881, in the immediate wake of the famous trial of an anthrax vaccine at Pouilly-le-Fort, and at Copenhagen in August 1884, when Pasteur announced that he was well on the way to a solution of the rabies problem.

But surely the two most glorious events in the last decade of Pasteur's life were the formal inauguration of the Institut Pasteur in November 1888, and the national celebration of his seventieth birthday on 27 December 1892. In the speeches he prepared for these two occasions, Pasteur produced some of his most stirring and memorable prose.

The gala official inauguration of the Institut Pasteur took place on 14 November 1888, when Pasteur was just a few weeks shy of his sixty-sixth birthday. With President Sardi Carnot in attendance, Pasteur was saluted above all for his discovery of the rabies vaccine, which had inspired an international flood of donations to establish a center for the treatment. The resulting fund, by then amounting to roughly 2.6 million francs, had made it possible to build, equip, and modestly endow the new institute. More than that, Pasteur himself became in effect its leading donor, for he pledged to donate to the institute named for him "the revenues from the sales in France of the vaccines discovered in [this] laboratory." His collaborators, Charles Chamberland and Emile Roux, joined him in this pledge.[8]

Pasteur had prepared a brief speech to conclude this ceremonial inauguration of the Institut Pasteur. But, reportedly overcome with emotion, he asked that his prepared remarks be read by his son, Jean-Baptiste, by then a junior member of the French diplomatic corps. Pasteur began by thanking the French state for all it had done in support of his own research, and for its crucial role in the recent educational renovation of France, "from village schools to the laboratories of advanced research [*hautes études*]." He then objected to the decision that "this Institut should carry my name." That, he said, was to "reserve to a man the homage due to a doctrine," by which he presumably meant the germ theory of disease. Yet Pasteur could not conceal his appreciation for this "excess of honor." "Never"—his son now read from the prepared text—"never has a Frenchman addressing himself to other Frenchmen been more profoundly moved than I am at this moment."[9]

But the most moving official occasion was yet to come. On 27 December 1892, in honor of his seventieth birthday, Pasteur was saluted in a famous celebration in the new grand amphitheater of the old Sorbonne. In a scene made familiar from the painting by Rixens, the now frail Pasteur was led into the amphitheater on the arm of President Carnot (see plate 13). The huge auditorium was filled to overflowing with young students from the French lycées and universities, with his own former pupils and assistants, with delegations from each of the major scientific schools and societies in France, and with government officials, foreign ambassadors, and assorted other dignitaries. Of the several distinguished speakers who honored his life and work, the English surgeon Joseph Lister was perhaps the most compelling, for he could testify to the direct influence of Pasteur's research on the surgical revolution represented by Lister's "antiseptic" techniques.

Weak of voice and fragile in health, Pasteur was unable to deliver his own brief speech of appreciation. Once again, he delegated the task to his only son, Jean-Baptiste. His prepared text counseled the young students in the audience to "live in the serene peace of laboratories and libraries," and he spoke to the foreign delegates of his "invincible belief that science and peace will triumph over ignorance and war, that nations will unite, not to destroy but to build, and that the future will belong to those who have done most for suffering humanity." Amid shouts of "vive Pasteur!" the president of the republic rose to offer Pasteur a congratulatory embrace.[10]

As it turned out, this jubilee celebration was Pasteur's last major public appearance, but far from his last honor. By the time he died three years later, on 28 September 1895, his name had been given to the collège in Arbois, to a village in Algeria, to a district in Canada, and to streets and schools throughout France and the world, not to mention the Pasteur institutes already proliferating beyond French borders.

In the case of Pasteur, then, there was no need to invent a posthumous hero. During the last two decades of his life, he had been festooned with honors, and his grand place in history was already secure. Yet the full apotheosis was yet to come. It is, of course, the French who find daily occasion to remember him. Everywhere in France, streets, schools, hospitals, and laboratories carry his name. In Paris, the Boulevard Pasteur is a major artery on the Left Bank, and the Station Pasteur is an important junction on the city's fabled Metro subway system. In the courtyard of the Sorbonne, Pasteur's statue faces that of his contemporary and counterpart in French letters, the great novelist Victor Hugo. A host of statues and other images of Pasteur have been erected elsewhere in France. Until quite recently his somber portrait was imprinted on the five-franc bill, making him the only French scientist to be honored on a note of fiscal exchange. Often portrayed as a pious Catholic and selfless benefactor of humanity, Pasteur remains the very model of a French hero. As recently as the 1960s, in an opinion poll asking

French schoolchildren which historical figure had done the most for France, Pasteur won in a landslide with 48 percent of the vote (St. Louis came in second with 20 percent; even Napoleon was a distant third with 12 percent).[11]

Finally, of course, there is the Institut Pasteur in Paris, which quickly acquired and still retains a reputation as one of the world's leading centers for biomedical research. Perhaps Pasteur's most important legacy to the country he loved so much, the Institut Pasteur is unique among major French research centers in that it is a private institution, as Pasteur insisted it should be from the start, although it has become increasingly dependent on state support as its patent revenues have declined. The centenary of the Institut Pasteur in 1988 generated considerable excitement, including an important conference and book on its history.[12] Its archives are now being organized for use by scholars, and we shall soon be learning vastly more about the history of this very special institution. Like the French language itself, the term "Institut Pasteur" even serves as a reminder of the colonial power that France once enjoyed. Pasteur institutes have long been a part of the French "civilizing mission" throughout the world, including especially its former colonies in Africa, Southeast Asia, and elsewhere. The disciples who first left Paris to spread Pasteur's message to Africa even gave the name Pastoria to a particularly beautiful area of that continent.[13]

Even today, the original building of the Institut Pasteur is quite literally a shrine. Outside the building is a statue depicting Jean-Baptiste Jupille's heroic struggle against a rabid dog. Inside, the Musée Pasteur gives its visitors a palpable, almost eery sense of the hero's presence, for his living quarters there have been preserved as they were at the time of his death. The visitor can look into his study, still furnished with his desk, personal library, and his early portraits of his parents. Also on view are Pasteur's bedroom and dining room and a small but important collection of scientific instruments, flasks, and other tangible relics of his career in science. A visitor who descends a few steps beneath the library of the original building can marvel at the "neo-Byzantine mausoleum of marble, gold, and mosaics in vivid colors" that contains the remains of Pasteur and his wife.[14] (See plate 18.) Until quite recently, all the workers at the Institut Pasteur assembled twice a year to commemorate the birth and death of its founder and namesake in a highly ritualized ceremony that has been vividly described and decoded by Nobel laureate François Jacob in his splendid recent autobiography, *The Statue Within*.

> Each year, at the end of September, everyone who worked at the Institute gathered to commemorate the death of its founder. . . . At the appointed

hour, I followed the crowd of people emerging from their laboratories and going to the garden toward the Institute's oldest building, where Pasteur was buried. When I arrived, the hall was already filled: young and old, department heads and cleaning ladies rubbing elbows, wearing smocks or city clothes. All were murmuring, greeting each other, gossiping in low voices. . . .

A sudden hush signaled the arrival of the dignitaries. . . . The director's brief address reminded the personnel of the virtues on which were founded "our house," its continuity and traditions.

Then, in silence, the descent into the crypt began, in Indian file, in hierarchical order: the director and the board; council; then the department heads, the eldest first; the heads of laboratories, their collaborators; then the technicians and assistants; finally, the cleaning women and lab boys. Each went slowly down some steps before passing in front of the tomb. . . . With a cupola, columns of poryphory, and arched vaults. At the entrance, over the whole of the vault, mosaics depicted, in the manner of scenes from the life of Christ, those from the life of Pasteur: sheep grazing, chickens pecking, garlands of hops, mulberry trees, grapevines, representing the treatment of anthrax, chicken cholera, the diseases of beer, of the vine, of the silkworm. And at the summit, the supreme image, the struggle of a child with a furious dog, to glorify the most decisive battle, that against rabies. In the center, on the cupola's pendentives, four angels with outspread wings: three representing the theological virtues of Faith, Charity, and Hope; the fourth, judged fitting by turn-of-the-century scientism, representing Science.[15]

THE CULT OF PASTEUR OUTSIDE OF FRANCE

The cult of Pasteur has obviously been promulgated most enthusiastically in France, but it is by no means confined to his native soil. We have as yet no systematic comparative study of Pasteur's international reputation, in his day or since. That is unfortunate, for such a study would likely reveal a great deal about the shifting relationship between political cultures and favored heroes or styles of science. The results would surely correlate to some degree with the larger history of political and cultural relations between France and other countries. In particular, it is hard to imagine that the Pastorian legend has had quite the same shape or power in Germany as in France, especially given Pasteur's own vocal hostility toward the "Prussian chancre" and his rivalry with such leading German scientists as Justus von Liebig and Robert Koch. True, even the Berlin institute headed by Koch sent a telegram of condolences upon hearing of Pasteur's death,[16] but civility under such circumstances is hardly the same thing as enthusiastic approval or long-term adulation. The cult of Pasteur has clearly not played as well in Germany as elsewhere.

But that "elsewhere" includes Russia, England, and the United States, to name but a few. The legend of Pasteur obviously gained no small part of its power from French nationalism and the "scientism" of the Third Republic, but it has also exerted a strong appeal well beyond French borders. Pasteur's name is familiar to schoolchildren everywhere, thanks especially to his rabies vaccine and the sterilizing techniques that have been known as "pasteurization" almost from their inception. His achievements have been celebrated throughout the world in song, verse, paintings, plays, posters, stamps, caricatures, films, and television.

In Russia, Pasteur was a hero from the day he treated Joseph Meister for rabies, if not long before. His immediate reputation there doubtless benefited from an emerging Franco-Russian *entente*, or political alliance, which Pasteur himself was eager to see realized. In fact, Russia is, to my knowledge, the only foreign country for which Pasteur ever expressed genuine admiration.[17] But it should also be remembered that a dozen or so Russian peasants from Smolensk who had been viciously bitten by a rabid wolf were among the first and most famous recipients of Pasteur's treatment for rabies. The czarist prince Alexandre the Third, who sent 100,000 francs, was among the most generous donors to the fund for the emerging Institut Pasteur.[18] Partly because of the distance and difficulties of transport between France and Russia, Russian scientists were among the first to establish foreign centers for the distribution of Pastorian vaccines.[19] Pasteur, who actively encouraged these efforts, eventually invited the Russian immunologist Ilya Metchnikoff to head up a section of the Institut Pasteur in Paris. In swift order, Metchnikoff's section became a virtual Russian colony within the Institut. The full story of the connections between the Institut Pasteur and Russian science is just beginning to be worked out by scholars, but it is already clear that Russia was an early and enthusiastic contributor to the cult of Pasteur.[20]

The cult also flourished in England and the United States. True, Pasteur was reviled in both countries by a few voluble antivivisectionists, but the vast majority of English and Americans saluted him as a hero. In fact, English and American scientists produced some of the more glowing tributes to Pasteur and his work. Forced to choose one brief summation of the classic legend of Pasteur, I can do no better than to evoke the words of the British scientist Stephen Paget, who played a leading role in the biomedical defense of vivisection in both England and the United States in the late nineteenth and early twentieth centuries. For Paget, as for the Canadian clinician William Osler after him, Pasteur was "the most perfect man who has ever entered the Kingdom of Science." Paget's remarkable tribute, published in the *Spectator* in 1910, continued as follows:

> Here was a life, within the limits of humanity, well-nigh perfect. He worked incessantly: he went through poverty, bereavement, ill-health, opposition: he lived to see his doctrines current over all the world, his facts enthroned, his methods applied to a thousand affairs of manufacture and agriculture, his science put in practice by all doctors and surgeons, his name praised and blessed by mankind: and the very animals, if they could speak, would say the same. Genius: that is the only word. When genius does come to earth, which is not so often as some clever people think, it chooses now and again strange tabernacles: but here was a man whose spiritual life was no less admirable than his scientific life. In brief nothing is too good to say of him.[21]

Americans of most stripes were no less enthusiastic about Pasteur. Indeed, Madame Pasteur, ordinarily so reserved and a bit suspicious of foreigners, was pleased to receive a visit from an American medical family after her husband's death, telling them that they represented the country that had first, most fully, and most deeply appreciated her late husband's genius.[22] If so, that warm American reception had less to do with Pasteur's theoretical concerns than it did with the practical consequences of his work. As early as the 1870s, Pasteur was awarded American patents for his methods of manufacturing and preserving beer and wine.[23] But it was above all his treatment for rabies that won Pasteur the enthusiastic attention of the American public at large, especially in the wake of the great Newark Dog Scare of December 1885. In that month, four children who had been bitten by a presumably rabid dog in Newark, New Jersey, were sent by ship to Paris to undergo the new Pastorian treatment. American newspapers provided breathless day-by-day coverage of the fate of these children, all of whom escaped rabies. Upon their return to the United States, they were even put on display, for a fee, at state fairs and carnivals.[24]

In fact, so appealing was the Pastorian myth in the United States that it survived, indeed was enhanced by, the muckracking journalist Paul De Kruif, scientist manqué and friend-collaborator of the prototypical American muckraker, Sinclair Lewis. In 1926 De Kruif included two chapters on Pasteur in his classic feat of scientific popularization, *The Microbe Hunters*, which became a phenomenal best-seller and remains in print to this day. No book did more to popularize Pasteur (and other microbiologists) in the English-speaking world. Yet despite De Kruif's best efforts to "humanize" Pasteur by criticizing his arrogance and reckless scientific style, the dominant impression of Pasteur that emerges from *The Microbe Hunters* is that of a scientific magus—in effect, a mythic hero, an impression De Kruif himself did much to create by his allusions to Phoenix, Zeus, and Prometheus. In a perhaps unintended tribute to Pasteur, De Kruif emphasized his entrepreneurial showmanship and went so far as to call him "a misplaced American."[25]

In the 1930s, even Hollywood was attracted to the drama of Pasteur's life and career. *The Story of Louis Pasteur*, produced by the fabled motion picture studio Metro-Goldwyn-Mayer featured the great actor Paul Muni, who deservedly won an Oscar in the title role. It portrays Pasteur as a dour and not always pleasant personality who was nonetheless a scientific magician, a sort of "great American success story." From time to time, *The Story of Louis Pasteur* still appears on late-night American television, where it enjoys the highest possible four-star rating from *TV Guide*. It remains the classic visual treatment of the scientist as hero. Even now, the dominant American image of Pasteur is that of a paragon of virtue, hard work, scientific genius, and technical virtuosity who was a "benefactor of humanity." With the possible exceptions of Galileo, Newton, Darwin, and Einstein, never has a scientist been so glorified, and the legend of Pasteur remains very much alive. There is no cause to regret this, as I shall insist at the end, but it is very instructive to examine the way in which the myth was created and sustained.

PASTEUR'S ROLE IN THE CONSTRUCTION OF HIS OWN MYTH

The first striking thing to notice about the Pastorian myth is the extent to which Pasteur himself participated in its construction. In a way that few scientists have the opportunity or talent to bring off, Pasteur laid the foundations of his own legend. The task would have been impossible, of course, in the absence of his real and widely acknowledged scientific achievements. But Pasteur also produced the first outlines of a saga that magnified his contributions to mythic proportions. It has been said that "a place in history is rarely attained without conscious image-molding in one's own time."[26] Pasteur was a master of this technique, as of so many others. It surely helped that he, like Freud and several other scientific heroes, came to believe early on in his future greatness. Recall, for example, that even before his thirtieth birthday he consoled his neglected wife by telling her that he would "lead her to posterity."[27] And so he did.

The first steps along that path are to be found in Pasteur's published papers. And here the reasons for the long reign of the Pastorian myth become entwined with the reasons for his success in his own time. The same practices that helped Pasteur win the support of the French scientific elite during his lifetime also help to explain his enduring fame. Like many scientists of his day, Pasteur often began his papers with a "historical" account of the problem at issue. In earlier chapters, we have already seen how effectively Pasteur used this convention for his own purposes, especially by turning his opponents into strawmen and by minimizing the contributions of others to positions he tended to present as his alone. The result was to exalt his own originality and his "revolutionary" ideas at the expense of others, including sometimes even his own collaborators, notably Emile Roux in the case of rabies.

In Chapter Three, for example, we saw in detail how Pasteur's retrospective accounts of his first major discovery of optical isomers erased the crucial influence of August Laurent on the research that pointed the way to that discovery. In other chapters, we saw how Pasteur's "histories" of prior studies of fermentation and spontaneous generation tended to obscure the importance of all of the work that had been done to clarify these problems before he arrived on the scene. His later published papers, including those on anthrax and rabies, pursued the same strategy, much to the annoyance of his critics then and since. What such critics fail to appreciate is that it was in fact Pasteur, however great his indebtedness to his predecessors, who managed to resolve so many of these problems with an experimental and rhetorical virtuosity that his alleged precursors lacked.

In no small part, Pasteur succeeded because of his flair for the dramatic gesture and his talent for self-advertisement, as we have already seen in the spectacular case of the public trial of his anthrax vaccine at Pouilly-le-Fort. He was also uncommonly skilled in rhetoric, in the old-fashioned sense of the art of persuasion. In this connection, we would do well to recall that rhetoric was part of the French curriculum as Pasteur experienced it, and that oratorical talent still played an important role in appointments and advancement in the French educational system, even for scientists. From the outset, Pasteur performed exceptionally well in this domain, as his high ranking in the *agrégation* competition suggests.[28] By the time he became a major force on the scientific scene, he had developed a refined sense of what sort of rhetorical devices would work best in particular contexts. He modified his tone and language according to the audience and purpose at hand. To take but one example, we should note his skillful deployment of "semantic stratagems" in the debates over fermentation and spontaneous generation. Despite the persistent efforts of philosophers of science to explain the triumph of the germ theory solely in terms of experimental facts and philosophical "realism,"[29] Pasteur's rhetorical talents were also a major factor in the success of his campaign.

So fully did Pasteur appreciate the subtleties of rhetoric that he knew how to deploy superficially *antirhetorical* language in appropriate contexts. In 1875, for example, during renewed debates over his theory of fermentation at the Académie des médecine, he explicitly complained about the baleful effects of "mere" rhetoric and eloquent language at the Académie:

> Several weeks ago, in some brilliant secret committees that I never left without being amazed by the talent of speech that I had heard deployed, you asked yourselves how the Académie could introduce, to a higher degree, the true scientific spirit in its works and discussions. Let me indicate one way, which will certainly not be a panacea but the efficacy of which inspires my full confidence. This way consists in a sort of moral pledge taken by each of us never to call this body a tribune, never to call a communication . . . a discourse, and never to call those who have just taken the floor or are about to take the floor orators. Let's leave these expressions to deliberating political assemblies, which discuss [*dissertent sur*] subjects where proof is often so difficult to give. These three words, *tribune*, *discours*, *orateur*, seem to me incompatible with scientific simplicity and rigor.[30]

Current students of the scientific enterprise will appreciate this ploy, not least because of its delicious irony. For it is by now a commonplace among historians and sociologists of science that, in the words of Steven Shapin, "Science, no less than any other form of culture, depends upon rhetoric." And the superficially antirhetorical language of most modern scientific discourse is itself but another rhetorical resource or strategy.[31] In the scientistic culture of Pasteur's day, however, such an antirhetorical strategy could be very effective. We should also not ignore the extent to which Pasteur's success depended on his rhetorical skills in the more limited sense of a dramatic and graceful prose style, epitomized perhaps by his famous memoir of 26 October 1885 on the stories of Joseph Meister and Jean-Baptiste Jupille.[32]

PASTEUR'S HEALTH AND THE CONSOLIDATION OF THE MYTH

A more surprising and even paradoxical factor in Pasteur's success was one he manifestly did not seek: his fragile health and prolonged physical decline. In a way, to scrutinize Pasteur's health is even more awkward than to analyze the laboratory notebooks he left behind. Both tasks verge on violations of his privacy. But if the state of his health is even more private in some sense than his notebooks, it was also more public in another, for his physical decline was visible to all who knew him. And its transforming effect may well have contributed to an ever more benign and valedictory assessment of his work and career. So long as Pasteur's combative powers were in full sway, he was less widely loved than he came to be as his declining health transformed him into a more sympathetic version of his former self.

One compelling account of Pasteur's reputation on the threshold of this transformation can be found in a manuscript recently brought to light by Richard Moreau. The account comes from the writer Maxime du Camp (1822–1894), a member of the Académie française from 1880, whose best-known work was his *Souvenirs littéraires*, but who also left behind a fascinating four-volume unpublished work that records his personal impressions of colleagues at the Académie française. Of Pasteur, his colleague in the Académie for more than three years, du Camp recorded the following in 1885: "It is said that he is brutal and despotic at the Académie des sciences, which doesn't surprise me. . . . His colleagues [there] fear him and hardly trouble themselves to please him. Of him, they openly say: 'He thinks he's a god.'"[33] These reflections were recorded in the immediate wake of Pasteur's early treatments of rabies in humans, and during the next two years he was embroiled in defenses of his treatment and conduct.

During these same years, Pasteur's health began visibly to decline, two decades after his recovery from a major stroke and a long period of stable health in the meantime. Beginning in November 1886, he displayed unmistakable signs of cardiac deficiency, and in October 1887 he suffered two small strokes. From then on, his health and strength ebbed away. Pasteur was obviously distressed by his physical decline, but it may have worked to the long-term advantage of his legend. His critics now fell respectfully silent, while the once gruff and aloof Pasteur displayed a more tender, sentimental, and even emotional side of his personality.

In Pasteur's surviving laboratory notebooks, the last entry—a note concerning still further trials of rabies vaccines in dogs—is dated 2 August 1887.[34] From that point on, Pasteur published very little, mostly brief speeches on public occasions or in defense of his rabies treatment. His energy for scientific work was spent, though he lingered on for seven more years. During this period of decline, he became more famous than ever, a living legend in fact, due partly to the inauguration of the Institut Pasteur in 1888 and the national celebration of his seventieth birthday in 1892, when more people saw him than ever before and the daily newspapers trumpeted his glory.

Mellowed by age and weakness, the once feared and "brutal" scientific conquistador took on the appearance of a frail, wise, and melancholy old sage. One eyewitness gave this striking verbal portrait of Pasteur's physical appearance toward the end of his life:

> Everyone knows that Pasteur is short, that since 1870 [*sic*, 1868] his leg and left arm, smitten by apoplexy [i.e., stroke], are somewhat stiff, and that he drags one foot much like a wounded veteran. Age, illness, the heavy labours of so many years, the bitterness of conflict, the intense passion for his work, and, lastly that prostration which follows triumph, have combined together to make a grand thing of his face.
>
> Weary, traversed with deep furrows, the skin and beard both white, his hair still thick, and nearly always covered with a black [skull] cap; the broad forehead wrinkled, seamed with the scars of genius, the mouth slightly drawn by paralysis, but full of kindness, all the more expressive of pity for the sufferings of others, as it appears lined by personal sorrow; and above all, the living thought which still flashes from the eyes beneath the deep shadow of the brow—this is Pasteur as he appeared to me: a conqueror, who will someday become a legend, whose glory is as incalculable as the good he has accomplished.[35]

Everyone who has left a description of Pasteur during these final years has referred to this "personal sorrow" or deep melancholy that could now be read in the lines of his face—"la mélancolie . . . du savoir," as Pierre Gascar has recently called it.[36] The verbal descriptions are in keeping with his late photographs. (See plates 14, 16.) Even outside his immediate entourage, Pasteur's melancholy was evident, not least on the honorific occasions that came to dominate his twilight years, when it became publicly clear that he no longer had full control of his emotions.[37]

At the official inauguration of the Institut Pasteur on 14 November 1888, for example, when Pasteur delegated the task of reading his speech of appreciation to his son Jean-Baptiste, the audience could hardly fail to see that Pasteur himself, seated nearby, was ill and weary. With this vision of a living but fading legend seated before them, the audience must have been moved when Jean-Baptiste read these words from his father's prepared text: "But alas! I have the poignant melancholy of entering [this great building, this house of work] as a man 'vanquished by time.'"[38]

Certainly by early 1889, when Pasteur and his family moved from the Ecole Normale into their elegant apartments within the completed Institut, the master's weakness and advancing cardiovascular ailments made it impossible for him to do any further research of his own. By a sad irony, then, Pasteur never really worked in the new Institut Pasteur, those large and well-appointed laboratories he had long sought and finally obtained. By some accounts, it was his inability to do research that made him so sad toward the end. And among the old problems that he most deeply wished he could take up once again was the relationship between optical activity and life.[39] Perhaps, in the end, Pasteur regretted most of all his failure to detect that "cosmic asymmetric force" that he had once seemed to regard as an argument for the existence of God.

During the last years of his life, Pasteur occupied himself by overseeing the construction and arrangements for the institute that bore his name, occasionally entering its laboratories to observe the work being produced. He also spent a lot of time sitting under the trees in the park at the Institut annex at Villeneuve l'Etang, surrounded by his family. He became ever more affectionate toward his inner circle and took great joy and consolation in his grandchildren. (See plates 14, 16.)

THE PASTEURS, THE PASTORIANS, AND THE MYTH

The construction of the Pastorian myth, already well advanced in the 1880s, shifted into high gear upon the master's death in 1895. Several of his collaborators and disciples eagerly joined in the cause. In a striking obituary notice, Pasteur's disciple and personal physician Joseph Grancher tried to convey his full sense of the goodness of the man.[40] In 1896, within a year of Pasteur's death, Emile Duclaux, his immediate successor as director of the Institut Pasteur, produced a brilliant full-length scientific biography of the master that remains a standard source for the legend. Given Pasteur's own preference in heroes, he might have preferred the brief obituary notice in which Duclaux found reason to exalt him even above Napoleon: "The only exact image [for Pasteur] is that of Napoleon dying triumphant in the midst of a pacified and fully conquered Europe. Yet even this vision, however grandiose it may be, is incomplete: Pasteur has conquered the world, and his glory has not cost a tear."[41]

Such contributions to the cult of Pasteur by his disciples might seem odd in a way, since they had lived in his shadow, often too little appreciated and often dismissed except as "Pastorians." The participation of Emile Roux in this project seems especially curious, given that he was sometimes sharply at odds with the master and allegedly wrote an unpublished autobiography that contained revealingly critical accounts of Pasteur and his relations with him. Yet in public, Roux spoke of Pasteur with unqualified praise, perhaps even more so than Duclaux, who was supposedly Pasteur's favorite.

No doubt a large part of the explanation for his disciples' enthusiastic public participation in the construction of Pasteur's legend can be found in their profound respect for what Pasteur had accomplished, and in their genuine affection for the more sympathetic version of the man that emerged most vividly toward the end. But some part of the explanation probably lies elsewhere. For in truth, Pasteur's immediate collaborators and disciples were not remarkably gifted scientists, even if several of them, notably Roux, were technically brilliant. Perhaps Pasteur preferred it that way. In any case, lacking major research talent or achievement themselves, the Pastorians in effect owed their high standing in French culture to their identification with the founding father of the institute that carried his name. In a way, to castigate Pasteur would have been to attack their own identity and to undermine such celebrity as they did enjoy. To be a Pastorian was—and is—to be something special in the scientific world and in elite French circles, and his disciples would have undermined their own position by challenging the cult that had grown up around him. In an important sense, Pasteur's collaborators and disciples had cast their own fate with Pasteur's posthumous reputation. Their own success was rooted in the success of the Pastorian program, and their present and future prospects would rise or fall with the fate of Pasteur's legend.

I have saved for last the most obvious force in the construction of the Pastorian myth: the careful stage management of the aging Pasteur and his immediate family. Particularly crucial assistance came from Pasteur's son-in-law, René Vallery-Radot, a popular writer of conservative cast and an unabashed enthusiast (and beneficiary) of everything his father-in-law accomplished and stood for.[42] In 1883, when its hero was sixty years old and very much alive, Vallery-Radot published a popular slim biography of Pasteur anonymously under the title *Histoire d'un savant par un ignorant*, which can perhaps be rendered into English as *The Story of a Scientist by a Layman*.

The book had its origins in a schoolboy speech of 1878 by Pasteur's nephew and later research assistant, Adrien Loir, who was then a sixteen-year-old student at the lycée in Lyons. Like all upperclassmen at the lycée, Adrien was obliged to deliver a talk on a topic of general interest to his classmates. He decided to devote his speech to his famous uncle, then fifty-five years old and already famous for his work on fermentation, wine, beer, and silkworms, among other things. Pasteur was given a sort of dress rehearsal of the speech when he came to spend the Easter holidays with his relatives in Lyons. He listened only briefly before interrupting his nephew with an offer to dictate the speech for him. The resulting notes for Adrien's speech, corrected in Pasteur's own hand, were soon passed on to René Vallery-Radot.[43] Under the direct supervision of his father-in-law, he expanded these notes into his *Histoire d'un savant*, the galley proofs of which Pasteur painstakingly corrected himself, as we know from an extant copy of the manuscript with Pasteur's handwritten revisions that is now deposited in the Burndy Library at the Dibner Center in Cambridge, Massachusetts. In effect, then, Vallery-Radot's *Histoire* was its hero's unofficial "autobiography." Similarly, his greatly enlarged two-volume *La vie de Pasteur* of 1900 can best be seen as its subject's own "authorized" biography. Certainly Pasteur, who had died five years before it appeared, would have approved, indeed applauded the result, both immediately and in its long-term effects.

René Vallery-Radot's standard biography has many virtues. It is still unsurpassed for its detail and extensive use of Pasteur's correspondence, including some that is still not published. It conveys a sense of human drama and excitement. It has had an enormous success across a wide range of audiences and is still in print in English.

But its chief function has been to transmit the image of Pasteur that he and his family preferred. The family long carefully protected that image. Until 1964, the immediate family retained possession of his private papers and correspondence, collecting much of that held elsewhere, and carefully managed what parts of the manuscripts and papers did see the light of day. René Vallery-Radot, in particular, even tried to control the publication of competing biographies. Consider, for example, the following remarkable passage from a letter of 16 June 1955 to "Doctor Larkey" from Peyton Rous of the Rockefeller Institute for Medical Research in New York City:

> In his endeavors for American science, [Christian] Herter determined to write a book for the general medical public about the deeds of Pasteur, and after this was well along he let Vallery-Radot know of it, thinking this but a due courtesy and having no idea that V-D [*sic*] could object; for the life was written simply and made no pretensions, as you will see on looking the manuscript through. Others tell me that even now the Vallery-Radot family consider Pasteur as their personal possession, and further that it is difficult to obtain pamphlets about Pasteur which show him as less than a hero. (One of the pamphlets now sent you tells, for example, that on the 'student' evenings held at his home he always went to sleep when the talk was not about science.) However all this may be, Vallery-Radot denounced Herter as having stolen from his book, and the shocked and wounded Herter never put pen to it more. I fear that now it may have little value.[44]

ADRIEN LOIR AND THE ULTIMATE IRONY: THE DECONSTRUCTION OF THE PASTORIAN MYTH FROM WITHIN

In their efforts to control the posthumous legend of their revered patriarch, the Pastorians were remarkably successful for a remarkably long time. For the half century after it was first published in 1900, René Vallery-Radot's hagiographic *Life of Pasteur* dominated the field. Even now it remains a popular introduction to the Pastorian saga. Beginning in 1950, however, some shelf space was made for René Dubos's more balanced and mildly critical biography, *Louis Pasteur: Free Lance of Science.*

In the meantime, from 1900 to 1950, serious students of Pasteur's career had precious little else to choose from. Only two published sources from that period are of current interest—and for very different reasons. In 1923, Ethel Douglas Hume wrote (or, more precisely, assembled from disparate sources) a book called *Béchamp or Pasteur?* that has since been reprinted (or reassembled) at least twice—in 1942 under the title *Pasteur Plagiarist. Imposter!* and in 1989 under the title *Pasteur Exposed: The False Foundations of Modern Medicine.*[45] What is interesting about this otherwise undistinguished book is its relentlessly hostile tone toward Pasteur and its lasting appeal to advocates of "alternative medicine," notably homeopaths. The book does reveal that Pasteur treated his sometime assistant Antoine Béchamp (1816–1908) very shabbily, but it does not persuade me that Pasteur "plagiarized" Béchamp's work and ideas in any meaningful sense of the term. In substance, if not in tone, I agree with the anonymous *Isis* reviewer who, in 1934, described Hume's effort as a "subsidized book of propaganda," in which "the animus throughout is to exalt Béchamp at the expense of Pasteur, and by . . . the inclusion of material not germane to the title to discredit vaccination, the use of serums, and animal experimentation." I agree even more fully with this reviewer's judgment that the book discusses scientific data and concepts with a sometimes "ludicrous incomprehension of their real relations."[46]

The second and vastly more important source is the series of essays published in 1937 and 1938 by Pasteur's nephew and sometime personal research assistant, Adrien Loir, who waited half a century before sharing his reminiscences with readers of the obsure journal *Le mouvement sanitaire.*[47] As already noted above, Loir's essays are anecdotal, personal, often vague about dates and details, and totally undocumented—qualities that are almost guaranteed to make historians wary and skeptical. Nonetheless, I want to insist that Loir's essays represent the most important, if often unrecognized, first step toward the deconstruction of the traditional myth of Pasteur. For it now seems clear to me that Loir's essays have played a crucial role in the revisionist accounts of the Pastorian story that have appeared with increasing frequency during the past two decades. If Loir's contributions have sometimes gone unacknowledged and have never been properly emphasized, that is due partly to the amazingly casual, nonchalant, even affectionate way in which he disclosed what were in fact profound revelations about his famous uncle's personality and scientific modus operandi. Perhaps it has also been hard to imagine that such iconoclasm could have emerged from within the Pastorian family itself. It is, in fact, a supreme irony that the myth of Pasteur should have first been "betrayed" by his own nephew—and, moreover, the man he had recruited as his personal research assistant precisely to guard the family secrets, so to speak.

It is not for me to say what first inspired each member of the expanding tribe of Pastorian revisionists. But I am struck by the extent to which each of them relies, more or less, on Loir's revelations. Surely Loir's testimony has played some part in the valuable work of such otherwise diverse critics as Jean Théodoridès, Philippe Decourt, and Donald Burke, all of whom have drawn attention to the neglected contributions of Pasteur's predecessors and "precursors," including not least Casimir Davaine and Henri Toussaint in the case of anthrax, Victor Galtier and Joseph-Alexandre Auzias-Turenne in the case of rabies, and Antoine Béchamp in the case of the silkworm diseases and almost every other domain of Pasteur's research.[48]

Speaking for myself, I can say that I was very much struck by Loir's revelations upon my first reading of his essays in the early 1970s—though I should quickly add that I did not really trust them at that time, as is clear from my 1974 monographic essay on Pasteur for the *Dictionary of Scientific Biography*, which cautiously and rather dismissively relegated Loir's essays to a list of other works described merely as "particularly informative with regard to Pasteur's personality and interaction with his assistants."[49] It took two very different sorts of experiences for me to begin to appreciate the full significance and basic accuracy of Loir's anecdotal essays. The first was watching the splendid 1974 BBC television series, *Microbes and Men*, which borrowed liberally from (although without explicit acknowledgment of) Loir's account at dramatic moments of the Pastorian work on anthrax and rabies vaccines.[50] I distinctly remember asking myself, in the midst of this TV program, "Could that really be true?" In particular, I began to wonder if Loir had been right after all about "the secret of Pouilly-le-Fort." Shortly thereafter, having been granted access to the Pasteur papers at the Bibliothèque Nationale in Paris, I quickly confirmed Loir's account through a close study of Pasteur's laboratory notebook records of that episode.

Loir's essays also inform the work of the other close student of Pasteur's laboratory notebooks, the Italian historian and philosopher of science Antonio Cadeddu, who has produced at least two major articles (in French in 1985 and 1987) and an important book on Pasteur (in Italian in 1991).[51] In the preface to his book, Cadeddu quite rightly says that his investigations of Pasteur's laboratory notes has culminated in thoroughly revisionist accounts of the Pastorian work on chicken cholera and anthrax. (Indeed, because I knew in advance of Cadeddu's research on the history of chicken cholera, I did not undertake a close study of Pasteur's laboratory notes on that disease and in this book merely refer the interested reader to Cadeddu's work.) As noted above in Chapter Six, Cadeddu and I agree on the basic facts of the matter in the Pouilly-le-Fort affair. That means that we also basically agree with Loir's version of the story. Where we may not agree is on what is the most interesting general point to take away from this famous episode in Pasteur's career. For Cadeddu, as I read him, the basic lesson is ultimately an epistemological one. For me, by contrast, the central point has to do with Pasteur's public presentation of self and is thus closely related to the question of the historical myth of Pasteur. In any case, the main goal I have had in mind here is to draw attention to Loir's crucial role in the deconstruction of the traditional Pastorian myth.

CONCLUDING REFLECTIONS: PASTEUR, MYTHS, AND HISTORY

In a standard encyclopedia of poetry and poetics, myth is defined as "a story or complex of story elements taken as expressing, and therefore as implicitly symbolizing, certain deep-lying aspects of human and transhuman existence."[52] For Erik Erikson, "a myth blends historical fact and significant fiction in such a way that it 'rings true' to an area or an era, causing pious wonderment and burning ambition."[53] However defined, myths are by no means "lies," as the word has much too casually come to be used.

Like all myths, the standard legend of Pasteur has served several useful functions. Especially in the form purveyed in René Vallery-Radot's *La vie* and the children's books derived from it, the legend served as a valuable reservoir of homilies for schoolteachers and French patriots, and as a source of inspiration for young would-be scientists. It has also provided a sense of human drama and excitement as opposed to the impersonal, collective sense of science about which so many complain today. Rarely has science been made so wonderfully simple, or so wonderfully grand and useful at once.

Furthermore, the myth of Pasteur, like all myths, embodies important elements of the truth. After all, Pasteur's scientific work was enormously important and fertile, and some of his principles continue to guide us today. As Bruno Latour and others have recently reminded us, it would be folly to deny the fruits of the Pastorian enterprise, and there was obviously something like a Pastorian "revolution," with consequences like the "pasteurization of France."[54]

More important, the fact is that a Pastorian legend was constructed, and that legend has itself had historical consequences, not all of which we would wish to deplore. In a strictly historical sense, the Pastorian myth cannot be undone. The myth of Pasteur cannot be taken away from the past. It is itself part of history.

But we also need a Pasteur for our times. This new story has yet to be fully written, but even in its current outlines it deserves its day in the sun. Without going so far as Philippe Decourt, who claims that "no serious book has been written about Pasteur,"[55] we can insist that it is time for another and even several more. The resulting transformation in the story of Pasteur may also serve useful functions for our time and beyond. After all, the deconstruction now underway is not merely the result of new scholarship, but partly of larger changes in our attitudes toward heroes, science, and technology. As in the case of the mythic Edison, whose legend has gone through several transformations in keeping with wider cultural and economic changes in American culture,[56] so too will each age get the Pasteur it deserves.

This book seeks to contribute to that larger project of revaluation—to deconsruct, as it were, the currently dominant image of Pasteur. That image was forged in a context that has lost much of its meaning for us—a context in which heroic biographies were used to transmit widely accepted moral verities and in which science was seen as straightforwardly useful and "positive" knowledge. Even in an age in need and search of heroes, we need no longer accept that image at face value. We need no longer perpetuate Pasteur's image of himself.

Appendixes

Appendix A

CHARBON. VACCINATION à MELUN.

26 avril [1881]. Projet de convention avec la *Société d'agriculture à Melun:*[1]

Le Société d'agriculture de Melun ayant proposé à M. Pasteur par l'organe de son président M. de la Rochette de se rendre compte par elle même, sous le rapport pratique, des résultats des expériences faites par M. Pasteur & Mm. Chamberland & Roux au sujet de l'affection charbonneux, il a été convenu ce qui suit:

1. La Société d'Agriculture de Melun met à la disposition de M. Pasteur 60 moutons.
2. 10 de ces moutons ne subiront aucun traitement & serviront comme témoins.
3. 25 de ces moutons subiront deux inoculations vaccinales à 12 ou 15 jous d'intervalle par le virus charbonneux atténué.
4. 25 de ces moutons seront en même temps que le 25 restant inoculés 12 ou 15 jours après par le charbon très virulent. Les 25 moutons non vaccinés périront tous, les 25 vacciné résisteront & on les comparera ultérieurement avec les 10 témoins réservés ci dessus afin de montrer que les vaccinations n'ont pas empeché [*sic*] les moutons de revenir après un certain temps, à un état normal.
5. Après l'inoculation du virus très virulent aux deux séries de 25 moutons vaccinés & non vaccinés, les 50 moutons resteront réunis dans le même étable; on distinguera une des séries de l'autre en faisant avec un emporte-pièce un trou à l'oreille des 25 moutons vaccinés.
6. Les 10 moutons témoins resteront toujours dans [*illegible*] bergerie à part afin qu'ils ne soient pas exposés à la contagion des moutons malades.
7. Tous les moutons qui mourront seront enfouis un à un dans des fosses distinctes, voisines les unes des autres & situées dans un enclos palaissadé.
8. Au mois de mai 1882 on fera parquer dans l'enclos dont il vient d'etre [. . .].

[1] Ce projet a été accepté par M. le President le Baron de la Rochette le 28 avril. Rendez-vous est pris avec lui pour la [*illegible*] à Melun le 5 mai à la gare de Lyon. On [vaccinera?] ce jour 25 moutons avec une bactéridie qui, affaiblé d'abord par le bichromate à Pot jusqu'à [ne plus tu?] que les souris a[reposé] alors par 3 souris successivement. 15 jours après, le 25 mai [*sic*], on vaccinera par une bactéridie. Puis 15 jours après, le 10 juin [*sic*], on passera à la bactéridie, dat de 4 ans, le plus virulent.

Translation

ANTHRAX, VACCINATION AT MELUN

26 April [1881]. Draft of agreement with the Agricultural Society of Melun.[1]

The Agricultural Society of Melun having proposed to M. Pasteur, through the offices of its president M. de la Rochette, of overseeing in a practical sense some results of experiments made by M. Pasteur and Mssrs. Chamberland and Roux on the subject of anthrax, the following has been agreed upon:

1. The Agricultural Society of Melun will put 60 sheep at M. Pasteur's disposal.
2. Ten of these sheep will be subjected to no treatment whatever and will serve as controls.
3. Twenty-five of these sheep will submit to two vaccinal inoculations with attenuated anthrax after an interval of 12 to 15 days.
4. Twelve or 15 days later, these 25 sheep will be inoculated with highly virulent anthrax at the same time as the remaining 25 sheep. The 25 unvaccinated sheep will all die; the 25 vaccinated sheep will resist and will later be compared with the 10 control sheep referred to above in order to show that the vaccinations have not impeded the [vaccinated] sheep from returning to a normal state after a certain time.
5. After the inoculation of the highly virulent virus into the two series of 25 sheep—vaccinated and unvaccinated—the 50 sheep will remain together in the same stable; one will distinguish the two series from each other by punching a hole in the ear of the 25 vaccinated sheep with a punch.
6. The 10 control sheep will remain apart in their own pen so that they will not be exposed to contagion from the sick sheep.
7. All the sheep that die will be buried one by one in separate graves, near one another and situated in a fenced enclosure.
8. In the month of May 1882, one will pen in the enclosure just referred to . . .

[1] This draft was accepted by the president, le Baron de la Rochette, on 28 April. A meeting is set with him for the [trip to the farm at Melun?] on 5 May at 11:55 at the Lyon railroad station. One [will vaccinate?] that day 25 sheep with an anthrax bacillus that, first weakened by pot[assium] bichromate until it kills only mice, has then been passed successively through three mice. Fifteen days later, on 25 May [*sic*], one will vaccinate by an anthrax bacillus [unspecified]. Then 15 days later, on 10 June [*sic*], one will pass to the most virulent anthrax bacillus, dating from 4 years.

Appendix B

SUITE DE PAGES 106 ET 107

Vaccination des moutons et des vaches près de Melun, à ferme de M. Rossignol, vétérinaire, commune de Pouilly-le-Fort

5 mai [1881].

Le bactéridie employée comme premier vaccin, ce 5 mai, en présence de MM. . . . et d'une foule d'autre personnes, vétérinaires et cultivateurs, etc., etc., a été une bactéridie atténué par Ch^{d} [i.e., Chamberland] par bichromate et qui ne tuant plus de tout aviat été renforcée par trois passages successifs dan trois souris.

Cette bactéridie est conservée en tube à 2 effilières ensemencée le 10 mai par moi et remise à l'étuve pour la conserver.

Je conservé aussi la tube a 2 effilières la bactéridie qui servira le 17 mai, après 13 jours depuis la vaccination du 5 mai. C'est une bactéridie issue directement de bichromate, après qques jours seulement. Une fois, elle a tué un mouton sur deux et à diverses reprises, employée sur des moutons après la bactéridie ci-dessous de 3ème souris elle a très bien vacciné pour la bactéridie de 4 ans, très virulente, inoculé en 3d lieu.

17 mai.

Le mardi 17 mai départ pour Melun à 11h 55′. On va inoculer la culture faite d'avant hier à aujourd'hui de la bactéridie ci-dessous qui a tué 2 moutons sur 4 (bactéridie B). Elle est en longs fils au peu grêles. Chaque jour à dater du 18, on va prendre leurs tempres. *Voir le tableau general du Tempres.*

28 mai.

On inocule par bact. de 4 ans (culture virulente) un mouton du 25 vaccinés et un mouton du 25 non vaccinés.

29 mai.

Le vacciné n'a presque pas changé de tempre. Il n'a augmente que de 0°, 1. Le non vacciné a augmenté de 2°, 3. Il est mort dans la nuit du 29 au 30 [mai] (ci-dessous). On réinocule un agneau vacciné [on 8 May] et un mouton du 25 non vaccinés.

30 mai.

Le vacciné a augmenté de 0°, 2. Le non vacciné a augmenté de 2°, 3. Il est mort dans le nuit du 30 au 31 mai (ci-dessous).

31 mai.

Réunion générale pour autopsie des deux morts et inoculation générale de 23 vaccinés et 23 non vaccinés par bactéridie de 4 ans.

Translation

CONTINUATION OF PAGES 106 AND 107

Vaccination of sheep and cows near Melun, at the farm of M. Rossignol, veterinarian, in the commune of Pouilly-le-Fort.

5 May.

The anthrax culture employed as the first vaccine, this 5th of May, in the presence of MM. . . . and a host of others, veterinarians, cultivators, etc., etc., was an anthrax culture attenuated by Ch[amberland] with [potassium] bichromate that, no longer lethal at all, had been reinforced by three successive passages through three mice.

This anthrax culture was preserved in a filed tube of 2 [?] sown by me on May 10th and put in the stove for preservation.

I also preserved in a filed tube of 2 [?] the anthrax culture that will be used on 17 May, 13 days after the vaccination of 5 May. It is an anthrax culture issuing directly from [potassium] bichromate, after only a few days [of exposure]. At one point it killed one sheep in two and at various repetitions, used on sheep in succession to the anthrax culture from the third mice (see below), it has fully vaccinated against the highly virulent anthrax of four years, inoculated third.

17 May.

On Tuesday, 17 May, departure for Melun at 11:55. One is going to inoculate the culture made the day before yesterday of the anthrax bacillus referred to below that has killed two sheep in four (anthrax culture B). It is of long and fairly slender threads. Each day, beginning on the 18th, one is going to take their [the sheep's] temp[eratures]. *See the general table of temp[eratures].*

28 May.

One inoculates one of the 25 vaccinated sheep and one of the 25 unvaccinated sheep with the anthrax bacillus of 4 years (virulent culture).

29 May.

The vaccinated [sheep] has hardly changed in temp[erature]; it has increased only by 0°, 1. The unvaccinated [sheep] has increased [in temperature] by 2°, 3. It died during the night of 29–30 [May] (see below). One reinoculates a lamb vaccinated [on 8 May] and one of the 25 unvaccinated sheep.

30 May.

The vaccinated [lamb] has increased by 0, 2. The unvaccinated [sheep] has increased by 2, 3 (see below). It died during the night of May 30th-31st.

31 May.

General meeting to autopsy the two dead [sheep] and [to do the] general inoculation of the [remaining] 23 vaccinated and 23 unvaccinated [sheep] with the anthrax bacillus of 4 years.

[*Note:* The results of this final injection into the vaccinated and unvaccinated sheep must have been recorded elsewhere; in any case, there can be no doubt that the outcome was that which Pasteur—and Rossignol—reported in their published accounts of the Pouilly-le-Fort trial.]

Appendix C

TOUSSAINT'S *PLI CACHETÉ*, DEPOSITED 12 JULY 1880, RELEASED TO THE ACADÉMIE DES SCIENCES, 2 AUGUST 1880

Procedure for the Vaccination of Sheep and Young Dogs

At first I used the filtration of anthrax blood coming from dogs, sheep, or rabbits. For that, I collected the blood of an inoculated animal at the moment when it was about to die or immediately after death. This blood was then defibrinated by churning, passed through a cloth, and filtered through ten or twelve sheets of paper. With this procedure, three three-month-old dogs and the first ewe were vaccinated. But it is a dangerous and not at all practical method, for the filters often allow the passage of some anthrax bacilli that are difficult to detect with the microscope because they are very rare, and one kills the animals that one wants to preserve.

In the face of these accidents, and being unable to procure any filter yielding the filtered matter in sufficient quantity, I had recourse to heat to kill the bacilli, and I heated the defibrinated blood to 55°[C] for ten minutes. The result was complete. Five sheep, inoculated with 3^{cc} of this blood, have since been inoculated with very active anthrax blood, and have felt no effects whatever from it.

But it is necessary, in order to assure complete innocuity, to make several inoculations. Thus after the first preventive inoculation, I inserted some rabbit anthrax blood and some bacilli spores under the skin of the ears of two sheep. One of them died with an immense quantity of bacilli in its blood. I then inoculated anew the four remaining sheep with the same blood from the dead sheep, after having carried it to 55°, and, since this period, each sheep has been inoculated twice with anthrax blood without suffering the least harm.

Not only are the animals refractory to anthrax, but the most bacilli-charged inoculations produce *no local inflammatory effect*; the wounds heal themselves like simple wounds, which leads me to believe that the obstacle to the development of anthrax is not only in the ganglions, but also in the blood or the lymph, in the liquids of the economy, which have become unsuitable to nourish the parasite.

The practical methods which will be able to serve for the inoculation of all the animals of a flock will immediately be sought. I hope that the difficulties will be easy to surmount, and that, a little time from now, I will be able to render public the method contained in this Note.

Appendix D

ROUX TO PASTEUR, 19 AUGUST 1880

I. Telegram of about 11:00 A.M.

Just seen Bouley. Toussaint has killed four of twenty sheep with anthrax at Alfort. Sending you today experiments which give an account of the thing. The temperature of 55° does not always kill the anthrax bacillus. Roux

II. Letter of the Same Day

19 August

Sir and Dear Master,

You will have understood by my telegram of today that Toussaint's virus did not keep all of its promises. Here's the report of things in detail. Bouley came this morning to the Laboratory. He began by speaking to me of your letter to the Académie [de médecine] and of the response that Roger addressed to you. But all of that in roundabout terms in such a fashion that I could see very well that he had not come to talk to me about these things. Finally he told me that he was about to speak to me of a recent experiment which modified that which Toussaint had announced in his note, and that [he was about to speak] under the seal of secrecy. Bouley had obtained from the minister [of agriculture] authorization to repeat Toussaint's experiments on a large scale at Alfort. Twenty sheep were then brought, which Toussaint "vaccinated" with a virus that he had brought from Toulouse and which had given the best results on sheep and even on rabbits. Of these twenty sheep, one died of anthrax the day after the inoculation, while three others died the next day and the surviving sixteen were so sick that a frightful mortality was anticipated. Nocard regarded them as lost, to the point that he wanted to sacrifice them in order to collect some blood before their death. Some blood was taken from the facial vein of one of these sheep, and this blood did not yet contain anthrax bacilli, say these gentlemen. These sixteen sheep are today on the way to recovery and evidently they are vaccinated. Toussaint was away for the Cambridge Congress. He returned yesterday, and it was after having talked with him, and with his authorization, that Bouley came to speak to us at the Laboratory of what he had just observed, and he will surely write you a letter on this subject. The experiments that I have carried out these last few days have given me the certitude that Toussaint did not completely destroy the anthrax bacilli in the blood, and so the case of his vaccine reverts to that of the chicken cholera [vaccine].

I then told Bouley that the experiments that you had instituted here and at Arbois authorized one to affirm that the case which astonished them today must inevitably have presented itself, that the thing was not new to us, that we have some [guinea?] pigs which have succumbed to the inoculation of anthrax blood heated for 10 minutes at 55°. I showed him a rabbit carrying a very large anthrax oedema that had received only anthrax blood heated according to Toussaint's method. Finally I had him look at some abundant cultures of anthrax bacilli obtained with blood heated at 55° for ten minutes. I told him that the experiments instituted by you at Arbois already left you no longer in doubt as to the true interpretation of [Toussaint's] experiment. Since he insisted on the fact that Toussaint's virus fell into the category of your attenuated virus of [chicken] cholera, I observed to him that there was between them this capital difference: that the [chicken] cholera vaccine cultivated itself [i.e., reproduced] in the vaccinal state, while everything leads to the conclusion that it is nothing of the sort for Toussaint's attenuated anthrax bacillus.

I do not know, Sir and dear Master, if you will approve of my conduct vis-à-vis Bouley, but here are the reasons that led me to act in such a manner. Evidently Toussaint has perceived that he had been very careless in this affair. Their Alfort experiment opened their eyes to the nature of things, and Bouley came immediately to establish that if Toussaint was at first mistaken, he was also the first to correct his error and to establish the true conditions of anthrax vaccination. They had no doubt whatever that you would have lost no time and that you would already have begun some control experiments, and they could find no better way of avoiding the lesson that threatened them than coming to disclose confidentially what had happened to them.

That's why I supposed it was better to tell Bouley everything that was known to us, that we have studied the facts much more closely than Toussaint, and that we have known for several days to what we ascribed them. I told him that you had in your hands all the information that could give Toussaint's experiment its true interpretation. Bouley seemed to me to be very much disappointed, and he told me that the Alfort experiment thus only confirmed our own, and that he would send you this confirmation by a letter. He seemed very surprised that you had already instituted experiments in the Jura.

Toussaint has never tried to culture the blood that he used. He is about to try it. One should expect a note from him promptly.

The virus that Toussaint brought from Toulouse had been obtained not by heat, but by the measured action of carbolic acid on anthrax blood. This blood had shown itself an excellent vaccine at Toulouse. To prepare the vaccines, Toussaint uses heat or carbolic acid, but in both cases the interpretation must be the same.

This last detail makes me think that all the enfeebled cultures of anthrax bacilli that Chamberland has obtained with gas vapors must be vaccines, at least for sheep.

So there, Sir and dear Master, is the story of what happened this morning at the Laboratory. I hope that you think that I acted not imprudently, but simply to reestablish in their proper place the situation of your laboratory and that of Toussaint.

Bouley's language, the precautions of requested secrecy, etc., seemed to me to have no other goal than that of protecting Toussaint against an adventure similar to that of Peter in the case of the "lepothrix puerpueralis." What I told him is, moreover, the absolute truth, as you see.

Receive . . . the assurance of my respectful affection. Roux

Appendix E

BOULEY TO PASTEUR, 19 AUGUST 1880

Paris, 19 August 1880

My dear Master,

I went this morning to the rue d'Ulm to get some news of you. Having learned that you were well, I thought I ought to transmit to you the preliminary results of our experiments at Alfort, which are of a sort to interest you, for they appear to me of such a nature as to do away with the "mysterious" character of the facts signaled by Toussaint.

You know that we inoculated 20 sheep on 8 August with Toussaint's "vaccinal liquid," having previously tried it on five sheep and on some rabbits, all of which resisted the effects of the inoculation. This vaccinal liquid was prepared, not *under the action of heat*, but by the action of carbolic action, which some trials had shown Toussaint to be very efficacious.

The following Tuesday, a sheep died "anthraxed"—that is to say, with anthrax bacilli "in mass" in the blood. On Thursday three others followed. I went to Alfort. All the lot was sick enough, but two sheep especially seemed even on the verge of succumbing (anal temperature 42°). I imagined a complete disaster. It was nothing of the sort. The loss was restricted to the first four sheep dead of anthrax. The other sixteen are in perfect health today and ready to be subjected to the counterproof of inoculation with anthrax.

What to conclude from this result? Evidently, that Toussaint does not vaccinate, as he believed, with a liquid devoid of anthrax bacilli, since he has given anthrax with this liquid, but that he uses a liquid in which the potency of the bacilli is reduced by a diminution in number and an attenuated activity. His vaccine is nothing other than anthrax liquid, the intensity of activity of which is weakened to the point of no longer being mortal for a certain number of "susceptible" individuals that receive it. But this would be a vaccine full of "treachery," since it would be capable of recovering its potency with time. The Alfort experiment makes it probable that the vaccine tested at Toulouse, and which showed itself inoffensive there, had acquired a greater intensity in the interval of the dozen days that elapsed before it was tested at Alfort—that the anthrax bacilli, temporarily anaesthetized by the carbolic acid, had had time to revive itself and to multiply in spite of this acid. As you see, dear Master, this experiment, which would give results that were not in Toussaint's program, clarifies the question in its true light and becomes a confirmation of that which you have established on viruses and vaccines. It is from this point of view that I believed I should communicate it to you, while asking you to be good enough not to make any public use of it yet; I would like it to be Toussaint himself who makes the required correction in the interpretation of his interesting discovery—which discovery remains, in spite of everything, for the sheep that he has vaccinated will be, I am convinced, immune to anthrax.

Your silence on the academic question [i.e., the controversy at the 27 July séance of the Académie de médecine] has led me to think that you have renounced any intention of making *an affair* of that.

Affectionately yours,
Bouley

Appendix F

PASTEUR TO BOULEY, 20 AUGUST 1880

Covering Letter

Dijon, 20 August 1880

My very dear Master,

Here I am at the Dijon station. I have four hours in front of me, waiting for my daughter who is arriving from Avallon with her mother-in-law, her husband being at Montauban beside M. de Freycinet. We will arrive this evening at Arbois, which I left early this morning.

I am going to use my four hours of waiting to compose a note that I ask you to present at the Académie [des sciences] on Monday! You will excuse me for the long delay that I have taken to thank you for your very obliging letter on the subject of the affair at the Académie de médecine. I am letting everything rest until a new day. My experiments have absorbed and are absorbing all my attention.

That's delivered me from a great burden, as you are about to see.

The Toussaint fact is going to be explained and is explained in the most natural fashion and is far from having the significance that was expected from it.

I am in a rush. I fear that I will not have time to write my note and eat lunch.

Again, a thousand thanks for your affectionate helpfulness.

Your very devoted colleague,

L. Pasteur

Appendix G

PASTEUR TO BOULEY, 22 AUGUST 1880

Arbois, 22 August 1880

Very dear colleague and Master,

You ought to have received from me yesterday morning a letter and a letter-note that I wrote you from the Dijon station, the day before, between two trains. On my return to Arbois, I found your letter [of 19 August] and I would have responded to you yesterday, if I had not had to pass the day near Lons-le-Saulnier, side-by-side with my experimental cows and sheep.

The facts in my letter from Dijon explain to you the results that you have obtained at Alfort. Why did Toussaint not try to culture his alleged vaccinal blood, and why call a vaccine something that has not been proved to reproduce itself indefinitely with vaccinal properties against the disease?

In your reinoculation of the sixteen sheep which you still have you will very probably have some more deaths. Use my note as you wish. Present it or hold it back at your discretion. If you present it, I ask you to please let me know. Depending on circumstances, I will or will not have to send you some other results.

Your very devoted colleague,

L. Pasteur

Appendix H

LETTER FROM ROUX TO PASTEUR, 22 AUGUST 1880

Sunday, 22 August 1880

Sir and dear Master:

I've just this instant received your telegram. I knew from Bouley, whom I saw yesterday, that you had sent him a note; I believe that he will present it tomorrow. Bouley and Toussaint came to the laboratory yesterday morning.

Toussaint did not deny that anthrax blood still contains living anthrax bacilli [*bactéridies*] after being heated for ten minutes at 55°. Since he sent his secret note [*pli secret*], it has too often happened that his vaccine killed animals by giving them anthrax for him to have any illusion in this respect. He has also since modified his modus operandi. He is no longer content to heat his blood at 55°. After having maintained it at this temperature for ten minutes, he adds to it some carbolic acid in a proportion of 1% to 1.5%. The flask of blood that he proposes to try at Alfort today was prepared in this way. He asked me to examine this blood and to investigate whether it still contains living anthrax bacilli. I examined it first under the microscope and was able to persuade myself that Toussaint operates in such a way as to have many impurities. In this blood one did not detect anthrax bacilli remaining in the form of globules; everything was resolved into extremely fine granules beside which one saw some mobile rods [*bâtonnets mobiles*] and some chains [*chapelets*]. I should tell you that this blood was in contact with carbolic acid for twelve days, that Toussaint keeps it in an ordinary large-necked flask with a ground stopper, and that he takes no precautions in the manipulation. I sowed this blood in neutral broth. In a few hours there developed numerous chains, double-points, and some small and agile anthrax bacilli. To extract the anthrax bacilli from his blood one proceeded [*il y aurait*] by inoculating a guinea pig with a heavy dose and trying to cultivate it by using some techniques that have sometimes succeeded in the search for anthrax bacilli in soils. I prepared a filed tube of this blood in this way. It was a very interesting trial. There were no anthrax bacilli in the blood and Toussaint will not vaccinate by injecting it, or if he did the vaccination would be no mystery. The matter was proposed to him and he made his own search for the vaccinal properties of his liquid, circumstances which add still more interest to the experiment. Unfortunately, I am leaving Paris tonight in order to attend my brother's wedding on Tuesday. If you think it useful to make some trials with this blood, Jean [the laboratory assistant] will send it to you—he knows where I put it—or else I can return to Paris to continue the work already begun.

Saturday morning Toussaint could not understand the death of these animals at

Alfort [that is, the deaths of four of the twenty sheep into which Toussaint had injected his alleged vaccine at Alfort], and he even persisted in the conclusion of his [published] notes: "The blood that cultivated the anthrax bacilli, deprived of all organisms, is the anthrax vaccine." I had to repeat the thing several times. But I saw him again that night and he no longer stuck to the same language at all. He said that he didn't know what to think and that he'd just made some reservations about his first opinion at the Scientific Association at Reims. The certain fact is that Toussaint is by no means the master of the conditions of his experiment. After having had some failures [*mécomptes*] with simple heating, he adds carbolic acid and is by no means sure of succeeding. He spoke incessantly in the course of our conversation of the numerous cultures he had made. Then, in the next instance, he conceded that he'd been unable to succeed at all because of the impurities which made his primitive seeding disappear. Thus he'd never been able to separate septicemia from anthrax, and when he received anthrax blood to study, his first inoculations yielded septicemia and anthrax mixed together, then septicemia alone. His method is therefore absolutely powerless to allow him to resolve this question: "Are there anthrax bacilli in the liquids that I inject in order to vaccinate?" He has heated up to 150cc of defibrated blood at once. In these conditions it is absolutely certain that he does not kill all the anthrax bacilli, that he even preserves a large number of them alive, for even with a vastly thinner layer they resist destruction in blood heated for ten minutes at 55°. Toussaint insists on the necessity of taking the [anthrax] blood at the moment when the animal has just expired. While he has succeeded in making a vaccine of blood extracted at the moment of death of the animal with anthrax, he had failed when taking blood from animals dead for several hours. However, the last experiment at Alfort, in which some blood—an excellent vaccine [in preliminary trials] at Toulouse—kills with anthrax eight days later at Paris, shows amply that not too much importance can be ascribed to the minute recommendations of an observer who is not the master of the essential conditions of his experiments. Toussaint seems convinced that in the organism, the anthrax bacilli can, under certain circumstances, give spores and that is how he explains the difficulty that presents itself of destroying [the bacilli] with a temperature of 55° in the blood of an animal dead for several hours.

Toussaint appeared very nervous and agitated about all that has happened. I think he regrets having opened the sealed envelope [*pli cacheté*]. He would undoubtedly have modified his initial conclusions.

Toussaint and Bouley were not averse to coming now and then to the laboratory these last couple of days to oversee the method of conducting the famous cultures that led them to failure. In this connection, I have easily evaded some questions, sometimes embarrassing ones, and to which it is not appropriate for me to respond. Beginning tomorrow, they will find themselves confronted by the discretion of a closed door.

I have received no sheep at all. I can't understand the behavior of the butcher, whom I have been unable to find at home. Bertrand tells me that one of his cousins is a livestock merchant, and that it will suffice to let him (Bertrand) know for him to furnish us through his cousin all [the sheep] that we want.

Here's my address:

ROUX Emile
chez M. de Fornet
à Larochefoucauld
Charentes

I will write to you upon my arrival at Larochefoucauld in connection with some of Toussaint's ideas about chicken cholera and septicemia. From our conversation, it was possible for me to learn how he conducts his experiments.

Accept, Sir and dear Master, the assurance of my respectful attention.

Roux

[*Source*: This letter is printed in full (in French) in Louis Nicol, *L'épopée pastorienne et la médecine vétérinaire* (Garches, 1974), pp. 336–339.]

Appendix I

LETTER FROM ROUX TO PASTEUR, 30 AUGUST 1880

Larochefoucauld 30 August

Sir and dear Master:

It has been difficult for me all this time, in the midst of small devilish nephews and nieces, to find enough peace and quiet to write you a little more fully. Please forgive me. I am now going to try to bring you up to date on several incidents from my interview with Bouley and Toussaint.

After Bouley had spoken to me on Saturday 21 August on the matter of the note that you had sent him, I thought that he would present it the following Monday. He had even said so in a manner that left me in no doubt. On reflection, however, I am not so astonished that he has acted as he has done. He was at least as upset as Toussaint, and his attitude was very embarrassed when I told him precisely what we knew of the effects of heating [anthrax bacilli]. Bouley embraces the interests of Toussaint with an extraordinary passion. He is himself much involved in the experiments made at this moment at Alfort and it is probable that a part of the honor will redound to him. He is also aware that Toussaint could be himself [led?] to the misadventures of the "vaccine against anthrax" and there is nothing he won't do to deflect anything that might diminish [Toussaint's] note of 2 August.

Bouley and Toussaint came to the laboratory on Saturday 21 August. They asked me for a culture of anthrax bacilli to inoculate into a rabbit. I gave them a culture that came from the blood of a rabbit dead of anthrax some days before. I also prepared a tube with an older culture, dating from the month of July, made by Chamberland under conditions unknown to me, and in which there had been many [anthrax] spores. Toussaint wanted a culture with spores. These cultures were very fine, as was immediately confirmed under the microscope. On leaving the laboratory Toussaint forgot one of the tubes on the table. I did not perceive this oversight until the evening and cannot say which of the two cultures he used at Alfort. That's the incident to which the passage from Bouley's letter refers. Why this culture did not show itself to be active, I have no idea. Could it be that in many of the flasks of anthrax cultures we have at the laboratory, the bacilli would show themselves benign for sheep and give them immunity, notably the cultures of the spores that Chamberland has left exposed for some time to gasoline vapors? Perhaps it was a culture of this sort that I gave to Toussaint without realizing it.

The same day, Saturday 21 August, Toussaint returned to the laboratory alone in the evening. He brought me the liquid mixed with carbolic acid about which I have told you and which contains some impurities to the extent that a trace put in a flask of bouillon has given an abundance of double points and chains within a few hours. It is this vaccinal blood that I have preserved at the laboratory in a tube. But you know these details . . .

Roux

[*Note:* The letter continues for several more pages, and though some of the passages omitted here are of some interest in other contexts, they do not bear directly on the competition between Pasteur and Toussaint in the quest for an effective anthrax vaccine.]

Appendix J

RESULTS OF EXPERIMENTS WITH CHAMBERLAND'S POTASSIUM BICHROMATE VACCINE BEFORE THE POUILLY-LE-FORT TRIAL

A. Bibliothèque Nationale. Papiers Pasteur, Oeuvres, *VI, "Maladies Virulentes," pp. 129–129v. [In Chamberland's hand; author's translation]*

129 *18 February 1881*

The secure results obtained by cultivating the anthrax bacillus in chicken bouillon containing potassium bichromate are the following:

10 [parts] of bouillon + 2 [parts] of bichromate at 1% [dilution] does not cultivate the bacillus.

10 bouillon + 1 [part] bichromate at 1% cultivates the bacillus, but poorly.

10 bouillon + 0.5 [part] bichromate at 1% cultivates the bacillus fairly well while giving spores.

All the experiments have therefore been made in flasks containing 10 [parts] bouillon + 1 [part] bichromate at 1%.

Weakening of the anthrax bacillus

After four days the flask kills guinea pigs, but only one sheep in two; the other [sheep] was very sick.

[After] 12–14 days the flask kills guinea pigs and no sheep; it vaccinates the latter.

[After] 29 days [the flask] sometimes kills guinea pigs and sometimes does not; does not kill sheep and vaccinates them.

[After] 40 days [the flask] no longer kills guinea pigs or sheep and does not vaccinate the latter. Successive cultures are harmless. There are, moreover, no germs.

Return of [virulence in] the anthrax bacillus by [passage through] guinea pigs

The culture of 40 days, inoculated into one- or two-day-old guinea pigs, kills them within a few days (five or six) and there are no anthrax bacilli in the blood.

After two or three inoculations [i.e., passages?] there are anthrax bacilli in the blood of guinea pigs and these bacilli kill large guinea pigs. These *first* cultures of return do nothing to sheep and do not vaccinate them.

129v *Return [of virulence] by [passages through] white mice*

The culture of 40 days, inoculated into white mice, kills them, and often, as with the small guinea pigs, there are no bacilli in the blood. After four or five successive inoculations [passages] in mice the bacilli still do nothing to large guinea pigs and to *vaccinated sheep.*

After eight or ten successive inoculations [passages] the blood from mice, inoculated into guinea pigs, produces a swelling that later disappears. One of these guinea pigs, which had had a fairly large swelling, recovered [after being] reinoculated with virulent anthrax. This [guinea pig] ought therefore to be vaccinated.

B. Papiers Pasteur, Correspondance, *XI, "Lettres adressés à Louis Pasteur, Pellerin à Susani", pp. 269–271. Excerpts from a letter from Roux to Pasteur, 17 April 1881. [Author's translation]*

At the time of your departure, there were 10 inoculated sheep:
4 with the culture of 7 February (inoculated on 30 March)
4 with the culture of 9 February (inoculated on 5 April)
2 with the culture of 5 March, series of 19 February (inoculated on 3 April)

13 April

— Two sheep of the series of 30 March received again the culture of 7 [February] to see if two preventive inoculations will protect them from the culture of 4 February.
— These sheep had an elevation of temperature of a few tenths of a degree after the inoculation; they are [now] returned to a normal state.
— The two other sheep of this series received the culture B of Chamberland. It is a culture that kills two sheep in four and served to test [the response of] Chamberland's sheep to the second inoculation on 18 March. They fully resisted [the effects of this second inoculation]. . . .

This experiment proves then that the culture of 7 [February], which does not vaccinate against the still highly virulent culture of four [years], vaccinates against a culture killing two sheep in four. In other words, if in the first experiment we had inoculated Chamberland's culture (B instead of the culture of 4 February), these sheep would not be dead and would be vaccinated. . . .

On 13 April Chamberland's sheep were reinoculated with the anthrax culture of four years [highly virulent anthrax]. . . . In sum, they fully resisted [the effects of this virulent injection]. . . .

It is thus determined through these experiments: (1) that Chamberland's sheep are vaccinated against the most virulent anthrax culture; (2) that the culture of 7 [February] is a good vaccine for an initial inoculation—that it, as fully as Chamberland's vaccine, preserves [sheep] from the [effects of] the culture killing two times in four. . . .

[*Note:* Most of these details, as well as the ultimate results of this trial of Chamberland's potassium bichromate vaccine, are recorded in Pasteur, *Cahier d'éxperiences 91,* fol. 102. The outcome persuaded Pasteur that his oxygen-attenuated vaccine was not a "secure" (*sûr*) vaccine, whereas Chamberland's antiseptic vaccine "should be secure."]

Appendix K

A REMARKABLE CASE HISTORY OF RABIES: THE CASE OF JOHN LINDSAY AS REPORTED BY SAMUEL ARGENT BARDSLEY IN 1807

John Lindsay, weaver at Fearn Gore near Bury, in the county of Lancaster, aged thirty-six, of middling stature, and spare habit of body, and of a temperament inclined to the melancholic, was brought into the Manchester Lunatic Hospital, on Friday May the sixteenth, 1794, about three o'clock in the afternoon. He was immediately visited by Dr. Le Sassier, who obligingly communicated to me the following particulars. The Patient expressed feelingly his sense of danger, from the persuasion that his disorder proceeded from the bite of a mad dog. He was desired to drink a little cold water, which on being presented to him he rejected, with every appearance of disgust and horror. Being again strongly urged to drink, he made the attempt, and with great exertion got down a small quantity of the liquid. He was perfectly rational, but appeared apprehensive of danger from the least noise, or approach of any person towards him. He expressed a desire to make water, and was quitting the room for that purpose; but no sooner had he approached the door, than he suddenly retreated, complaining of an unpleasant sensation he felt from the cold air, and particularly that it produced a convulsive twitching, about his throat. To screen him from the effects of the air, when conveyed from the examining room into the Hospital, an umbrella was held over his head, and his body closely muffled up in a wrapping cloak. As soon as he had gone into his apartment, he ate some bread and cheese, but with difficulty; and requested to be allowed to drink some buttermilk. He attempted to swallow this liquid, and in part succeeded; but not without the most violent struggling efforts, attended with distortions of his countenance, which remained slightly convulsed for some time afterwards.

. . . I saw the Patient, in company with the other Physicians, about six o'clock the same evening; and we found him very willing, and sufficiently composed, to give a distinct account of the circumstances preceding the disease, and to describe his sufferings since its attack. The following particulars were collected. He had been industrious, sober, and regular in his mode of living; but subject to low spirits from the difficulty he found, at times, of maintaining a wife and six young children. His exertions, however, were in general proportionate to his difficulties. But of late, from the depreciation of labor, he found, that the most rigid oeconomy and indefatigable industry were not sufficient to ward off, from himself and family, the calamities of hunger, debt, and most abject poverty. The anxiety of his mind now became almost insupportable. As the last refuge for his distress, he applied a few days previous to the attack of his complaint, to the Overseers of his Parish for their assistance to pay his rent, and thereby prevent the seizure of his goods; but obtained no relief. Overwhelmed with grief and disappointment, he yielded to despair, resigning himself and family to their wretched fate. He was soon roused from this state of fancied apathy, by the piercing cries of his children demanding bread. In a paroxysm of rage and tenderness, he sat down to his loom on the Monday morning, and worked night and day, seldom quitting his seat, till early on the ensuing Wednesday morning. During this period of bodily fatigue and mental anxiety, he was entirely supported by hasty draughts of cold buttermilk, sparingly taken. Nor did he quit the loom, until his strength was completely exhausted. He then threw himself upon his bed, and slept a few hours. On waking, he complained of giddiness and confusion in his head, and a general sense of weariness over his body. He walked five miles that morning, in order to receive his wages, for the completion of his work; and, on his return, felt much fatigued, and troubled with a pain in his head. During the night, his sleep was interrupted by involuntary and deep sighs—slight twitchings in the arms—and a sense of weight and constriction at the breast. He complained of much uneasiness at the light of a candle, that was burning the room. On evacuating his urine, he was obliged to turn aside his head from the vessel, as he could not bear the sight of the fluid without great uneasiness. Being rather thirsty, he wished for a balm tea to drink; but was unable to swallow it from a sense of pain and tightness, which he experienced about the throat, when the liquid was presented to him. He suddenly exclaimed, on perceiving this last symptom, "Good God. It is all over with me." and immediately recalled to his wife's recollection, the circumstance of his having been bitten, twelve years ago, by a large dog apparently mad. . . .

During the whole of Thursday, his abhorrence of fluids increased; and he now began to feel an uneasy sensation of being exposed to the air. The slight twitchings of his arms were also increased to sudden startings; attended with a violent agitation of his whole body. He had suffered much from his journey, being brought eight miles in an open cart. I perceived at this time (half past six, Friday evening) that his countenance expressed the utmost anxiety; his breathing was laborious and interrupted; and he complained of a dull pain, shooting from the arms toward the praecordia and region of the stomach. A livid paleness overspread his face; the features were much contracted; and the temples moistened with a clammy sweat. He suffered greatly from excessive thirst, and dryness of the mouth and fauces.

An unusual flow of viscid saliva occasioned him to spit out frequently. He complained of a remarkably fetid taste in his mouth, and a loathsome smell in his nostrils. He ate some bread and butter, at his own request, but with great difficulty, as he was obliged to throw the head backward, in order to favor the descent of the morsel down the gullet. He was requested to wash down this solid food, with some liquid; and he expressed a readiness to make the trial. On receiving a bason of buttermilk, he hastily applied it, with a determined countenance, to his lips; when he was instantly seized with so severe a spasm and rigidity of the muscles of the neck, that he was compelled, in an agony, to desist from drinking. Shortly after, he raised himself upon his knees in bed, took the bowl again into his hands, and by forcibly stretching his neck forward, at the moment he received the liquid into his mouth, and then violently throwing his head backwards, he succeeded in swallowing a small portion. He appeared highly gratified with the success of this effort, and the fortitude he had exhibited; and exultingly demanded another draught of the buttermilk, as he now thought he could conquer the difficulty he had hitherto experienced. But a violent return of the spasms in the throat and neck checked this attempt. These convulsions were terminated by the stomach discharging the liquid previously swallowed, highly tinged with bile. I perceived that he had conveyed a piece of orange, under the bed cloaths, which at intervals he applied to his mouth by stealth, and as it were unperceived by himself; for he constantly hurried it to his lips, when his attention appeared to be engaged on other objects. This stratagem did not succeed. No sooner had the morsel touched his mouth, than he was seized with convulsions about the throat, and a stricture in the breast. I saw him again, in consultation, at eight o'clock this evening. . . . He appeared rather more composed, but expressed great anxiety at the idea of being left alone. He courted eagerly the conversation of those around him; apparently from the motive of withdrawing his mind from the contemplation of his miserable state. The repugnance he felt at swallowing liquids, and the uneasiness occasioned by the attempt, he now considered as his chief complaints; and was determined to conquer the first by perseverance, and an undaunted resolution. His spasms seemed to be somewhat mitigated, as he got down a little milk-porridge with less difficulty than usual. . . . At nine o'clock the next morning (Saturday) he was visited again; and we learned that he had passed the night without a moment's rest, frequently shouting out with looks of horror, and sometimes wailing in broken and confused murmurs; but, on being spoken to, he always returned rational answers. He was now alarmed to a degree of distraction at being left alone. He examined every object with a timid and suspicious eye; and, upon the least noise of a footstep in the gallery, he begged, in the most piteous accents, to be protected from harm. He had never offered the least violence to any one, since the commencement of the disease; and, even now, when the encreased secretion of saliva occasioned him to spit out very frequently, he apologized to the bystanders, and always desired them to move out of the way. I observed, he frequently fixed his eyes, with horror and affright, on some ideal object; and then, with a sudden and violent motion, buried his head underneath the bed-cloaths. The last time I saw him repeat this action, I was induced to enquire into the cause of his terror.—He eagerly asked, if I had not heard howlings and scratchings. On being answered in the negative, he suddenly threw himself upon his knees, extending his arms in a defensive posture, and forcibly throwing back his head and body. The muscles of the face were agitated by various spasmodic contortions;—his eye balls glared, and seemed ready to start from their sockets;—and at that moment, when crying out in an agonizing tone:—"Do you not see that black dog." his countenance and attitude exhibited the most dreadful picture of complicated horror, distress and rage, that words can describe, or imagination paint;—The irritability of the whole system was now becoming excessive. He discovered the highest degree of impatience on the least motion of the air. Every action was accompanied with that hurry and inquietude, which marks an apprehension of danger from surrounding objects. The oppression of the praecordia was evidently encreased; and, when he gasped for breath, the whole body was writhed with convulsions. His speech was interrupted by convulsive sobs. The pulse was tremulous and intermitting; and, at some times, so hurried as not to be counted. He had frequent retchings, and brought up occasionally small quantities of a yellow liquid. Solids were now swallowed with excessive difficulty; and the attempt always produced strong spasms about the neck and breast. . . . At four o'clock the same day, the consultation was renewed. We found the patient had been able to swallow his boluses without much difficulty, and had drank several times with infinitely more ease than usual; but, the fluid had been immediately rejected by the stomach, and had come up, deeply tinged with yellow. His countenance exhibited a cadaverous aspect. His voice was hoarse, indistinct, and faltering. He complained of a fixed pain at the region of the stomach; which he had felt, more or less, during the disease. The pulse was feeble and scarcely perceptible. He swallowed some tea with less difficulty, than had been observed since his entrance into the Hospital. His dissolution was apparently drawing near. . . . His mental faculties at this period suffered very little derangement; for although, when not attending to external objects, he could utter some incoherent sentences; yet, the moment he was spoken to, he was perfectly collected, and returned rational answers. At half past four o'clock, he submitted willingly to have his body rubbed with the oil, and for that purpose sat down upon the side of the bed; when he was seized with an instantaneous convulsion, threw himself backward—and expired without a groan!

[*Source*: Samuel Argent Bardsley, *Medical Reports of Cases and Experiments, with Observations, chiefly derived from Hospital Practice: to which are added, An Enquiry into the Origin of Canine Madness; and Thoughts on a Plan for its Extirpation from the British Isles* (London, 1807).]

Author's Note on Notes and Sources

PASTEUR'S PUBLISHED WORKS

Virtually every word that Pasteur published during his lifetime, including all of his books and scientific papers, is reprinted in the monumental and magnificent *Oeuvres de Pasteur*, ed. Pasteur Vallery-Radot, 7 vols. (Paris: Masson et Cie, 1922–1939). This work also contains a significant number of letters, notes, lectures, and manuscripts that were not published during Pasteur's lifetime, as well as documents by others relating to his work, including several reports by commissions of the Académie des sciences.

Each volume of the *Oeuvres de Pasteur* has a brief introduction by Pasteur Vallery-Radot, who adds helpful editorial notes and comments throughout. The volumes are organized topically as follows: I, Molecular asymmetry; II, Fermentation and spontaneous generation; III, Studies on vinegar and wine; IV, Studies on the silkworm diseases; V, Studies on beer; VI, Infectious diseases, virus vaccines, and rabies prophylaxis; and VII, Scientific and literary miscellania. Volume VII also contains a complete index of names cited in all of the volumes and a masterful "analytic and synthetic" subject index. In addition, it provides a complete chronological bibliography of Pasteur's publications ("Table chronique de l'ouevre de Pasteur": Vol. VII, pp. 473–512).

In this book, I have used the following convention in citing Pasteur's published works: *Oeuvres*, volume number in roman numerals, and immediately pertinent page numbers, often those with cited quotations. Thus, for example, *Oeuvres*, I, pp. 344–345, refers to *Oeuvres de Pasteur*, vol. I (Molecular asymmetry), pages 344–345. To recover the full title, date, and original place of publication for any citation, one need only refer to the "Table chronique," which indicates the location of the citation both in the *Oeuvres* and in the original publication.

PUBLISHED CORRESPONDENCE

A significant portion of Pasteur's vast correspondence was assembled and published in Pasteur, *Correspondance*, ed. Pasteur Vallery-Radot, 4 vols. (Paris: Flammarion, 1940–1951). Arranged chronologically over the period 1840–1895, these letters provide a detailed account of Pasteur's activities and illuminate every aspect of his life and career. Pasteur's own letters dominate the collection, but many letters to him and many by members of his family and collaborators are also included. For published versions of scores of other letters to or by Pasteur, see *Pages illustres de Pasteur*, ed. Pasteur Vallery-Radot (Paris: Hachette, 1968); *Correspondence of Pasteur and Thuillier Concerning Anthrax and Swine Fever Vaccination*, translated and edited by Robert M. Frank and Denise Wrotnowska, with a preface by Pasteur Vallery-Radot (Tuscaloosa: University of Alabama Press, 1968); and Louis Nicol, *L'épopée pastorienne et la médecine vétérinaire* (Garches: Chez l'Auteur, 1968).

MANUSCRIPT SOURCES: THE PAPIERS PASTEUR AT THE BIBLIOTHÈQUE NATIONALE IN PARIS

Pasteur's grandson, Dr. Pasteur Vallery-Radot, devoted much of his life to collecting his grandfather's letters, manuscripts, and other unpublished materials. In 1964 he gave most of the collection to the Bibliothèque Nationale in Paris, but it was not made generally available to scholars until the 1970s. A printed inventory of this vast and rich collection did not appear until 1985. See Bibliothèque Nationale, *Nouvelles acquisitions latines et françaises du département des manuscrits pendant les années 1977–1982: Inventaire sommaire* (Paris, 1985), pp. 106–128 (N.a.fr. 17923–18112). The general heading for all of these materials is *Papiers Pasteur*.

I began my research in this archival collection before the printed inventory appeared. In the notes that follow, I have used my original conventions, but it is easy enough to convert them into the now-printed codes. In the case of Pasteur's laboratory notebooks, for example, what I cite as Pasteur, *Cahier 94* conforms to acquisition number 18019 in the printed inventory. *Cahier 77* is acquisition number 18002, while *Cahiers 91*, *92*, and *93* are equivalent to acquisition numbers 18016, 18017, and 18018, respectively. For the other items in this archive, I have used the following convention: Papiers Pasteur, followed by, for example, *Correspondance* or *Oeuvres*. In the notes that follow, these archival sources are distinguished from Pasteur's *published Correspondance* and *Oeuvres* by the prefix "Papiers Pasteur."

ON PASTEUR, "NOTES DIVERS"

As indicated in Chapter Three, Pasteur's "Notes divers" refers to his first laboratory notebook, which is not among the Papiers Pasteur at the Bibliothèque Nationale. In fact, the current location of the original version of this notebook is unknown, but Professor Seymour Mauskopf of Duke University has a microfiche of the notebook, one copy of which he also generously donated to Firestone Library at Princeton University.

OTHER MANUSCRIPT COLLECTIONS, MUSEUMS AND MISCELLANIA

There are several other fairly substantial and interesting collections of Pasteur manuscripts, in France and elsewhere, although the Papiers Pasteur at the Bibliothèque Nationale in Paris is vastly more important. A number of official and administrative documents by and concerning Pasteur are deposited in the Académie des sciences and the Archives Nationales in Paris and in other French national and provincial archives. Several such documents were extracted or otherwise put to use by Denise Wrotnowska (see Bibliography).

Outside of France, the three most extensive and significant collections are to be found at the Wellcome Institute for the History of Medicine in London; the National Library of Medicine in Bethesda, Maryland; and the Burndy Library of the Dibner Institute for the History of Science and Technology in Cambridge, Massachusetts.

The original building of the Institut Pasteur is the site of the Musée Pasteur, which includes the following: Pasteur's personal apartment, preserved as it was when he lived there; Pasteur's personal library, including annotated volumes of his communications to the Académie des sciences; about one thousand pieces of Pasteur's laboratory instruments and equipment, including microscopes, wooden models of crystals, flasks, and bottles; Pasteur's medals, diplomas, and other personal souvenirs; several of the portraits and pastel drawings he did as a youth (including two splendid portraits of his parents); an iconography of about five thousand photographs, drawings, and portraits of Pasteur, his disciples, and the Institut Pasteur. The archives of the Institut Pasteur itself are now being organized and catalogued for scholarly purposes.

Pasteur museums also exist in Arbois, Dole, and Strasbourg. The Wellcome Institute in London also has artifacts and instruments, some of dubious authenticity.

TRANSLATIONS AND BRACKETED DATES

Except where the notes indicate otherwise, I am responsible for all translations from the French. I have used brackets around dates in the Bibliography to indicate the original date of publication of later or translated editions to which my notes refer. For example, Duclaux [1896], 1920 translation, p. 14, is meant to indicate that my pagination is taken from the 1920 translation of a book that Duclaux originally published in 1896.

Notes

CHAPTER ONE
LABORATORY NOTEBOOKS

1. Prévost 1977, p. 101.

2. See Bibliothèque Nationale 1964.

3. See Bibliothèque Nationale 1985. See also "Author's Note on the Notes and Sources."

4. Pasteur, *Correspondance*. A small but important selection of previously unpublished manuscripts can also be found scattered throughout Pasteur, *Oeuvres*.

5. Gillispie 1981, p. 76; Piel 1986, p. 201.

6. Tyndall 1885, p. xiii.

7. Morson 1985, p. 32. The quotations are from Morson rather than Bakhtin himself. More generally, see, e.g., Bakhtin 1981; Morson and Emerson 1989, 1990.

8. On Darwin's "repression" of his private thoughts, see Gruber 1974.

9. Gould 1988.

10. On Pasteur's "Olympian silence," see Duclaux [1896], 1920 translation, p. 147. For one example of Pasteur's acknowledgment that he could not always keep his thoughts secret from his collaborators, see Papiers Pasteur, *Cahier 94*, fol. 9–9v.

11. Roux to Pasteur, 29 July 1880, Papiers Pasteur, *Correspondance*, XI, fols. 225–229, esp. 226v.

12. Loir 1937–1938.

13. Loir 1937–1938, *15*, p. 347.

14. Roux to Pasteur, 14 and 29 July 1883. Papiers Pasteur, *Correspondance*, XI, fols. 275–275v, 285–286.

15. Roux, "Medical Work of Pasteur" [1896], 1925 translation, pp. 389ff.

16. For a flawed but useful journalistic survey of alleged fraud or deceit in science, see Broad and Wade 1982. A more scholarly analysis of scientific fraud is given in Zuckerman 1977. For a superb critical review of the book by Broad and Wade, see Joravsky 1983. And for a sophisticated, if not fully convincing, attempt to argue that definitions of scientific "truth" and "fraud" are simply matters of social negotiation, see Sapp 1987, 1990.

17. Grmek 1973; Holmes 1974.

18. See Holton on the Millikan-Ehrenhaft dispute in Holton 1978a, pp. 25–83, and Gooding 1989, 1990, 1992.

19. Holmes 1974, 1985, 1991a, 1993. For Holmes's very important historiographic essays on laboratory notebooks, see, e.g., Holmes 1984, 1986, 1987, 1988, 1990, 1991b, 1993a, 1993b.

20. These are two of a set of similar sentiments quoted by Merton 1968, and Merton 1978, pp. 119–120n.31.

21. Merton 1978, p. 119n.31.

22. Holton 1978a, pp. 25–83.

23. Holmes 1974, 1985, 1993.

24. Merton 1978. The history of the scientific paper is one of the great underdeveloped topics in the history of science. For preliminary forays, see Porter 1964, Barnes 1934, and McKie 1948.

25. Holmes 1991b.

26. Medawar 1964.

27. See, e.g., Collins 1992; Latour 1986a, 1987.

28. Pasteur, *Oeuvres*, II, pp. 497–585.

29. De Kruif 1959, pp. 95–98.

30. Pasteur, *Oeuvres*, II, pp. 559, 567, 588–589, and esp. pp. 596–597, 608.

31. Pasteur, *Oeuvres*, II, pp. 527–535, 559–562.

32. Prévost 1977, pp. 101–102.

33. Pasteur, *Oeuvres*, II, pp. 549, 556. This phrase was one of Pasteur's favorite rhetorical ploys in controversy; for other examples of his use of it, see Pasteur, *Oeuvres*, V, p. 39; VI, pp. 102, 192.

34. Pasteur, *Oeuvres*, II, pp. 595–596, 600–604, 608–615.

35. Pasteur, *Oeuvres*, VI, pp. 93, 101–102; VII, pp. 33–35.

36. Pasteur, *Oeuvres*, VII, pp. 160–165.

37. Pasteur, *Oeuvres*, VII, pp. 271–281.

38. Pasteur, *Oeuvres*, VII, p. 162.

CHAPTER TWO
PASTEUR IN BRIEF

1. This chapter draws liberally on scattered portions of Geison 1974 and Geison 1985.

2. For a detailed account of Pasteur's early years and of his later visits to his native region, see Ledoux 1941.

3. Ledoux 1941, p. 33.

4. See Pasteur, *Correspondance*, I, pp. 119–155 passim.

5. Pasteur, *Correspondance*, I, p. 170.

6. Pasteur, *Correspondance*, I, pp. 173–181.

7. Pasteur, *Correspondance*, I, pp. 182–185.

8. Pasteur, *Correspondance*, I, pp. 189–190.

9. Pasteur, *Oeuvres*, VII, pp. 129–130.

10. Pasteur, *Correspondance*, VII, p. 131.

11. Pasteur, *Correspondance*, I, pp. 371–374, 382–383, 400–406.

12. Pasteur, *Correspondance*, I, pp. 428–429.

13. René Vallery-Radot 1926, p. 84.

14. Duclaux [1895], pp. 449–450.

15. See Wrotnowska 1958.

16. See Wrotnowska 1959.

17. Pasteur, *Oeuvres*, VII, pp. 156–159.

18. Pasteur, *Oeuvres*, VII, pp. 147–155, 160–163.

19. Pasteur, *Correspondance*, II, p. 348.

20. Glachant 1938, p. 97.

21. Pasteur Vallery-Radot 1954, pp. 36–37.

22. See Pasteur Vallery-Radot 1954, pp. 37–41; and Pasteur, *Correspondance*, II, pp. 124–125.

23. See Pasteur, *Correpondance*, II, pp. 136–142, 162, 179–180.

24. Pasteur, *Correspondance*, II, pp. 333–334.

25. See Pasteur, *Correspondance*, II, pp. 332–339; and Pasteur Vallery-Radot 1954, pp. 49–57.

26. Pasteur, *Correspondance*, II, pp. 340, 345–351.

27. Pasteur, *Oeuvres*, V, p. 5.

28. Pasteur, *Correspondance*, II, pp. 551–552, 565–567, 573–574.

29. Pasteur, *Correspondance*, III, pp. 121–122, 136–138, 190–191.

30. Pasteur, *Correspondance*, III, p. 271n.2.

31. See Pasteur, *Correspondance*, III, pp. 421, 425–428, 441–445; and René Vallery-Radot 1926, pp. 398, 406, 410–411.

32. See Institut Pasteur, 1888.

33. Pasteur, *Oeuvres*, I, p. 376.

34. See Dubos 1950, pp. 359–362, 377–384.

35. See Chapter Five.

36. Pasteur, *Correspondance*, I, p. 228.

37. See Pasteur, *Correspondance*, III, pp. 173ff; and Pasteur, *Oeuvres*, VI, p. 489.

38. For a valuable but rather uncritical survey of Pasteur's involvement in controversies, see Pasteur Vallery-Radot 1954, pp. 62–131.

39. See, e.g., Roux [1896], 1925 translation, pp. 372–373, 389; and Loir 1937–1938, *14*, pp. 278–279.

40. See Roux [1896], 1925 translation, p. 384; Loir 1937–1938, *14*, pp. 278–279; and Chauvois et al. 1966, pp. 197ff.

41. See Partington 1964, IV, pp. 751–752; and Ihde [1964], pp. 322–323.

42. See Duclaux [1896], 1920 translation, pp. 280–281. Cf. Dubos 1950, p. 327; and Lagrange 1954, pp. 38–39.

43. Cadeddu 1985.

44. See René Vallery-Radot 1883, where there is no allusion to a "lucky" discovery of the chicken cholera story. For evidence that this little book amounted to Pasteur's own autobiography, see Chapter Nine.

45. First used by Pasteur in his 1854 inaugural address at Lille. See Pasteur, *Oeuvres*, VII, p. 131. He later used it again at least twice in print. See ibid., VI, p. 348; VII, p. 215. More generally on luck and chance in Pasteur's work, see Dubos 1950, pp. 100–101, 219, 327, 340, 342.

46. Pasteur, *Correspondance*, II, p. 183.

47. On Pasteur's relations with the imperial house, see Pasteur, *Correspondance*, II, pp. 62, 215–235, 245–246, 268, 286–287, 297, 345–346, 355, 385, 387–388, 407–408, 451, 459, 461–463, 471, 484–485, 489, 586, 627; and his correspondence with Colonel Fave, Napoleon III's aide-de-camp, ibid., pp. 98–100, 110–111, 120–121, 125–126, 146–148, 160–161, 236–238.

48. Pasteur's annual governmental research budget was about 50,000 francs in the early 1880s (see Pasteur, *Correspondance*, III, pp. 190–191, 285–286, 317), at which time the matériel expenditures for *all* scientific research in France were apparently less than 500,000 francs (see Forman et al. 1975, esp. p. 66).

49. See Pasteur, *Correspondance*, IV, pp. 227–229, 231–232.

50. See Pasteur, *Correspondance*, IV, pp. 229–231.

51. See Institut national de la propiété industrielle, *Catalogue des brevets d'inventions*, 1857, p. 21 (procedure for alcoholic fermentation); 1861, pp. 201, 385 (manufacture of acetic acid); 1865, pp. 114, 167, 358 (procedure for preserving wines); 1871, pp. 48, 196, 198 (procedure for manufacturing beer); 1871, p. 99 (manufacture and preservation of beer); 1872, pp. 58, 313; and 1873, p. 78 (procedure for manufacturing unalterable beer). See also Pasteur, *Correspondance*, II, p. 190; Pasteur, *Oeuvres*, III, pp. 13n.3, 410n.1; V, pp. 346–357; and Frederico 1937.

52. See Lutaud 1887b, pp. 405–431, esp. p. 412.

53. Pasteur, *Correspondance*, III, p. 271n.2.

54. Pasteur, *Correspondance*, IV, p. 365.

55. For the average salaries of professors and councillors of state in France in c. 1870, see Weisz 1983, pp. 57n.5, 58 (table 2.1); for the income of sales clerks in a department store at the time, see Miller 1981, pp. 81, 91–93.

56. For assessments of Pasteur's attitudes toward money and toward the commercial exploitation of his discoveries, see Cuny 1965, pp. 15–18; Loir 1937–1938 *14*, pp. 188–192; Maurice Vallery-Radot 1985, pp. 181–223; and the fiercely critical Lutaud 1887b. Lutaud often distorts and misuses the documents he adduces in support of his claims, but those documents are sufficiently compelling to warrant a more dispassionate reexamination. For some of Pasteur's own statements about the commercial exploitation of scientific discoveries, see Pasteur, *Correspondance*, II, pp. 236–238; and Pasteur Vallery-Radot 1968, pp. 26–28.

57. On Pasteur's general philosophical and religious attitudes, see Pasteur Vallery-Radot 1954, pp. 221–238; Dubos 1950, pp. 385–400; George 1958; and especially Maurice Vallery-Radot 1985, pp. 267–295, which I consider the fullest and most convincing of the writings on Pasteur's religious attitudes.

58. Pasteur, *Correspondance*, II, pp. 213–214.

59. Pasteur, *Oeuvres*, VII, p. 326.

60. Pasteur, *Oeuvres*, II, pp. 328–346.

61. On Pasteur's general political views, see Cuny 1965, pp. 19–25; Pasteur Vallery-Radot 1954, pp. 175–220; Pasteur, *Correspondance*, II, pp. 355, 459, 461–463, 489, 523, 534, 593, 600, 611–630; III, pp. 424, 436–437; IV, pp. 268–270, 300; and Maurice Vallery-Radot 1982.

62. Pasteur, *Correspondance*, I, pp. 228, 230.

63. Pasteur, *Correspondance*, II, pp. 216–236. On Pasteur's more general relations with the imperial household, see the sources cited in n.49 above.

64. See Pasteur Vallery-Radot 1968, p. 8.

65. Pasteur, *Correspondance*, II, p. 612. More generally on this campaign and the election, see ibid., pp. 611–630; and Pasteur Vallery-Radot 1954, pp. 203–215.

66. See Ledoux 1941, pp. 55ff; and Pasteur, *Correspondance*, IV, pp. 268–270.

67. Pasteur, *Correspondance*, IV, p. 340.

68. Pasteur, *Correspondance*, II, pp. 502–503, 511–514, 517–519.

69. Pasteur, *Correspondance*, II, pp. 491–492.

70. See [Anonymous] 1873.

71. Pasteur, *Correspondance*, IV, pp. 286–287.

72. On Pasteur's stroke and subsequent health, see Pasteur, *Correspondance*, II, pp. 406, 428–429, 440, 464ff, 466–468, 481–483, 589 (where he notes that paralysis is "cruel for a chemist"); IV, pp. 185, 200, 205–206, 224, 230, 232, 305, 329, 336, 341, 357n.1. For secondary accounts and "retrospective" diagnoses, see Friedlander 1967, and especially Heron 1982, which is the fullest and most persuasive discussion known to me.

73. Pasteur, *Correspondance*, III, p. 418.

74. Loir 1937–1938, *14*, pp. 135, 620.

75. Loir 1937–1938, *14*, pp. 189–190; and Roux [1896], 1925 translation, p. 389.

76. Roux [1896], 1925 translation, p. 389. See also Roux [1910], pp. 102–104; and René Vallery-Radot 1914.

77. See Pasteur, *Correspondance*, III, p. 91.

78. Duclaux [1896], p. 147. See also Loir 1937–1938, *14*, pp. 144, 619–621.

79. Duclaux [1896], p. 178. See also Loir 1937–1938, *14*, pp. 139–140.

80. Loir 1937–1938, *15*, p. 508.

81. Loir 1937–1938, *14*, pp. 278–280; Roux [1896], 1925 translation, p. 385; and Lagrange 1954, pp. 38–40.

82. See Pasteur, *Oeuvres*, VI, p. 167.

83. Loir 1937–1938, *14*, pp. 135–138.

84. For a striking account of Fermi's school of physics in Rome as a state-protected "family enterprise," see Holton 1978b, esp. pp. 193–196.

85. See Delaunay 1967, esp. p. 52, where he describes a succession of laboratory assistants from Eugène Viala to his nephew and then to the latter's son as a "beautiful example of the familial tradition" of the Institut Pasteur.

86. This paternalism is the central theme of Miller 1981.

87. Loir 1937–1938, *14*, p. 85.

CHAPTER THREE
THE EMERGENCE OF A SCIENTIST

1. René Vallery-Radot 1926, p. 39.
2. For a sketch of Mitscherlich, see Szabadváry 1974.
3. See Partington 1970, vol. 4, pp. 256–259, 750–751.
4. Biot 1844. Cf. Pasteur, *Oeuvres*, I, p. 323.
5. See Pasteur, *Oeuvres*, I, p. 323.
6. Pasteur, "Notes divers," 1/5/6 (see n. 10). See also "Author's Note on the Notes and Sources." The quoted phrase, later crossed out, is part of the introduction to Pasteur's first draft account of his first major discovery, written between 4 May 1848 and 15 May 1848.
7. See René Vallery-Radot 1883. The galley proofs of this little book, corrected in Pasteur's own hand, were once deposited at the Burndy Library in Norwalk, Connecticut, whose founder and director, the late Bern Dibner, very kindly allowed me to study them and to cite them in my publications. The Burndy Library is now part of the Dibner Institute for the History of Science and Technology, Cambridge, Mass.
8. Pasteur, *Oeuvres*, I, pp. 314–328, quoting from p. 326.
9. Pasteur *Oeuvres*, I, pp. 325–326.
10. See Bernal 1953, pp. 181–219, esp. pp. 195–206; and Mauskopf 1976, esp. pp. 75–77.

This first of Pasteur's laboratory notebooks, entitled "Notes divers," was once in the possession of Professor Wyatt of the Laboratoire de Mineralogie-Cristallographie, Université de Paris, who showed it to Bernal and gave a microfilm copy of it to Mauskopf. Mauskopf, in turn, made a microfiche copy of the microfilm. Mauskopf's copy is on two microfiches, and the references to "Notes divers" indicate fiche number/row number/column number. For example, "Notes divers," 1/5/6, refers to the first fiche, fifth row, sixth column. Pasteur did not himself paginate the notebook.

Bernal also offered an analysis of Pasteur's notebook, but his examination of it was limited to one day (see Bernal 1953, p. 195, n. 2), and his interpretation is sometimes dubious or incomplete. Unfortunately, Mauskopf's otherwise penetrating analysis merely refers the reader to Bernal's account of several crucial pages in the notebook. See Mauskopf 1976, esp. p. 77 and n. 49.

11. See Mauskopf 1976, pp. 79–80, quoting from p. 80.
12. On the tendency to "telescope" historical accounts of the process of discovery, see Kuhn 1970, esp. pp. 54–56. The problematic nature of discovery accounts is also emphasized in Brannigan 1981, Schaffer 1986 and Gooding 1982. For a programmatic call for rational analyses of the "context of discovery," see Nickles 1978, pp. 1–60. For an update and review, see Nickles 1985.
13. For a survey of Laurent's life and work, see Kapoor 1973. For an extensive

and excellent account of the French crystallographic tradition stemming from Haüy, see Mauskopf 1976.

14. For Laurent's theories, see Mauskopf 1976, pp. 45–51, and Kapoor 1973. Laurent's ideas evolved considerably over time; the intent here is to present them as they stood when Pasteur was working with him in 1848.

15. For a sketch of Delafosse's life and career, see Taylor 1978. For Delafosse's crystallographic concepts and their influence on Pasteur, see Mauskopf 1976, pp. 51–55, 69–70, 76. Although Pasteur's general debts to Delafosse are unmistakable, Mauskopf may somewhat exaggerate the extent and depth of Delafosse's importance for the actual discovery. For example, Mauskopf argues that Delafosse's discussion of the relation between optical activity and molecular structure in the case of quartz "led on most directly" to Pasteur's discovery (Mauskopf 1976, p. 52). But as shown later in this chapter, there is no evidence from the notebook that the issue of optical activity raised by Delafosse played much of a role until the very end of Pasteur's discovery of optical isomers in the tartrates.

A decade later, in a letter of 1857, Pasteur gave an uncharitable but arguably accurate assessment of Delafosse's work and its impact on him. He did concede—and it is a crucial concession—that he might not have recognized hemihedrism in the tartrates had Delafosse not given such "particular development and special attention" to hemihedrism in his lectures at the Ecole Normale and the Sorbonne, but "apart from that," Pasteur continued, "my researches had had nothing to do with those of my excellent professor." The rest of Pasteur's letter casts doubt on the importance and originality of Delafosse's research. It should be noted, however, that Pasteur was at the time seeking election to the mineralogy section of the Académie des sciences and therefore had reason to stress the independence and originality of his own work. See Pasteur, *Correspondance*, IV, pp. 384–386.

On Mitscherlich's use of waters of crystallization as a guide to structure, see Melhado 1980, pp. 113–120.

16. Pasteur, *Correspondance*, I, pp. 144–146, 152–153, quoting from p. 152.

17. Pasteur, *Oeuvres*, I, pp. 1–18, quoting from pp. 8, 15. In the same thesis Pasteur also wrote: "Guided by the kindly advice of M. Laurent, beside whom I had the good fortune to work, for too short a time, in the chemistry laboratory of the Ecole Normale, I undertook to prove one of the points of his theory of *acides amidés*" (ibid., p. 3), and he described the first part of his thesis as "rather the work of M. Laurent than my own" (p. 8).

18. Pasteur, *Oeuvres*, I, pp. 19–30, quoting from pp. 19–20.

19. Mauskopf 1976, p. 73.

20. Pasteur, *Oeuvres*, I, pp. 20–21, 27.

21. Pasteur, *Correspondance*, I, pp. 154–155.

22. Laurent 1847. Cf. Pasteur, *Oeuvres*, I, p. 20, and Mauskopf 1976, p. 72.

23. Pasteur, *Oeuvres*, I, pp. 35–37.

24. Pasteur, *Oeuvres*, I, p. 37.

25. Pasteur, *Oeuvres*, I, pp. 38–58.

26. Pasteur, "Notes divers," 2/5/2. Deciphering the exact order in which the pages of this notebook were written is somewhat problematic. Particularly puzzling is the fact that the rough drafts of the extended essay on dimorphism, along with other early subjects, come after the work on the enantiomorphism of the paratartrate. Thus 2/5/2, which comes *after* the dimorphism material, is almost at the end of the notebook. If the notebook had been filled front-to-back in chronological order, this page on the isomorphism of the tartrates would actually belong to a period after the famous experiments of April 1848. However, this seems highly unlikely, for the questions Pasteur sets out on this page suggest that he had not yet examined the tartrates at any great length, as he was soon to do. Internal evidence thus suggests that this passage was written at some point before April 1848.

27. Pasteur, "Notes divers," 2/5/2. Pasteur's reference is to Kopp 1840. For an easily accessible summary of Kopp's theory, see Leicester 1973.

28. Pasteur, "Notes divers," 1/1/2 through 1/2/2. "April 1848" is written on 1/2/1. The first page of this sequence comes four pages earlier than the series described by Bernal 1953, p. 195, or by Mauskopf 1976, p. 75. It is true, however, that the page with which they began their analyses ["Tartrates (questions à resoudre)," 1/3/1] is the first point at which Pasteur sets up an organized, *systematic* attack on the tartrate problem.

29. Pasteur, "Notes divers," 1/2/8–9 (sodium tartrate); 1/2/10 (paired compounds); and 1/2/11–12 (calculations). It is impossible to know for sure why Pasteur chose to start with the neutral tartrate of sodium, but it was the only one generally believed to have four waters of crystallization; the other tartrates were believed to have either one or eight waters of crystallization. The paired compounds were the sodium-ammonium tartrate (which forms a right rectangular prism) and simple ammonium tartrate (oblique rectangular prism); when mixed together, they yielded crystals of the bitartrate of ammonium (right rectangular prism). Pasteur wrote "à revoir" at the end of this page, and returned to the topic in the same notebook at 1/3/1.

30. Pasteur, "Notes divers," 1/3/1.

31. See Geison and Secord 1988, p. 18n.30.

32. Pasteur, "Notes divers," 1/3/1, transcribed in Mauskopf 1976, p. 75, n. 41. As Mauskopf notes, the projected new experiments were indicated by a marginal "x."

33. Pasteur, "Notes divers," 1/3/2.

34. Pasteur, "Notes divers," 1/3/2. Actually Pasteur listed five mixtures, but the fourth was the same as the first (namely, ammonium tartrate with sodium-ammonium tartrate).

35. Pasteur, "Notes divers," 1/3/2; transcribed in Mauskopf 1976, p. 76, n. 44.

36. Pasteur, "Notes divers," 1/3/2; transcribed in Mauskopf 1976, p. 76, n. 44. As Mauskopf notes, Bernal (1953, p. 197) partly misread this passage.

37. Pasteur, "Notes divers," 1/3/3, transcribed in Mauskopf 1976, p. 76, n. 45.

38. Pasteur, "Notes divers," 1/3/3; transcribed in Mauskopf 1976, p. 76, nn. 46–

47. Pasteur cited Hankel 1843, p. 135. On de la Provostaye and hemihedrism in ammonium tartrate, see de la Provostaye 1841, fig. 6. Bernal 1953, p. 198, mistranslated Pasteur's passage on hemihedrism in ammonium tartrate, giving it precisely the opposite meaning.

39. Pasteur, "Notes divers," 1/2/11–12, 1/3/5. See also the pertinent entries in table 1 in the text, esp. note b.

40. Pasteur, "Notes divers," 1/3/4, transcribed in Mauskopf 1976, p. 77, n. 48.

41. Pasteur, *Oeuvres*, I, p. 37.

42. Pasteur does mention Laurent explicitly at least once in his "Notes divers," referring with some skepticism to an article of 1845 in which Laurent reported that two chlorinated compounds could crystallize together in all proportions and were isomorphic.

43. Pasteur, "Notes divers," 1/3/5.

44. Pasteur, "Notes divers," 1/3/1 (for list of chemical formulas). See also table 1 in the text, above, esp. note b. In their rush to get to the material dealing with the sodium-ammonium tartrate and paratartrate, both Bernal and Mauskopf skip lightly over "Notes divers," 1/3/5 and 1/3/6, but these pages of the notebook are important because they show that Pasteur was still working out his Laurentian research plan.

45. Pasteur, "Notes divers," 1/3/7–9. The first page in this sequence is reproduced is photographically in Bernal 1953, p. 201. In his accompanying text (p. 200), Bernal mistakenly identifies the sodium-ammonium tartrate as Seignette salt, which is in fact sodium-*potassium* tartrate.

46. Pasteur, "Notes divers," 1/3/10. Bernal 1953, p. 204, reproduces this page, but neither he nor Mauskopf offers any explanation for Pasteur's return to the sodium-ammonium tartrate. In fact, however, it seems clear that Pasteur was now focusing on the precise nature of hemihedrism in the tartrates. He was by no means the first crystallographer to orient hemihedral crystals according to a consistent convention, but he had now come to recognize the importance of doing so for furthur insight into this specific problem in the tartrates.

47. Pasteur, "Notes divers," 1/3/10. See Bernal 1953, p. 203, for quotations from this same notebook page and for a strikingly different interpretation of what Pasteur had in mind at this point. In particular, Bernal assumes that Pasteur decided immediately to cross out the phrase "and sometimes all the faces repeat themselves according to the laws of symmetry"—presumably because the empirical evidence led him to see at once that he had been briefly "on the wrong track" (Bernal 1953, p. 203, n. 2). If so, Pasteur's subsequent pronouncement, "Therein lies the difference between the two salts," would refer to the difference between the right-handed hemihedry of the sodium-ammonium tartrate and the simultaneous right- *and* left-handed hemihedry of the corresponding paratartrate. There is no decisive evidence against Bernal's interpretation, since we simply do not know exactly when Pasteur crossed out the phrase in question or exactly when he wrote, "C'est là qu'est la différence des deux sels." But, in the full context of Pasteur's first major research program, as revealed by the whole of his first notebook, it seems more likely that

Pasteur only somewhat later came to doubt the symmetry that he here supposed he had "often" seen in the paratartrate crystals. If so, the difference to which he referred in the quoted statement would have had to do initially with the contrast between the right-handed hemihedry of the tartrate and the crystalline *symmetry* of the paratartrate. In either case, as already noted, Pasteur could have felt satisfied that he had resolved the third of his Laurentian anomalies; for in either case he had found a consistent crystallographic distinction between the tartrate and the paratartrate. The central point, for current purposes, is that *either* interpretation—whether Bernal's or the one preferred here—can be traced to Pasteur's brief but crucial apprenticeship under Laurent.

48. Pasteur, *Oeuvres*, I, p. 324.

49. See n. 22 above and the paragraph in the text that precedes it.

50. For a similar sort of argument in the case of the discovery of diamagnetism, see Gooding 1982.

51. I borrow the phrase "privileged material" for the tartrates from Salomon-Bayet 1986. The Salomon-Bayet "Postface" occupies pp. 255–281; my quotations of the phrase "privileged material" come from pp. 266 and 269. See also, in the same collection, the superb preface by Jean Jacques (pp. 7–45), which gives a wonderfully lucid and insightful account of past and current developments in the study of molecular asymmetry. On Pasteur's "luck" in the discovery of optical isomers in the tartrates, see esp. pp. 27 and 43. Happily, Salomon-Bayet also expresses her indebtedness (which I share) to the work of François Dagognet on Pasteur, including most pertinently here his allusion to Pasteur's work on "the production of new bodies." See Dagognet 1967 and more specifically Dagognet 1985, p. 220.

52. See Salomon-Bayet 1986, esp. the preface by Jean Jacques.

53. Pasteur, "Notes divers," 1/3/10. In reprinting and commenting on this notebook page, Bernal (1953, pp. 204–206) and even Mauskopf (1976, p. 77) make it sound as if Pasteur's discovery was now complete in every respect. But Pasteur had not yet recorded any measurement of optical activity in the second hemihedral form in the paratartrate nor established its identity with the naturally occurring tartrate.

54. Pasteur, *Oeuvres*, I, pp. 20–21. See Mauskopf 1976, pp. 55–68, for an extensive discussion of Biot's conception of the relation between optical activity and molecular constitution.

55. Pasteur, *Oeuvres*, I, p. 64.

56. Pasteur, "Notes divers," 1/4/12; see also ibid., 1/3/11–1/5/6.

57. For a list of intended projects, see Pasteur, "Notes divers," 1/5/1. Among them was "Transformation des tartrates en paratartrates (chaleur-acid sulfurique)," indicating that even at this early date Pasteur was considering the possibility of transforming an asymmetric compound into its racemate. For more on this issue, see Pasteur, *Oeuvres*, I, pp. 258–262. See also Kottler 1978, esp. pp. 70–79.

58. Compare Mauskopf 1976, pp. 77–78, with Pasteur, *Oeuvres*, I, pp. 61–64, 77–80. For the sudden decline in Pasteur's expressions of indebtedness to Laurent, see the index in ibid., VII, p. 461, where Laurent's name appears in thirteen entries

before the discovery and only once in relevant material published during the rest of Pasteur's life.

59. Pasteur, *Correspondance*, I, p. 152.

60. Pasteur, *Correspondance*, I, p. 236.

61. On the contrast between the theories of Dumas and Laurent, see Mauskopf 1976, pp. 44–48.

62. For a sketch of Dumas's work and career, see Kapoor 1971. For a more extensive account, with special attention given to Dumas's students, see Klosterman 1985. More generally on the social system of nineteenth-century French science, see Fox 1976 and Fox and Weisz 1980, and (for a slightly earlier period) Outram 1984.

63. On Laurent, see Kapoor 1973. On Pasteur's political views, see, e.g., Pasteur, *Correspondance*, I, pp. 228–230; II, pp. 216–236, 345–351, 484–489, 567–568, 611–630; Pasteur Vallery-Radot 1954, pp. 203–215; Maurice Vallery-Radot 1982, and Farley and Geison 1974, esp. pp. 186–188.

CHAPTER FOUR
FROM CRYSTALS TO LIFE

1. Pasteur, *Oeuvres*, II, pp. 3–13.

2. See Geison 1974, p. 362. For a fascinating "partial semiotic analysis" of Pasteur's memoir on lactic acid, the burden of which is to make the "lactic ferment" no less an "actor" than Pasteur, see Latour 1992.

3. René Vallery-Radot 1926, p. 83.

4. See Pasteur, *Oeuvres*, I, p. 376, where he states that in moving from crystallography through fermentation to disease, he had been "enchained . . . by the almost inflexible logic of my studies."

5. Pasteur, *Oeuvres*, VII, pp. 129–132.

6. René Vallery-Radot 1926, p. 79.

7. De Kruif 1959, p. 59.

8. Bernal 1953, p. 82.

9. Cf. Dubos 1988, pp. 41–42, with Dubos 1950, pp. 41–42, where the name Bigo does not appear.

10. Paul 1985, pp. 141–142.

11. See Pasteur, *Cahier 10*, fols. 31–41 passim, 61, 70, 85 et seq. (all in BN pagination).

12. In this more limited sense, Bernal is right to insist on the industrial context for Pasteur's early work. Bernal 1953, p. 187.

13. Pasteur, *Oeuvres*, II, p. 3.

14. Pasteur, *Oeuvres*, II, pp. 4, 25–28.

15. Pasteur, *Oeuvres*, I, pp. 275–279; and *Cahier 3*, fols. 77–66, esp. fols. 74v, 72v (using BN pagination, which inverts the chronological order).

16. Pasteur, *Oeuvres*, I, pp. 275–279.

17. See Pasteur, *Cahier 7*, which covers all of 1855 and is entitled "Amy Alcohol"; and Papiers Pasteur, *Oeuvres*, I, fols. 92–117, which is a draft of the paper published in August 1855. Particularly noteworthy is this passage (fol. 96v), which was deleted from the published version of the paper: "The simplicity of the results that I have the honor of communicating to the Académie [des sciences] will conceal from everyone the difficulties that I encountered in the course of my work. I arrived at the procedure indicated here only through innumerable trials that had all been fruitless."

18. Pasteur, *Oeuvres*, I, pp. 284–288.

19. Pasteur, *Cahier 1*, fol. 77v. My emphasis.

20. For example, Pasteur, *Cahier 1*, fol. 7; *Cahier 2*, fols. 19–20.

21. For example, Pasteur, *Cahier 1*, fols. 7–7v.

22. For example, Pasteur, *Cahier 1*, fols. 14–16.

23. For example, Pasteur, *Cahier 2*, fols. 9, 11v, 14; *Cahier 2*, fols. 19, 22.

24. For example, hydrochloric acid (Pasteur, *Cahier 1*, fols. 7v et passim); nitric acid (*Cahier 1*, fol. 19v; *Cahier 2*, fol. 2); boric acid, *Cahier 1*, fol. 18v; *Cahier 2*, fols. 24, 27, 63.

25. For example, Pasteur, *Cahier 2*, fol. 27v.

26. For example, *Cahier 1*, fol. 13v; *Cahier 2*, fols. 29v, 35, 41, 50–52, 55.

27. Pasteur, *Oeuvres*, I, pp. 284–288.

28. See, for example, the rapid decline of crystal drawings and angle measurements in Pasteur, *Cahier 6* and thereafter.

29. Duclaux [1896], 1920 translation, pp. 67–69.

30. See Geison and Secord 1988; Mauskopf 1976.

31. Pasteur, *Cahier 2*, fol. 64 using the BN pagination; fol. 62 in Pasteur's handwriting.

32. Pasteur, *Cahier 2*, fol. 65 (BN) or 63 (Pasteur).

33. Pasteur, *Oeuvres*, I, pp. 160–188.

34. Pasteur, *Oeuvres*, I, pp. 198–202.

35. Pasteur, *Oeuvres*, I, pp. 334–336. But see Huber 1969, esp. pp. 40–58.

36. Pasteur, *Oeuvres*, I, pp. 160–188.

37. Pasteur, *Oeuvres*, II, pp. 18–22.

38. Pasteur, *Oeuvres*, II, pp. 25–28.

39. Pasteur, *Oeuvres*, I, pp. 314–344, but also Salomon-Bayet 1986b.

40. Pasteur, *Oeuvres*, I, pp. 314–344.

41. Pasteur, *Oeuvres*, I, p. 341.

42. Pasteur, *Oeuvres*, I, p. 337.

43. Pasteur, *Oeuvres*, I, pp. 333–334.

44. Pasteur, *Oeuvres*, I, p. 343.

45. For a valuable and deeply informed background, see Fruton 1972.

46. Duclaux [1896], 1920 translation, p. 73.

47. Pasteur, *Oeuvres*, II, pp. 71–72.

48. Pasteur, *Oeuvres*, II, esp. pp. 102–113.

49. Geison 1981. For critiques, see Temple 1986 and Latour 1992.

50. See Fruton 1972, and Kohler 1971.
51. Fruton 1972, esp. p. 58.
52. Pasteur, *Oeuvres*, II, p. 224.

CHAPTER FIVE
CREATING LIFE IN NINETEENTH-CENTURY FRANCE

Important parts of this chapter are drawn from a controversial article that John Farley and I published twenty years ago (see Farley and Geison 1974). In the meantime, that article has been the object of considerable commentary and criticism, both positive and negative, both published and unpublished. For examples of generally positive commentary, see Hesse 1980, esp. pp. 34–35; Collins 1981; and Shapin 1982, esp. pp. 190–192. The critiques, sometimes severe, include Kottler 1978; Roll-Hansen 1979, 1983; Gálvez 1988; and Latour 1989, 1992.

This is not the place to respond in detail to the critics. It is, however, worth noting that Roll-Hansen's critiques, which he offers in defense of traditional scientific method and "rationalism," amount to a rehash of the debate as Pasteur and the Académie des sciences saw it. Roll-Hansen offers no novel evidence or documents in support of his position. Indeed, I would go so far as to say that Roll-Hansen and I do not disagree on a single fact of consequence. We simply disagree about which claims and issues deserve attention, credence, and emphasis. What Roll-Hansen's critiques actually and inadvertently show is that logic and the bare "facts of the matter" do no more to decide between rationalist and relativist interpretations of the Pasteur-Pouchet debate than they did to determine the outcome of that debate itself. Gálvez, for his part, joins Roll-Hansen in defense of the "objectivity" of the Académie des sciences commissions on spontaneous generation, mainly by insisting that the Académie had long been concerned with the issue of plant and animal generation and that the Pasteur-Pouchet debate should be seen in that context. I find his argument interesting and suggestive but ultimately unconvincing.

Collectively, however, the critics have persuaded me that the original Farley-Geison interpretation was rather too crudely "externalist" in form and asymmetrically tilted in Pouchet's favor. We tended to overlook or excuse Pouchet's own violations of the Scientific Method and to minimize the role of "external" factors in his case, while emphasizing both in the case of Pasteur. Gálvez's pointed critique of Pouchet's work on embryology and fertilization is particularly pertinent, and it has been reinforced through my additional research in the Pasteur collection at the Bibliothèque Nationale, which includes a lengthy correspondence between Pouchet and his collaborators Joly and Musset. A small but significant portion of that correspondence was published long ago, I now realize, by Pasteur Vallery-Radot (see Pasteur Vallery-Radot 1954, pp. 66–69). This correspondence is the opposite of helpful to the cause of Pouchet and his collaborators, for it reveals an increasingly desperate, even pathetic, trio of woolly-headed zealots who were at least as nasty toward Pasteur as he was toward them. Indeed, I am now almost tempted to agree with Pasteur Vallery-Radot's judgment that this collection of letters suggests that Pouchet was more than a bit paranoid. All of this evidence, together with a re-reading of Emile Duclaux's typically fair assessment of the debate (see Duclaux [1896], pp. 104–111), underscores the extent to which Pouchet was not in the same scientific league as Pasteur, at least as an experimentalist.

The tone of this chapter is therefore quite different from that of the Farley-Geison article of 1974. In particular, Pouchet is assigned a distinctly less prominent role and is less vigorously defended for his now discredited stand. This is basically a story about Pasteur, not Pouchet. None of this is meant to suggest, however, that I here restore Pasteur to the traditional "internalist" throne on which Roll-Hansen tends to place him. My interpretation of the debate remains fundamentally contextual, and Pasteur's religious and political commitments are still very much in play.

1. For the text of this lecture, see Pasteur, *Oeuvres*, II, pp. 328–346.

2. On Pouchet's life and career, see Farley and Geison 1974; and esp. Gálvez 1988.

3. Pouchet 1847, Law 2.

4. Pouchet 1853.

5. Pouchet 1859b.

6. Pouchet 1859b, pp. 7–9.

7. Pouchet 1859b, pp. 97–98.

8. Pouchet 1859b, pp. 127–128.

9. Pouchet 1858.

10. Pasteur, *Oeuvres*, II, pp. 34–36.

11. Pasteur, *Oeuvres*, II, pp. 628–630.

12. Pasteur, *Oeuvres*, II, p. 246.

13. Pasteur, *Oeuvres*, II, pp. 187–191.

14. Pasteur, *Oeuvres*, II, p. 191.

15. Duclaux [1896], 1920 translation, p. 107; and Pasteur, *Oeuvres*, II, p. 190.

16. Pasteur, *Oeuvres*, II, pp. 192–196.

17. Cf. Pasteur, *Oeuvres*, II, p. 343.

18. Pasteur, *Oeuvres*, pp. 197–201, 202–205.

19. See Farley 1977, pp. 116–117; and Geison 1991.

20. Pasteur, *Oeuvres*, II, pp. 197–201, 295–317.

21. Pasteur, *Oeuvres*, II, p. 337. My emphasis.

22. Pasteur, *Oeuvres*, II, p. 342.

23. Pasteur, *Correspondance*, II, p. 134.

24. Pasteur, *Oeuvres*, II, pp. 345–346.

25. Pasteur, *Oeuvres*, II, p. 346.

26. For details of this period, see Farley 1972.

27. See, e.g., Coleman 1964.

28. Appel 1987.

29. Williams 1953.

30. For examples of Geoffroy's attempts to defend himself from charges of alle-

giance to *Naturphilosophie*, materialism, and impiety, see *Notions synthetiques*, pp. 26, 33, 82, 110; *Comptes rendus* 5 (1837): 183–194; ibid., 7 (1839): 489–491; and "Hérésies panthéiestiques," *Dictionaire de la conversation et de la lecture* 31 (1836): 481ff.

31. The naturalistic basis of French geology by the mid-nineteenth century is discussed in Rudwick 1972, chap. 3.

32. This brief summary of the politico-theological issues during the Second Empire is based mainly on Charlton 1963, Dansette 1961, and Wright 1966.

33. Quoted in Dansette 1961, p. 311.

34. Guérard 1920.

35. Faivre 1860, p. 172.

36. Guizot 1862, p. 18.

37. Royer 1862. Details of the French Darwinian debate and its association with the issues of spontaneous generation are given in Farley 1974.

38. Owen 1868, p. 814.

39. Pennetier 1907, p. 10.

40. Pennetier 1907, p. 10.

41. On coverage of the Pasteur-Pouchet debate by the French press, both "popular" and "scientific," see Galérant 1974; Diara 1984; and Bensaude-Vincent 1991.

42. Pouchet, Joly, and Musset 1863.

43. M.J.P. Flourens, *Comptes rendus* 57 (1863): p. 845.

44. Milne-Edwards 1859, p. 24.

45. On this episode, see Bulloch 1938, pp. 103–105; Duclaux 1896, pp. 104–109; Pennetier 107, pp. 10–12; and Pasteur, *Oeuvres*, II, pp. 321–327, 637–647.

46. Pennetier 1907, p. 12.

47. Flourens 1864, p. 170.

48. See Farley 1974.

49. Diara 1984, pp. 203–204.

50. Farley 1977, esp. chap. 7. In addition, my student, James Strick, is engaged in doctoral research on Bastian and the spontaneous generation debates of the 1870s.

51. As quoted by Farley 1977, p. 137.

52. As quoted by Farley 1977, p. 140.

53. Farley 1977, pp. 138–140.

54. Duclaux [1896], 1920 translation, p. 141.

55. Pasteur, *Oeuvres*, II, pp. 345–346.

56. Pasteur, *Oeuvres*, II, pp. 321–323. For an attempt to defend Pasteur's claim, see Roll-Hansen 1979.

57. See Pasteur, *Oeuvres*, II, pp. 637–647.

58. See Pasteur, *Oeuvres*, II, p. 459; VI, pp. 25n.1, 41, 54, quote on p. 54.

59. See Duclaux [1896], 1920 translation, pp. 109–111.

60. See Vandervliet 1971, pp. 43–54. On the exact relation between Cohn's work and Tyndall's, see Geison 1971, p. 340n.3.

61. Pasteur, *Oeuvres*, II, pp. 253–259.

62. Pasteur, *Oeuvres*, II, p. 337.

63. Pasteur, *Oeuvres*, II, pp. 321–323.

64. Owen 1868, p. 814.

65. Latour 1984, 1988, 1989, 1992.

66. Latour 1988.

67. See, e.g., Duclaux [1896], 1920 translation, p. 85.

68. Pasteur's acknowledgment of and appreciation for the power of preconceived ideas is particularly evident in the "ghost-written autobiography" by his son-in-law; see René Vallery-Radot 1883.

69. Duclaux [1896], 1920 translation, esp. pp. 86–87.

70. Pasteur, *Cahier 3*, fol. 3v.

71. Pasteur, *Oeuvres*, I, p. 155.

72. See Pasteur, *Oeuvres*, I, pp. 345–387.

73. Pasteur, *Oeuvres*, I, pp. 364–365; Pasteur's italics.

74. Pasteur, *Oeuvres*, I, p. 386.

75. Pasteur, *Oeuvres*, I, p. 376.

76. Dubos 1950, pp. 112–114, quote on. p. 114.

77. Pasteur, *Correspondance*, I, p. 227.

78. Pasteur, *Correspondance*, I, p. 324.

79. Pasteur, *Correspondance*, I, p. 326.

80. See Pasteur Vallery-Radot 1968, pp. 10–11. I have also benefited here and for the next couple of paragraphs from an unpublished paper by my student Robert Root-Bernstein, who has made the closest study of these early notebooks known to me. See Root-Bernstein 1979. He made use of a small part of this research in Root-Bernstein 1989. I have also examined the early notebooks, as has my student James Strick. I thank him and Root-Bernstein for sharing their research and ideas with me.

81. Pasteur, *Cahier 6*, fols. 17, 22–22v.

82. See Pasteur, *Oeuvres*, I, p. 292.

83. Pasteur, *Cahier 6*, fol. 37.

84. Pasteur, *Oeuvres*, VII, p. 23.

85. Pasteur, *Oeuvres*, I, p. 362.

86. Pasteur, *Oeuvres*, I, pp. 364–365.

87. Pasteur, *Oeuvres*, I, pp. 377–378. My emphasis.

88. Pasteur, *Ouevres*, VI, pp. 26–28.

89. Pasteur, *Oeuvres*, VI, pp. 37–58.

CHAPTER SIX
THE SECRET OF POUILLY-LE-FORT

1. *The Times* (London), 3 July 1880, p. 5.

2. Pasteur, *Oeuvres*, VI, p. 371.

3. For two extensive accounts of the Pouilly-le-Fort trial in the secondary literature, see René Vallery-Radot 1926, pp. 313–325; and Nicol 1974, pp. 365–389. A

quasi-official and exquisitely detailed account was published by the sponsoring Agricultural Society of Melun; see Rossignol 1881. A substantial part of this now rare brochure of ninety-five pages is conveniently reprinted in Pasteur, *Oeuvres*, VI, pp. 697–720. An original copy of the complete brochure was deposited at the Burndy Library in Norwalk, Connecticut, since relocated to the Dibner Institute for the History of Science and Technology, Cambridge, Mass. I am grateful to the late Bern Dibner for granting me access to this important document.

4. See Pasteur, *Correspondance*, III, pp. 196–199.

5. Nicolle 1932, pp. 62–65; cf. René Vallery-Radot 1926, pp. 320–321.

6. Roux [1896], 1925 translation, p. 379.

7. Among other things, the protocol stipulated that ten additional sheep were to be kept aside as controls, to be compared with the twenty-five vaccinated sheep at the end of the experiments. At the request of the Agricultural Society of Melun, Pasteur also ultimately agreed to substitute two goats for two of the fifty experimental sheep and to extend the trial to ten cows, of which six were to be vaccinated. Although somewhat less confident of the outcome in the case of these cows, Pasteur did predict that the six vaccinated cows would remain healthy when injected with the virulent anthrax culture, while the four unvaccinated cows would die or at least become very ill. In addition, the protocol called for a subsequent experiment (to take place a year later) in which healthy sheep would be kept in an enclosure above the buried carcasses of the dead (unvaccinated) sheep, "in order to prove that the new sheep will become spontaneously infected by anthrax germs which will have been carried to the surface of the soil by earthworms." See Pasteur, *Oeuvres*, VI, pp. 346–351; and Rossignol 1881. As we shall see, Pasteur and his collaborators also departed from the signed protocol by injecting four of the experimental sheep with virulent anthrax in advance of the scheduled date for such injections.

8. Pasteur, *Oeuvres*, VI, p. 348.

9. Pasteur, *Oeuvres*, VI, pp. 349–350. Much of the discussion that followed Pasteur's address at the Académie des sciences concerned the death and autopsy of the vaccinated and pregnant ewe. See ibid., VI, pp. 351–357.

10. Pasteur, *Oeuvres*, VI, pp. 348–349.

11. Pasteur, *Oeuvres*, VI, pp. 350–351. My emphasis.

12. Pasteur, *Oeuvres*, VI, pp. 323–330, esp. pp. 328–329. For more of Pasteur's speculations on natural epidemics, see ibid., VI, pp. 337–338.

13. Pasteur, *Oeuvres*, VI, pp. 332–338. Among other things, Pasteur insisted here (p. 332) that "we have applied all our efforts to search the possible generalization of the action of atmospheric oxygen in the attenuation of viruses" and described this vaccinal culture of anthrax bacilli as having been cultivated at 42°–43° C. in contact with "pure air" (p. 333). He gave the same description of his method a month later, on 21 March 1881, just six weeks before the Pouilly-le-Fort trial began. See ibid., VI, p. 343.

14. Pasteur, *Oeuvres*, VI, pp. 358–369.

15. Pasteur, *Oeuvres*, VI, p. 368: "ce microbe [i.e., the microbe of saliva] s'attenué

également par l'action de l'oxygène de l'air." Earlier in this memoir, Pasteur wrote explicitly of "applying the method of which I just spoke, the influence of the oxygen of the air, to the anthrax parasite" (p. 362). For more on the "microbe of saliva," see Chapter Seven of this book.

16. Hippolyte Rossignol himself clearly assumed that an oxygen-attenuated vaccine had been used in the famous experiments at his farm. See Rossignol 1881, esp. pp. 10–11. Among the secondary sources that adopt the same assumption is my own monographic essay in the *Dictionary of Scientific Biography* on Pasteur; see Geison 1974, esp. pp. 392–395. That essay was published before I began my study of Pasteur's laboratory notebooks. For the few exceptions that have challenged this standard version of the story of Pouilly-le-Fort, see note 18 below.

17. Loir 1937–1938, *14*, pp. 91–92.

18. Until the investigations of Pasteur's laboratory notebooks by Antonio Cadeddu (see below) and me, there were, to my knowledge, only five authors who adopted Loir's claim that an antiseptic vaccine was used at Pouilly-le-Fort: (1) Lagrange 1954, pp. 43–48; (2) Ramon 1962; (3) Théodoridès 1968, pp. 119–120, and then again in Théodoridès 1977; (4) Decourt 1974b; and (5) Reid 1975. Of these works, Théodoridès's book on Davaine is the most scholarly. It cites Loir 1937–1938, *14*, pp. 91–92, as the original source for this "revisionist" account. Reid's book, explicitly directed at a popular audience, contains no bibliography and no footnotes whatever. Its main interest is that is was the "companion volume" for the superb television series *Microbes and Men* by the British Broadcasting Corporation, originally broadcast in 1974. For my recent review of this video series, see Geison 1993.

19. Dubos 1950.

20. Roux [1896], 1925 translation, pp. 377–379. Admittedly Roux did not say in so many words, "The Pouilly-le-Fort vaccine was oxygen-attenuated," but that is the unmistakable impression that he (and Pasteur) conveyed. This impression emerges even more clearly—almost explicitly—in Roux's Croonian Lecture of 1890. See Roux 1890a, esp. p. 161. Lagrange claims that Roux in 1890 "published the 'secret of Pouilly-le-Fort,' the 'asporogenous anthrax bacillus'—where he showed how the addition of potassium bichromate or carbolic acid to cultures renders them *definitively* asporogenous" (quote from Lagrange 1954, p. 48). In this second paper of 1890 Roux does indeed discuss this effect of carbolic acid (though not potassium bichromate) on anthrax cultures, but he conspicuously omits any reference to the Pouilly-le-Fort trial. Far from publishing "the secret of Pouilly-le-Fort" here or elsewhere, Roux continued to keep it. See Roux 1890b.

21. For the first published verification of Loir's account, based on Pasteur's laboratory notes, see Cadeddu 1987. By the mid-1970s, when I began to disclose my findings in public lectures, I too had confirmed Loir's account from my analysis of the same notebook entries. This chapter does, however, represent the first published account of my analysis and interpretation of the secret of Pouilly-le-Fort. Cadeddu and I are in full agreement on this basic fact of the matter: the vaccine Pasteur used

at Pouilly-le-Fort had been attenuated by potassium bichromate rather than oxygen, as Pasteur implied in his publications. We do, however, put this finding to rather different uses. Cadeddu's central aim is philosophical or epistemological, with special attention given to the complexities of the process of discovery. My aim is more historical; specifically, I seek to explain Pasteur's deception in terms of his competition with Toussaint in the race for a safe and effective anthrax vaccine. Cadeddu mentions the Toussaint affair, but only in passing. For a separate account, in Chamberland's own hand, of the major results of his experiments with potassium bichromate as of 18 February 1881, see Papiers Pasteur, *Oeuvres*, VI, "Maladies Virulentes," pp. 129–129v, translated in Appendix J at the back of this book.

22. Pasteur, *Cahier 91*, fols. 106v, 107 (according to Pasteur's handwritten pagination; fols. 108–109 in BN stamped pagination).

23. Pasteur, *Correspondance*, III, pp. 196–197 (on the "foretaste" of success); Nicolle 1932, pp. 62–65 (on the anxious final moments before the triumph at Pouilly-le-Fort).

24. Loir 1937–1938, *14*, p. 92.

25. See Pasteur, *Cahier 91*, fol. 113 (in Pasteur's pagination; fol. 115 in BN staff pagination).

26. Pasteur's belief that Chamberland's vaccines were produced by oxygen attenuation is clear from an interesting episode described by Loir 1937–1938, *14*, pp. 88–90. During one of Pasteur's rare visits to the annex on the rue Vauquelin—while Chamberland and his research assistant were both away on holiday—Loir prepared the commercial vaccines according to Chamberland's instructions. As Pasteur watched Loir do so, he suddenly realized that Chamberland had been adding a "foreign" culture of the hay bacillus (*bacillus subtilis*) to his attenuated anthrax vaccines. It seems that Pasteur's initial surprise and possible irritation at this alteration in technique—which Chamberland had introduced without his knowledge—evaporated when he conceived of an explanation for it that was in keeping with his oxygen theory of attenuation. Specifically, Pasteur suggested that the hay bacillus might absorb any free oxygen that remained in the sealed tubes above the vaccinal liquid and thus help to maintain a fixed level of attenuation in the anthrax culture. As Pasteur now interpreted it, Chamberland's alteration in technique was a justifiable "precaution." But Dubos has suggested that the addition of the hay bacillus to Pasteur's commercial anthrax vaccines may well have been responsible for complaints about their "impurity" and may even have led to the occasional failure of these vaccines to produce immunity in sheep. See Dubos 1950, pp. 341–342.

27. On the criticisms, occasional failures, and overall success of Pasteur's commercial anthrax vaccines, see Geison 1974, pp. 395–398.

28. This possibility was suggested to me by my Princeton colleague, philosopher David Lewis, when I presented an earlier version of this chapter in the colloquium series of the Program in History and Philosophy of Science at Princeton University on 1 December 1979.

29. Chamberland and Roux 1883, pp. 1088–1091, 1401–1412.

30. I do not mean to suggest that Toussaint has been entirely ignored. Indeed, two extensive and valuable accounts of the Pasteur-Toussaint rivalry appeared a decade ago: Nicol 1974, esp. pp. 174–176, 214–224, 277–281, 291–389; and Wrotnowska 1975. Nicol's detailed account consists largely of extensive quotes from the correspondence and published papers of the participants in the story, including notably the monthly commentaries of Henri Bouley in the *Recueil de médecine vétérinaire*. It makes no use of Pasteur's laboratory notebooks and, despite occasional resort to Loir 1937–1938, does nothing to dispute—indeed, it tends rather to confirm—the standard story that the Pouilly-le-Fort vaccine was produced by oxygen attenuation. More generally, Nicol's book lacks an adequate scholarly apparatus and often fails to distinguish between published and unpublished sources. For my review, see Geison 1977.

Curiously, Wrotnowska's article also ignores Loir's version of the Pouilly-le-Fort vaccine, despite her careful examination of Pasteur's laboratory notebooks on anthrax, which she cites and quotes with some frequency. She refers to the Pouilly-le-Fort trial only in passing and without conveying the contents of the crucial laboratory pages reproduced and analyzed here. More generally, despite the valuable new information she provides on Toussaint and his relations with Pasteur and on Pasteur's early work on anthrax, Wrotnowska's account reflects her long-standing tendency to celebrate uncritically everything about Pasteur and his work.

In short, neither Nicol nor Wrotnowska establish or even discuss the "secret" of Pouilly-le-Fort. On the other hand, they do cover some of the same background, as I have tried to acknowledge in my frequent citations of their work (especially Nicol's) below.

31. Pasteur, *Oeuvres*, VI, pp. 291–303.

32. Pasteur, *Oeuvres*, VI, pp. 359, 495.

33. Pasteur, *Oeuvres*, VI, pp. 289–303 (esp. p. 303n.3), 358, 495; VII, pp. 48–51. See also Nicol 1974, pp. 277–281; and Wrotnowska 1978, pp. 268–275.

34. Pasteur, *Oeuvres*, VI, p. 298.

35. See Pasteur, *Oeuvres*, VI, pp. 303–312, esp. pp. 304–305.

36. Nicol 1974 focuses on Bouley and gives a detailed account of his relations with Toussaint. On Bouley's visit to Toulouse in early July 1880, see Nicol 1974, p. 297.

37. See Nicol 1974, p. 297n.l.

38. Colin 1880, pp. 650–670, with subsequent discussion on pp. 671–677. For Bouley's reference to Toussaint's new vaccine and Colin's response, see pp. 674–676. See also Nicol 1974, p. 298.

39. Toussaint 1880a; Nicol 1974, pp. 298–299.

40. "Observations à l'occasion du procès-verbal," *Bull. Acad. méd.*, 2d series, 9 (1880): 753–756. For a fuller account of this meeting, based mainly on a long and previously unpublished letter from Bouley to Pasteur, see Nicol 1974, pp. 299–308.

41. See Pasteur, *Correspondance*, III, pp. 145–155, 158–162; and Nicol 1974, pp. 313–323.

42. Nicol 1974, pp. 308–309.

43. Toussaint 1880b.

44. Pasteur, *Correspondance*, III, pp. 158–160.

45. Pasteur, *Oeuvres*, VI, pp. 290–291.

46. For a brief account of Chauveau's influence and his own work on anthrax, see Nicol 1974, pp. 291–294, 437–441. For Toussaint's initial conclusion that "a substance secreted by a parasite . . . would be vaccinal for the disease provoked by that parasite itself," see Toussaint 1881, p. 1023. For the stark contrast that Pasteur drew between this theory and his own "biological" conception of vaccines, see Pasteur, *Oeuvres*, VI, p. 340.

47. Nicol 1974, p. 334 et passim.

48. Nicol 1974, pp. 327–335.

49. Roux's long and previously unpublished letter of 19 August 1880 is printed in Nicol 1974, pp. 331–333, and translated in Appendix D at the back of this book.

50. See the letters of Bouley, Pasteur, and Roux printed in Nicol 1974, pp. 334–353.

51. Toussaint 1881, p. 1025.

52. The statutes and rules of the French Association for the Advancement of Science at the time Toussaint gave his address at Reims are printed in its *Compte rendu* of the session, Reims 1880 (Paris, 1881), pp. iii–xiv. Articles 60–63 of the rules (pp. xiii–xiv) concern the submission of manuscripts for the annual published volume of Association meetings, which was to appear "ten months or more after the session to which it corresponds," while the deadline for submission of manuscripts was 1 December.

53. Toussaint 1881, pp. 1024–1025, refers to the results of experiments extending from 22 August into late October or early November 1880.

54. Pasteur, *Oeuvres*, VI, p. 342, including n. 1.

55. Wrotnowska 1978 discusses the relations between Pasteur and Toussaint at considerable length but in a way that essentially repeats Pasteur's claims of priority and superiority over Toussaint. While Pasteur's (and thus Wrotnowska's) version of the relation between his and Toussaint's sometimes competing work on chicken cholera and anthrax does seem generally persuasive, it should nonetheless be compared with the more balanced accounts that emerge from Théodoridès 1973 and 1977, and the documents and spare commentary printed in Nicol 1974, esp. pp. 214–224, 277–281, 334–353. For Pasteur's private and wholly negative assessments of Toussaint's priority claims, see Pasteur, *Cahier* 92, fols. 7–8, 20–21.

56. Pasteur, *Oeuvres*, VI, pp. 323–330.

57. Pasteur, *Oeuvres*, VI, p. 327.

58. Cadeddu 1985.

59. Pasteur, *Oeuvres*, VI, pp. 332–338. In 1882 and 1883, Auguste Chauveau investigated the respective roles of oxygen and heat in attenuating the anthrax bacillus, seeking (with mixed success) to produce an effective and stable vaccine by modifying Toussaint's procedure of heating anthrax blood. See Nicol 1974, pp. 439–

441. Remarkably enough, Pasteur himself later undertook similar experiments. In 1888, he reported occasional success in attempts to vaccinate rabbits by heating anthrax blood at temperatures between 40° and 45° C for periods of two to nine days. Almost astonishingly, given his earlier rejection of Chauveau's and Toussaint's chemical theory of immunity, Pasteur now wrote that the "attentive reader" of these results would be left in no doubt that "the anthrax parasite . . . is associated with a vaccinal chemical matter in the anthrax blood." In all of this Pasteur was using procedures and drawing conclusions very similar to Toussaint's earliest work, though he declined to mention his sometime rival. Pasteur, *Oeuvres*, VI, pp. 462–466, quote on p. 464.

60. Pasteur, *Oeuvres*, VI, p. 335.

61. Pasteur, *Oeuvres*, VI, p. 342.

62. Roux [1896], 1925 translation, p. 378.

63. Théodoridès 1968, esp. pp. 109–115. See also note 65 below.

64. See Wrotnowska 1978, p. 276.

65. Davaine had used antiseptics to treat anthrax blood in vitro and cases of clinical anthrax in vivo, but had not sought to develop an antiseptic *vaccine* against anthrax. Nor had he sought a vaccine when claiming still earlier (in 1863 and 1864) that anthrax blood could be rendered "inoffensive" by heating at 55° C for ten minutes. He seems to have had mainly therapeutic rather than preventive goals in mind. See Théodoridès 1968. Toussaint's claim of August 1880 was thus a novel one, as was his use of carbolic acid, which had not been included among the antiseptics tested by Davaine. Toussaint nonetheless cited Davaine's work as the inspiration for the procedures he was now deploying toward a different goal. See Toussaint 1881, pp. 1023–1024.

66. These two letters, deposited among the Papiers Pasteur at the Bibliothèque Nationale, are printed in full in Nicol 1974, pp. 331–333 and 343–349. Partly because Nicol's book is not readily available, and partly to increase their accessibility to English-language readers, I have provided translations of Roux's letters in Appendixes D, H, and I at the back of this book.

67. Bouley's letter of 27 August 1880 is printed in Nicol 1974, pp. 350–351; and, with some deletions, in Pasteur, *Correspondance*, III, p. 168n.l. A week later, in a letter of 3 September, Bouley informed Pasteur of new experiments showing that the surviving sheep at Alfort had indeed been rendered immune to anthrax. Nicol 1974, pp. 350–351.

68. See Appendix D.

69. See Appendix I.

70. Nicol 1974 suggests in passing (p. 345n.l and p. 353) that Toussaint's experiments with the inactive culture given him by Roux may have put Pasteur "on the path to success" in his search for an effective anthrax vaccine. But Nicol fails to note that the most likely effect of this episode was to intensify and redirect Chamberland's experiments with agents other than atmospheric oxygen. It is hard to see how Roux's suggestion that Chamberland may have created vaccinal cultures by using

gasoline vapors could have given impetus to Pasteur's ongoing search for an oxygen-attenuated vaccine, as Nicol implies. It is vastly more likely that Toussaint's results with Chamberland's "inactive" culture put Pasteur's collaborators on a very different "path to success" that led them toward the potassium bichromate vaccine used at Pouilly-le-Fort.

71. See Loir 1937–1938, *14* (1937), p. 92.

72. Pasteur, *Oeuvres*, VI, p. 335.

73. See Nicol 1974, pp. 184–185, 365–368.

74. See esp. Pasteur, *Cahier 91*, fols. 34v-35, 42–42f.

75. Pasteur, *Cahier 91*, fol. 54.

76. See esp. Pasteur, *Cahier 91*, fols. 78–78v.

77. Pasteur, *Cahier 91*, fols. 102–103.

78. Loir 1937–1938, *14* (1937), p. 91.

79. Loir 1937–1938, *14* (1937), pp. 90–91; Lagrange 1954, pp. 43–48; and Rossignol 1881, passim.

80. See the sources in note 79, and especially Nicolle 1932, pp. 62–65.

81. Pasteur, *Correspondance*, III, pp. 158–164.

82. The letter is quoted in full on pp. 162–163.

83. Pasteur, *Oeuvres*, VI, pp. 298–299, 350–351.

84. In August 1880, the same month in which Toussaint disclosed the *modus fasciendi* of his heat-produced anthrax vaccine, Pasteur and his collaborators tested the procedure on six sheep as follows: two sheep were inoculated with *three* cubic centimeters of anthrax blood heated at 55° C. for ten minutes; two sheep were inoculated with *five* cubic centimeters of anthrax blood heated at 55° for ten minutes; and the last two sheep were inoculated with *six* cubic centimeters of anthrax blood heated at 55° for twenty minutes. The results were as follows: both of the sheep inoculated with three cubic centimeters of heated anthrax blood survived the initial injection; both inoculated with five cubic centimeters of blood died; and one of the two sheep injected with six cubic centimeters of blood heated for twenty minutes survived. All three of the surviving sheep had been quite sick. Subsequent inoculation with virulent anthrax showed that they had been rendered immune to anthrax. Almost needless to say, these results, which could have been used in support of the efficacy of Toussaint's vaccine under certain conditions (specifically doses of three cubic centimeters), were not disclosed outside of the immediate Pastorian circle. They are recorded in a manuscript deposited at the Bibliothèque Nationale: Papiers Pasteur, *Oeuvres*, VI, "Maladies virulentes," p. 129. If Pasteur and his collaborators undertook other tests of Toussaint's procedures, I have thus far failed to locate any record of them.

85. Toussaint apparently used only heat and carbolic acid in his attempts to produce an effective anthrax vaccine. There is no evidence of his testing the effects of potassium bichromate. And though Davaine exposed the anthrax bacillus to a very wide range of antiseptics, potassium bichromate was apparently not among them. See Théodoridès 1968, pp. 108–115.

86. See Wrotnowska 1978, pp. 284–290, for a complete list of Toussaint's publications. The statement that Toussaint's "mind gave way" in 1882 is quoted in ibid., p. 265, from a speech at Toussaint's funeral by the director of the Toulouse Medical School.

87. See Appendix H, esp. p. 295.

88. Pasteur, *Oeuvres*, VI, p. 350.

89. The letter is paraphrased in Pasteur, *Correspondance*, III, p. 271n.2.

90. See Loir 1937–1938, *14* (1937), pp. 189–191.

91. See Lutaud 1887b, pp. 405–431, esp. p. 412.

92. See Wrotnowska 1978, esp. pp. 265–280.

93. See Appendix D, esp. p. 288.

94. See Pasteur, *Oeuvres*, VI, p. 342.

95. For an insightful discussion of the inherent tension between the institutionalized norms and reward structure of science, which can lead scientists to "deviant" behavior, including even fraud, see Merton 1973, pp. 286–324, as well as the other essays in part 4 of that volume.

96. For a useful journalistic survey of examples of scientific "fraud" see Broad and Wade 1982. But their examples of "fraud" are by no means equivalent in form or in degree of "turpitude." Sometimes, especially in the cases taken from the more distant past, Broad and Wade misleadingly use the word "fraud" to describe what are really quite typical manipulations—rather than "inventions"—of data. For a properly critical review of their book, see Joravsky 1983.

97. Koch 1882. For Pasteur's spirited, indeed sarcastic response, see Pasteur, *Oeuvres*, VI, pp. 418–440.

CHAPTER SEVEN
FROM BOYHOOD ENCOUNTER TO "PRIVATE PATIENTS"

1. The rabid wolf attack of 1831 is mentioned in René Vallery-Radot 1926; Ledoux 1941, pp. 16–17; and Dubos 1950, p. 332. The most extensive account remains unpublished; see note 2 immediately below.

2. This entire story, including transcriptions of the letters exchanged between Pasteur and Mayor Perrot as well as Perrot's report on the rabid wolf attack of 1831, can be found in Papiers Pasteur, *Oeuvres*, VIII, "Notes et documents concernant la rage," fols. 86–103. Perrot sent a copy of his report and of the correspondence to René Vallery-Radot in 1909. In a covering letter of that date, Perrot indicated that Madame Pasteur had not agreed to the publication of these documents on the grounds that to do so would present unstated "inconveniences" to the Pasteur family. See ibid., pp. 86–87.

3. Pasteur, *Correspondance*, IV, pp. 257–259, quote on p. 258.

4. Pasteur, *Oeuvres*, IV, pp. 590–602, quote on p. 591.

5. According to Pasteur himself, sixty people died of rabies in Paris hospitals

between 1880 and 1885, leading to an annual mortality rate, for Paris, of twelve; Pasteur, *Oeuvres*, VI, pp. 630–631. For France as a whole, the average annual mortality rate for the period 1850 to 1885 was estimated at twenty-five to thirty; see Lutaud 1887a, pp. 8–9. Cf. also Pasteur, *Oeuvres*, VI, pp. 809–810.

6. Ritvo 1987, pp. 169–170.

7. See Kete 1988, pp. 92–95; Ritvo 1987, esp. p. 181; Nicol 1974, p. 104.

8. Kete 1988, esp. p. 90; and Ritvo 1987, pp. 167–202.

9. I should emphasize that my discussion of rabies, here and elsewhere, depends importantly on Lépine 1948, a classic contribution to clinical and scientific knowledge of the disease; and Théodoridès 1986, which is the most comprehensive and systematic study of the history of rabies to date, with valuable citations to the leading primary sources from antiquity into the twentieth century.

10. Over the full sweep of history, according to the U.S. Center for Disease Control, there were as of 1977 only three "well-documented, non-fatal cases of [symptomatic] rabies in humans"—and all of those since 1970. See U.S. Public Health Service, *Morbidity and Mortality Weekly Report 26* (1977): 275.

11. See Rosenblatt 1993. In his vivid way, Thomas spoke as follows: "There's really no such thing as the agony of dying. . . . A lot of people fear death because they think that so overwhelming an experience has to be painful, but I've never known anyone to undergo anything like agony. . . . You see, something happens when the body knows it's about to go. . . . Peptide hormones are released by cells in the hypothalamus and pituitary gland. Endorphins. They attach themselves to the cells responsible for feeling pain. The exception was a patient in a charity hospital in New Orleans. . . . He'd been bitten by a rabid squirrel, and he kept repeating, raving, that he was dying. He couldn't stop talking about his symptoms. And he was heaving in pain. People thought he'd gone crazy. He died that afternoon. I wondered if the rabies hadn't knocked out some center in his brain stem designed to prevent that kind of thing. On the whole, though, I believe in the kindness of nature at the time of death." Lewis Thomas died less than a month after this interview.

12. See, e.g., Kete 1988, p. 90, citing Ernest Renan's comment in the speech that marked Pasteur's election to the Académie française in 1882: "Humanity will recognize you as its deliverer."

13. Pasteur, *Cahier 77*, fol. 3v.

14. On the book by Enaux and Chaussier, see Théodoridès 1986, pp. 136–137.

15. Pasteur, *Cahier 77*, fol. 3v.; on Davaine as "precursor" of Pasteur, see Théodoridès 1968.

16. See Pasteur, *Oeuvres*, VI, pp. 553–566; and, for corroboration from the laboratory notes, Pasteur, *Cahier 91*, fols 15–15v.

17. Pasteur, *Oeuvres*, VI, pp. 559–566.

18. See, e.g., Pasteur, *Cahier 91*, fols. 25–29.

19. Pasteur, *Oeuvres*, VI, pp. 564–565.

20. Pasteur, *Oeuvres*, VI, pp. 570–571.

21. Pasteur, *Oeuvres*, VI, p. 399; Pasteur, *Cahier 91*, fols. 79–86.

22. Pasteur, *Oeuvres*, VI, pp. 367–368. Cf. ibid., pp. 398–401, 570–571.

23. See, e.g., Dubos 1950, pp. 263, 375.

24. Pasteur, *Cahier 91*, fol. 67.

25. See, e.g., Pasteur, *Cahier 92*, fols. 27v, 49–50, 151v, 152, 159v; cf. Pasteur, *Oeuvres*, VI, pp. 599–600.

26. See, e.g., Pasteur, *Oeuvres*, VI, pp. 586–602, passim.

27. For an extensive account of Galtier's work, see Théodoridès 1986, esp. pp. 189–199. For Théodoridès, Galtier is "the great forgotten figure [*le grand oublié*] in the history of rabies." Ibid., p. 189. See also Nicol 1974, pp. 555–563.

28. Galtier 1879.

29. Galtier 1881a,b; cf. Théodoridès 1986, pp. 189–195.

30. See Pasteur, *Oeuvres*, VI, pp. 577–578, esp. p. 578n.1.

31. See Théodoridès 1986, p. 197.

32. See Théodoridès 1986, pp. 196–199, where Galtier's later publications are discussed; for the most part, they rehearse the results already reached by 1881.

33. My thanks to Dr. Donald Burke for emphasizing the importance of this point to me. See also figure 7.1.

34. See Bulloch 1938, p. 226.

35. See Pasteur, *Oeuvres*, VI, pp. 335–336; Duclaux [1896], 1920 translation, pp. 306–314.

36. Pasteur, *Oeuvres*, VI, pp. 586–589.

37. See, e.g., Pasteur, *Cahier 92*, fols. 49–50, 80 (anthrax), 119 et seq., 132 (oxygenated water).

38. Pasteur, *Cahier 92*, fols. 159–160.

39. See Loir 1937–1938, *14* (1937), p. 270; and, more important, Roux 1883.

40. Pasteur, *Oeuvres*, VI, pp. 573–574.

41. Pasteur, *Oeuvres*, VI, pp. 575–579. For laboratory notes on "recovery" from rabies, see *Cahier 92*, fols. 83, 145v, 152.

42. See, e.g., *Cahier 92*, fols. 105–105v.

43. Pasteur, *Oeuvres*, VI, pp. 579–586.

44. Pasteur, *Oeuvres*, VI, pp. 586–589.

45. Pasteur, *Oeuvres*, VI, pp. 391–411.

46. Pasteur, *Oeuvres*, VI, pp. 527–534.

47. Pasteur, *Oeuvres*, VI, pp. 586–589.

48. Pasteur, *Oeuvres*, VI, p. 589.

49. See Pasteur, *Oeuvres*, VI, pp. 753–758, for printed version of the French committee's report.

50. See Pasteur, *Oeuvres*, VI, p. 757.

51. Pasteur, *Oeuvres*, VI, pp. 590–602.

52. See Pasteur, *Cahier 94*, fol. 62 (Pasteur's pagination; fol. 66 of BN stamped pagination). Except where otherwise noted, all of the rest of the story of Girard is based on this one page in Pasteur's laboratory notebook.

53. On Dujardin-Beaumetz, see Pasteur, *Oeuvres*, VI, pp. 630, 736; and Pasteur, *Correspondance*, IV, p. 96.

54. To repeat, this entire story, including records of correspondence between Pasteur and Dujardin-Beaumetz, is drawn from that same notebook page cited in note 52 above, i.e., Pasteur, *Cahier 94*, fol. 62.

55. Like the story of Girard, this story of Julie-Antoinette Poughon is based on another single page in Pasteur's laboratory notebook: Pasteur, *Cahier 94*, fol. 79 (Pasteur's pagination; fol. 83 in BN pagination).

56. Elsewhere in the Pasteur manuscripts at the Bibliothèque Nationale, the name Girard is included on a list of four men whom Dujardin-Beaumetz reported as having died of rabies. If this was the same Girard—the name is not especially rare in France—it seems odd that Pasteur merely recorded his name along with the other three, without comment or expression of surprise. Stranger still, Pasteur failed to add this crucial piece of information to his laboratory notes on Girard, as was his usual practice when later results affected the findings recorded in his laboratory notebook. And given the brief clinical course of rabies, it is almost impossible to understand how Girard could have died of the disease so long after his admittance to the hospital.

57. See, yet again, Pasteur, *Cahier 94*, fol. 62.

58. Pasteur, *Oeuvres*, VI, pp. 665–666.

59. Pasteur, *Oeuvres*, VI, p. 666. For interesting accounts of other cases of "false rabies," see Kete 1988 and Ritvo 1987, chap. 4.

60. On the 1887 debates over the safety and efficacy of Pasteur's rabies vaccine, see Chapter Eight.

61. See U.S. Center for Disease Control, *Morbidity and Mortality Weekly Report 28* (1979): 109–111.

62. Pasteur, *Cahier 94*, fol. 62. Pasteur's pagination here and in subsequent notes.

63. Pasteur, *Cahier 94*, fol. 62.

64. Pasteur, *Cahier 94*, fol. 79.

65. Pasteur, *Cahier 94*, fol. 65.

66. See, e.g., Bernard [1865], 1957 translation, pp. 101–108.

CHAPTER EIGHT
PUBLIC TRIUMPHS AND FORGOTTEN CRITICS

1. The Joseph Meister story is ubiquitous in the literature on Pasteur. For Pasteur's famous first published paper on Meister, see Pasteur, *Oeuvres*, VI, pp. 603–610. For the classic secondary account, see René Vallery-Radot 1926, pp. 414–418. In 1889, Pasteur himself published a popular account of his work on rabies, translated into English with the simple title "Rabies," in the inaugural volume of *The New*

Review (1889), I, pp. 505–512, 619–630 (see pp. 619–621 on Meister and Jupille). A somewhat revised version of this popular article also appeared in French: see Pasteur, *Oeuvres*, VI, pp. 672–688. See also Huas 1985.

2. Pasteur, *Oeuvres*, VI, p. 606. Interestingly, in his original draft of this famous paper, Pasteur wrote that he decided to treat Meister "with the approval of Vulpian and Grancher [*avec l'approbation de M.Vulpian et du M. Grancher*]," but then crossed out the quoted phrase. Papiers Pasteur, *Oeuvres*, VII, "Rage I," fol. 96. It is thus conceivable that Vulpian and Grancher did not wholly and immediately approve of Pasteur's decision to treat Meister, but the evidence is ambiguous.

3. For the Meister story as recorded in Pasteur's laboratory notes, see Pasteur, *Cahier 94*, fols. 83–83v (Pasteur's pagination; stamped BN pagination is fols. 87–87v).

4. Part of the Jupille story is told in René Vallery-Radot 1926, pp. 421–423. A full and fascinating account is to be found in Papiers Pasteur, *Oeuvres*, VIII, "Notes et documents concernant la rage," fols. 86–103. On the circumstances of the Jupille family, see ibid., fols. 101v-102.

5. Papiers Pasteur, *Oeuvres*, VIII, "Notes . . . ," fols. 93–93v.

6. Ibid. Papiers Pasteur, *Oeuvres*, VIII, "Notes . . . ," fols. 93–93v.

7. For the Jupille story as recorded in Pasteur's laboratory notes, see Pasteur, *Cahier 94*, fols. 109–109v. (Pasteur's pagination; stamped BN pagination is 107–107v).

8. See Pasteur, *Oeuvres*, VI, pp. 603–610, quotes from p. 603.

9. Pasteur, *Oeuvres*, VI, p. 603.

10. Pasteur, *Oeuvres*, VI, p. 604.

11. Pasteur, *Oeuvres*, VI, p. 604.

12. Pasteur, *Oeuvres*, VI, pp. 604–605.

13. Pasteur, *Oeuvres*, VI, pp. 606–607. For the details of Meister's treatment as recorded in Pasteur's laboratory notes, see Pasteur, *Cahier 94*, fols. 83–83v (Pasteur's pagination).

14. Pasteur, *Oeuvres*, VI, pp. 609–610.

15. The comments of the three speakers who rose to praise Pasteur and his paper are conveniently printed in Pasteur, *Oeuvres*, VI, pp. 610–611. For Vulpian's comment, see ibid., p. 610.

16. See Pasteur, *Oeuvres*, VI, pp. 610–611. That Jupille was in fact awarded a prize of 1,000 francs is clear from two of Pasteur's letters to Mayor Perrot of Villers-Farlay. See Pasteur, *Correspondance*, IV, pp. 44–48.

17. See Pasteur, *Oeuvres*, VI, p. 611.

18. Pasteur, *Oeuvres*, VI, pp. 628–629.

19. See Duclaux [1896], 1920 translation, pp. 298–299.

20. On the English critics, see Ritvo 1987, chap. 4.

21. See "Prof. Huxley and M. Pasteur on Hydrophobia," *Nature 40* (1889): 224–226, on 225.

22. See [Anonymous] 1873; Pasteur, *Oeuvres*, II, p. 379; VI, pp. 446–447, 450.

23. See Bernadette Bensaude-Vincent, "Louis Pasteur face à la presse scientifique," in Morange 1991, pp. 75–88. Bensaude-Vincent makes it clear that most of the criticism Pasteur did receive in the popular scientific press concerned the debate over spontaneous generation. In the case of rabies, however, criticism was virtually absent; indeed, he was by then very "popular" in the popular scientific press.

24. Lutaud 1887b, p. 6.

25. See Brouardel, meeting of 12 July 1887. *Bull. Acad. de méd. 18*:51–64.

26. In sum, between January and July 1887, there were three mini-discussions upon Peter's presenting negative works (1 February, pp. 138–140; 8 February, pp. 162–164, and 22 February, pp. 206–209), for a total of 10 pages; two mini-debates (4 January, pp. 16–23; and 5 July, pp. 6–11), for a total of 14 pages; and three prolonged discussions (11 January, pp. 28–66; 18 January, pp. 72–120; and 12 July, pp. 37–68), for a total of 120 pages in the *Bulletin* of the Académie de médecine. Throughout the debates, Peter was the sole public critic; and though the first salvos were fired in Pasteur's absence for the sake of his health, the Pastorian cause was amply represented by Grancher (through the perpetual secretary, Dr. Beclard), Dujardin-Beaumetz, Brouardel, and Vulpian. At a meeting of 22 February, the Académie voted to suspend all discussion on Pasteur's vaccine until he could return to defend himself against the attacks. Pasteur returned on 10 May (p. 530), but no discussion took place until 5 July, when Pasteur himself took the opportunity to present a copy of the report of the official English Commission on Rabies. When Peter complained, this led to a final major discussion of 12 July.

27. On Michel Peter, see Loir 1937–1938, *14*, pp. 43–47.

28. See, e.g, the meeting of 22 February 1887. *Bull. Acad. de méd. 17*: 206–209.

29. See, e.g., the meeting of 11 January 1887. *Bull. Acad. de méd. 17*:56.

30. See Brouardel and Charcot, meeting of 12 July 1887. *Bull. Acad. de méd. 18*:64, 68.

31. See Vulpian, meeting of 18 January 1887. *Bull. Acad. de méd. 17*:93.

32. Peter, meeting of 12 July 1887. *Bull. Acad. de méd. 18*:39.

33. See esp. Latour 1988 and Latour 1992.

34. Pasteur, *Correspondance*, IV, pp. 75–76.

35. See Peter and Brouardel, meetings of 8 February 1887 and 22 February 1887, *Bull. Acad. de méd. 17*:162–164, 206–209.

36. Von Frisch 1887.

37. See, e.g., Papiers Pasteur, *Correspondance* IV, fols. 295, 374–376, 379; and Pasteur, *Correspondance*, IV, pp. 140–143, 161–166, 176–179.

38. Von Frisch 1887, pp. 109, 121–127.

39. Von Frisch 1887, pp. 35–107.

40. Von Frisch 1887, pp. 106–107.

41. More generally on this issue, see Farley and Geison 1974; and Collins, 1981, 1992.

42. For a conveniently accessible printed version of the report of the English Commission on Rabies, see Pasteur, *Oeuvres*, VI, pp. 870–883.

43. See Papiers Pasteur, *Correspondance*, IV, pp. 264–266; and Pasteur, *Correspondance*, IV, letters to Victor Horsley (passim).

44. See Report of English Commission in Pasteur, *Oeuvres*, VI, p. 877.

45. See Ritvo 1987, Walton 1979, and Geison 1979.

46. See Pasteur meeting of 5 July 1887. *Bull. Acad. de méd. 18*:6–11.

47. See Peter, meeting of 18 January 1887. *Bull. Acad. de méd. 17*:78–92; and Peter, meeting of 12 July 1887. *Bull. Acad. de méd. 18*:37–49.

48. See, e.g., Peter, meeting of 18 January 1887. *Bull. Acad. de méd. 17*:79–91.

49. Pasteur, *Correspondance*, IV, pp. 75–76.

50. See Peter, meeting of 12 July 1887. *Bull. Acad. de méd. 18*:44–49.

51. Brouardel as quoted in Pasteur, *Oeuvres*, VI, p. 856.

52. See Peter, meeting of 11 January 1887. *Bull. Acad. de méd. 17*:31.

53. See Pasteur, *Oeuvres*, pp. 350–351, 358–360, 363–364.

54. See Peter, meeting of 18 January 1887. *Bull. Acad. de méd. 17*:79. See also Lutaud 1887a,b.

55. See Peter, meeting of 12 July 1887. *Bull. Acad. de méd. 18*:36–49.

56. See, e.g., Brouardel, meeting of 12 July 1887. *Bull. Acad. de méd. 18*:49–64.

57. United States Center for Disease Control, 1977, *Morbidity and Mortality Weekly Report 26*:275.

58. Pasteur, *Oeuvres*, VI, p. 619.

59. For a thoughtful critique by one of Pasteur's contemporaries, see von Frisch 1887; and, for a much more recent and even more critical assessment, Webster 1942.

60. See Webster 1942, Lépine 1948, and Decourt 1974.

61. Pasteur, *Oeuvres*, VII, pp. 363–371.

62. Pasteur, *Oeuvres*, VI, pp. 588, 591.

63. Pasteur, *Correspondance*, III, pp. 445–446.

64. Pasteur, *Correspondance*, IV, p. 21.

CHAPTER NINE
PRIVATE DOUBTS AND ETHICAL DILEMMAS

1. Loir 1937–1938, *14*, p. 329.

2. Loir 1937–1938, *14*, p. 330. In his important recent book, Antonio Cadeddu argues that Loir's account is misleading in several respects, especially when compared to the documentary evidence in Pasteur's laboratory notebooks on rabies. In general, Cadeddu suggests, Loir simplifies the process of discovery and exalts Roux's contributions at the expense of Pasteur's. Cadeddu's claim is plausible enough, but it seems to be based entirely on the assumption that Loir's account was referring to events in 1885, an assumption for which there is no direct evidence in Loir's text. See Cadeddu 1991, pp. 173–286.

3. Loir 1937–1938, *14*, p. 331.

4. Loir 1937–1938, *14*, p. 331.

5. See, in addition to Loir 1937–1938, Cressac 1950; Nicolle 1932, esp. p. 60; Lagrange 1954, esp. p. 8; and Gascar 1986, passim.

6. I have previously explored this divide in Geison 1979, pp. 67–90. See also, in the same volume, Maulitz 1979. There is a vast body of other pertinent literature, including Feinstein 1987.

7. Gascar 1986, p. 51, specifically compares Roux to Don Quixote. For one among a host of examples comparing Pasteur with Napoleon, see Jacob 1988, p. 248.

8. Lagrange 1954, esp. pp. 15–22, 33–38.

9. Lagrange 1954, pp. 33–38.

10. Lagrange 1954, p. 38, where Lagrange reports that Pasteur withdrew an offer he had already made to a young veterinarian in favor of Roux, who was hired "en qualité d'inoculateur."

11. Loir certainly implies as much, though he does not quite put it in these words. See Loir 1937–1938, *14*, pp. 338–340, 347–348.

12. Loir 1937–1938, *14*, p. 335. For more on Grancher, see Roussillat 1964.

13. Pasteur, *Oeuvres*, VI, p. 606.

14. Pasteur, *Cahier 93*, fols. 148ff [late June 1884].

15. Pasteur, *Cahier 93*, fols. 70–71, 75, 116, 125–129; *Cahier 94*, fols. 7–8; 22, 33, 36ff.

16. Pasteur, *Cahier 93*, fols. 125–129; *Cahier 94*, fols. 33–37.

17. Pasteur, *Oeuvres*, VI, pp. 603–607.

18. See Pasteur, *Cahiers 93 and 94*, folios cited in notes 15 and 16 above.

19. Pasteur, *Oeuvres*, VI, pp. 586–589.

20. See Pasteur, *Cahier 93*, fols. 116, 125–129.

21. See Pasteur, *Cahier 94*, fols. 7–8, 22, 33–37.

22. See Pasteur, *Cahier 94*, fols. 60–73 passim.

23. Pasteur, *Oeuvres*, VI, pp. 290–291.

24. Pasteur, *Oeuvres*, VI, p. 463.

25. From this point through note 28, much of what follows is based on Pasteur, *Cahier 94*, fols. 7–7v.

26. Pasteur, *Cahier 94*, fol. 7.

27. Pasteur, *Oeuvres*, VI, pp. 462–466.

28. Pasteur, *Oeuvres*, VI, pp. 607–609, 637–652, esp. pp. 644–650.

29. See Pasteur, *Cahier 94*, fols. 1–70 passim.

30. Pasteur, *Cahier 94*, fols. 7–7v.

31. Pasteur, *Cahier 94*, fols. 60–70.

32. Pasteur, *Cahier 94*, fols. 10, 27, 43, 54v, 71. Cf. Cadeddu 1991, pp. 278–282.

33. See especially Roux 1883.

34. Pasteur, *Oeuvres*, VI, p. 338.

35. Pasteur, *Cahier 94*, fols. 73–103.

36. Roussillat 1964, p. 68.

37. Papiers Pasteur, *Correspondance*, XI, pp. 322–323.
38. Loir 1937–1938, *14*, pp. 343–348.
39. Pasteur, *Correspondance*, IV, pp. 189–190.
40. Roussillat 1964, p. 69. My emphasis.

CHAPTER TEN
THE MYTH OF PASTEUR

1. See Lumet 1923, p. 166; George 1958; and especially Maurice Vallery-Radot 1985, chap. 6.
2. Lumet 1923, p. 168.
3. Lumet 1923, p. 167.
4. Lumet 1923, pp. 168–170.
5. See Lumet 1923, p. 167; and Geison 1974, p. 408.
6. See Geison 1974, pp. 352, 408.
7. For the printed text of Pasteur's tribute to his parents, see Pasteur, *Oeuvres*, VII, pp. 360–361.
8. See Institut Pasteur 1888, p. 25.
9. See Pasteur, *Oeuvres*, VII, pp. 417–420.
10. See *Jubilée de M. Pasteur* 1893; and Pasteur, *Oeuvres*, VII, pp. 426–428.
11. Zeldin 1977, vol. 2, p. 390.
12. See Morange 1991.
13. See Simmonet 1942; Delaunay 1962; Morange 1991, pp. 239–319; and Gascar 1986, esp. chaps. 11–20.
14. Jacob 1988, p. 246.
15. Jacob 1988, pp. 244–247.
16. Lumet 1923, p. 166.
17. See, e.g., Papiers Pasteur, *Correspondance*, IV, pp. 392ff (to a Russian scientist: "You know my sympathy for your country"), and p. 435 ("Vivent Russie et France unies"); and Pasteur, *Correspondance*, IV, pp. 251, 316.
18. Pasteur, *Correspondance*, IV, p. 108.
19. See, e.g., *Correspondance*, IV, pp. 75–76; and his extensive correspondence with Dr. Gamaleia in ibid., IV, passim.
20. See, e.g., Gascar 1986, passim; and Hutchinson 1988.
21. Paget 1910. For further evidence of the cult of Pasteur in England, see Priestley 1897, 1908, and Ward 1994.
22. See the letters exchanged between René Vallery-Radot, Dr. Chrisian Herter and Dr. Herter's sister in the McChesney Archives, the Johns Hopkins University Medical School.
23. See Frederico 1937.
24. See Hansen 1991.
25. De Kruif 1959, p. 97.

26. See Jacoby 1986.

27. Pasteur, *Correspondance*, I, p. 228.

28. See Fox 1976; Geison 1974, p. 356; and Maurice Vallery-Radot 1985.

29. Cf. Farley and Geison 1974, Geison 1981, Roll-Hansen 1979, 1983, and Temple 1986.

30. Pasteur, *Oeuvres*, VI, p. 37. See also, for Pasteur's "antirhetoric" rhetoric, *Oeuvres*, II, p. 163.

31. Shapin 1984, pp. 127–128. Further on science and rhetoric, see Dear 1985, 1991.

32. This point is emphasized by Bensaude-Vincent 1991. On Pasteur as prose stylist, see also Maurice Vallery-Radot 1985, pp. 59–60.

33. Moreau 1986, esp. pp. 271–273.

34. Pasteur, *Cahier 95*, final entry.

35. See Frankland 1898, pp. 209–210.

36. Gascar 1986, p. 205.

37. One especially striking description of Pasteur's health toward the end of his life comes from the man who painted the most famous portrait of him—Edelfelt. See Weisberg 1995.

38. Pasteur Vallery-Radot 1954, pp. 68–69.

39. Grancher 1895.

40. See Lumet 1923, p. 170.

41. Lagrange 1954. Cf. Nicolle 1932.

42. Long ago, Decourt made the same point; Decourt 1974.

43. Hemphill 1977.

44. See the correspondence of Richard Herter in the McChesney Archives, The Johns Hopkins University Medical School.

45. See Hume 1923 and the later reprintings.

46. [Anonymous] 1934. *Isis 21*:404–405.

47. Loir 1937–1938.

48. See Théodoridès 1968, Decourt 1989, Burke 1993a, 1993b, 1993c.

49. Geison 1974, p. 415.

50. For my review of the TV series, see Geison 1993.

51. Cadeddu 1985, 1987, 1991.

52. See Alex Preminger, ed., *Princeton Encyclopedia of Poetry and Poetics*, enlarged edition (Princeton, New Jersey: Princeton University Press, 1974), p. 538.

53. Erikson 1965, p. 318.

54. See Latour 1984, 1988, and Salomon-Bayet 1986.

55. See Decourt 1974, 1989, pp. 300.

56. See Wachhorst 1981.

Acknowledgments

This book owes much to a few funding agencies, several institutions, and a host of individuals, many of whom I shall, alas, be unable to acknowledge. Early on, a generous grant from the National Endowment for the Humanities allowed me to spend an academic year in Paris, just as the Papiers Pasteur at the Bibliothèque National were becoming generally available to scholars. Thanks to equally generous support from the National Science Foundation, I was able to return to Paris for further archival research during subsequent summers. The Howard Foundation also contributed to this project, as did the National Institutes of Health, during a semester I spent (in 1987–88) at the National Library of Medicine as Visiting Senior Historical Scholar. None of this external support would have been possible but for the internal advantages of my position at Princeton University and its History Department, which several times allowed me to take leave or released time from my usual teaching and administrative responsibilities to work on this book.

It is hard to separate my indebtedness to institutions from the people who make them work. To speak only of libraries, friends, and colleagues in Paris, I owe a very special debt to Marie-Laure Prévost, curator of the Papiers Pasteur at the Bibliothèque Nationale, who was unfailingly helpful to me and remarkably tolerant of my halting French. Warm thanks also to Dr. Anne-Marie Moulin for her willingness to share with me her extensive knowledge of the history of immunology and the Institut Pasteur. She was also kind enough to read and comment upon major portions of this manuscript. Over the last decade or so, my good friend Bruno Latour has done everything he could to encourage and improve this book, with what he will regard as mixed results at best, I fear.

Closer to home, I want to thank my colleagues and students in the Program in History of Science at Princeton, who have offered valuable criticism and steadfast support for many years. Richard Weisberg of New York University graciously shared with me his knowledge of the iconography of Louis Pasteur, including especially the famous painting by Edelfelt. William H. Helfand has been remarkably generous in allowing me to reproduce—without charge!—several Pastorian images from his private collection.

In the Preface, I have already thanked my great good friend and mentor, Professor Larry Holmes of Yale. There are at least three other friends, all at Princeton University Press, without whom this book would still not exist. Emily Wilkinson has spent years trying to prod me toward this result. Alice Calaprice has been a supportive and spectacularly efficient editor. And then there's Kevin Downing, my good friend, former student, and gentle, genial, but persistent goad. In the final months and days that led to this product, Kevin was the single most important force of all.

Bibliography

Adam, Alison Evelyn. 1988. "Spontaneous Generation in the 1870s: Victorian Scientific Naturalism and Its Relationship to Medicine." Ph.D. thesis, Department of Historical and Critical Studies, Sheffield City Polytechnic, Sheffield, England.

[Anonymous]. 1873. "Fabrication de la bière," *Journal de pharmacie 17*:330–331.

[Anonymous]. 1889. "Prof. Huxley and M. Pasteur on Hydrophobia." *Nature 40*: 224–226.

Appel, Toby. 1987. *The Cuvier-Geoffroy Debate: French Biology in the Decades before Darwin*. New York: Oxford University Press.

Arcieri, G. P. 1938. *Agostino Bassi and Pasteur*. New York: Vigo Press.

Auzias-Turenne, Joseph-Alexandre. 1878. *La syphilisation*. Publication de l'oeuvre du Docteur Auzias-Turenne fait par les soins de ses amis. Paris: Librairie Baillière.

Bachelard, Gaston. 1975. *La formation de l'esprit scientifique: Contribution à une psychanalsye de la connaissance objective*. 9th ed. Paris: Librairie Philosophique J. Vrin.

Bakhtin, M. M. 1981. *The Dialogic Imagination: Four Essays*. Ed. Michael Holquist, trans. Caryl Emerson and Michael Holquist. Austin: University of Texas Press.

Bardsley, Samuel Argent. 1807. *Medical Reports of Cases and Experiments, with Observations, chiefly derived from Hospital Practice: to which are added, An Enquiry into the Origin of Canine Madness; and Thoughts on a Plan for its Extirpation from the British Isles*. London.

Barnes, S. B. 1934. "The Scientific Journal, 1665–1730." *Scientific Monthly 38*:257–260.

Bastian, Henry Charlton. 1877. "The Commission of the French Academy and the Pasteur-Bastian Experiments." *Nature 16*:277–279.

Bazerman, Charles. 1988. *Shaping Written Knowledge: The Genre and Activity of the Experimental Article in Science*. Madison: University of Wisconsin Press.

Bean, William B. 1977. "Walter Reed and the Ordeal of Human Experiments." *Bull. Hist. Med. 51*:75–92.

Béchamp, Antoine. [1883]. *Les microzymas*. Réédité par le Centre Internationale d'Études A. Béchamp, Paris. 1990.

Benichou, H. 1991. "Pasteur, Pasteurization, and Medicine." In R. Schofield, D. S. Reher, and A. Bide, eds., *The Decline of Mortality in Europe*, pp. 220–232. Oxford: Clarendon Press.

Bensaude-Vincent, Bernadette. 1991. "Louis Pasteur face à la presse scientifique." In Michel Morange, ed., *L'Institut Pasteur*, pp. 75–88. Paris: Editions la Découverte.

Bernal, J. D. 1953. *Science and Industry in the Nineteenth Century*. London: Routledge.

Bernard, Claude. [1865]. *An Introduction to the Study of Experimental Medicine*. Trans. H. C. Greene. Reprint, New York: Dover, 1957.

Bibliothèque Nationale. 1964. *Pasteur. Donation du Professeur Pasteur Vallery-Radot.* Paris.

Bibliothèque Nationale. 1985. *Nouvelles acquisitions latines et françaises du département des manuscrits pendant les années 1977–1982: Inventaire sommaire.* Paris. 17923-18112. Louis Pasteur, Papiers. See also "Author's Note on the Notes and Sources."

Bichet, Robert. 1989. *Pasteur et le Sénate.* Besançon, France: Cêtre.

Biot, Jean-Baptiste. 1844. "Communication d'une note de M. Mitscherlich." *Comptes rendus de l'Acad. des sci. 19*:720.

Bird, Christopher. 1991. *The Persecution and Trial of Gaston Naessens: The True Story of the Efforts to Suppress an Alternative Treatment for Cancer, AIDS, and Other Immunologically Based Diseases.* Tiburon, Calif.: H. J. Kramer.

Blaisdell, J. D. 1992. "Louis Pasteur, Alexandre Liautard, and the Riverdale Dog Case." *Vet. Herit. 15*:2–15.

Blaringhem, Louis. 1923. *Pasteur et le transformisme.* Paris: Masson et cie.

Bliss, Michael. 1982. *The Discovery of Insulin.* Chicago: University of Chicago Press.

Bouley, Henri. 1883. "Les découvertes de M. Pasteur devant la médecine." *Revue scientifique*, 3d ser., *31*:439–443.

Bowers, Edgar. 1989. *For Louis Pasteur.* Princeton, N.J.: Princeton University Press.

Bowman, Ida A. 1978. "The Pasteur Collection at UTMB Library." *Bookman 5* (5): 1–6.

Brannigan, Augustine. 1981. *The Social Basis of Scientific Discoveries.* Cambridge, U.K.: Cambridge University Press.

Broad, William, and Nicholas Wade. 1982. *Betrayers of the Truth: Fraud and Deceit in the Halls of Science.* New York: Simon and Schuster.

Brock, Thomas D. 1988. *Robert Koch: A Life in Medicine and Bacteriology.* Madison, Wis.: Science Tech.

Bulloch, William. 1938. *The History of Bacteriology.* London: Oxford University Press.

Burke, Donald S. 1993a. "Of Postulates and Peccadilloes: Robert Koch and Vaccine (Tuberculin) Therapy for Tuberculosis." *Vaccine 11*:795–804.

Burke, Donald S. 1993b. "Vaccine Therapy for HIV: A Historical Review of the Treatment of Infectious Diseases by Active Specific Immunization with Microbe-Derived Antigens." *Vaccine 11*:883–891.

Burke, Donald S. 1993c. "Joseph-Alexandre Auzias-Turenne, and Early Concepts of Virulence, Attenuation and Vaccination." Unpublished ms. of 45 pages.

Cadeddu, Antonio. 1985. "Pasteur et le choléra des poules: Révision critique d'un récit historique." *Hist. Phil. Life Sci. 7*:87–104.

Cadeddu, Antonio. 1987. "Pasteur et la vaccination contre le charbon: Une analyse historique et critique." *Hist. Phil. Life Sci. 9*:255–276.

Cadeddu, Antonio. 1991. *Dal mito alla storia: Biologia e medicina in Pasteur.* Milan: Franco Angeli.

Callon, Michel, and Bruno Latour, eds. 1982. *La sciences telle qu'elle se fait: Anthologie de la sociologie des sciences de langue anglaise.* [Paris]: Pandore. (No. spécial.)

Canguilhem, George, et al. 1984. *Anatomie d'un épistémologue: François Dagognet.* Paris: Librairie Philosophique J. Vrin.

Carter, K. Codell. 1982. "Nineteenth-Century Treatments for Rabies as Reported in the Lancet." *Med. Hist.* *26*:67–78.

Carter, K. Codell. 1988. "The Koch-Pasteur Dispute on Establishing the Cause of Anthrax." *Bull. Hist. Med.* *62*:42–57.

Carter, K. Codell. 1991. "The Development of Pasteur's Concept of Disease Causation and the Emergence of Specific Causes in Nineteenth-Century Medicine." *Bull. Hist. Med.* *65*:528–548.

Chamberland, Charles, and Emile Roux. 1883. "Sur l'attenuation de la virulence de la bactéridie charbonneuses, sous l'influence des substances antiseptiques." *Comptes rendus de l'Acad. des sci.* *96*:1088–1091, 1401–1412.

Charlton, D. G. 1963. *Secular Religion in France, 1815–1850.* Oxford: Oxford University Press.

Chauveau, Jean-Baptiste Auguste. 1880. "Du renforcement de l'immunité des moutons algériens, à l'égard du sang du rate, par les inoculations préventives." *Comptes rendus de l'Acad. des sci.* *93*:150.

Chauvois, Louis, Denise Wrotnowska, and E. Perrin. 1966. "L'optique de Pasteur." *Rev. d'optique théor. et instr.* *45*:197–213.

Cohn, Ferdinand. 1875. "Untersuchungen über Bacterien, II." *Beiträge zur Biol. der Pflanzen* *1*:141–207.

Coleman, William. 1964. *Georges Cuvier: Zoologist.* Cambridge, Mass.: Harvard University Press.

Coleman, William, and Frederic L. Holmes, eds. 1988. *The Investigative Enterprise: Experimental Physiology in Nineteenth-Century Medicine.* Berkeley: University of California Press.

Colin, [Gabriel]. 1880. "Analyse éxperimentale de la pustule maligne et de l'oedème charbonneux; détermination de leurs formes variées et de leurs degrés de virulence." *Bull. de l'Acad. de méd.*, 2d ser., *9*:650–670.

Collins, Harry M. 1981. "The Place of the 'Core-Set' in Modern Science: Social Contingency with Methodological Propriety in Science." *History of Science* *19*:6–19.

Collins, Harry. 1992. *Changing Order: Replication and Induction in Scientific Practice.* With a new afterword Chicago: University of Chicago Press.

Conant, James Bryant. 1957a. "Pasteur's Study of Fermentation." In *Harvard Case Histories in Experimental Science*, vol. 2, pp. 437–485. Cambridge, Mass.: Harvard University Press.

Conant, James Bryant. 1957b. "Pasteur's and Tyndall's Study of Spontaneous Generation." In *Harvard Case Histories in Experimental Science*, vol. 2, pp. 487–539. Cambridge, Mass.: Harvard University Press.

Coulter, Harris L. 1993. "Pasteur, Metchnikoff, Behring, and the Therapeutic of

Similars." In Coulter, *Divided Legacy*, vol. 4, *The Bacteriological Era*. Berkeley, Calif.: North Atlantic Books.

Crellin, John K. 1966a. "Airborne Particles and the Germ Theory: 1860–1880." *Annals of Science* 22:49–60.

Crellin, John K. 1966b. "The Problem of Heat Resistance of Microorganisms in the British Spontaneous Generation Controversy of 1860–1880." *Medical History* *10*:50–59.

Cressac, Mary. 1950. *Le Docteur Roux: Mon oncle*. Paris: L'Arche.

Cuny, Hilaire, 1963b. *Louis Pasteur: Choix de textes*. Bibliographie, portraits, facsimilés. Lingugé, France: Seghers.

Cuny, Hilaire. 1965. *Louis Pasteur: The Man and His Theories*. Trans. by Patrick Evans. London: Souvenir Press.

Dagognet, François. 1967. *Méthodes et doctrines dans l'oeuvre de Pasteur*. Paris: Presses universitaires.

Dagognet, François. 1985. *Rematérialiser: Matières et matérialismes*. Paris: Librairie Philosophique J. Vrin.

D'Andey, P. 1878. "Auzias-Turenne: Sa vie, son oeuvre." In J.-A. Auzias-Turenne, *Syphilisation*, pp. v–xxiii. Paris: Librairie Baillière.

Dansette, A. 1961. *Religious History of Modern France*. New York: Herder and Herder.

Dear, Peter. 1985. "*Totius in verba:* Rhetoric and Authority in the Early Royal Society." *Isis* *76*:145–161.

Dear, Peter, ed. 1991. *The Literary Structure of Scientific Argument: Historical Studies*. Philadelphia: University of Pennsylvania Press.

Decourt, Philippe. 1971. "Béchamp et Pasteur." *Arch. internat. Claude Bernard* *1*:[182–185].

Decourt, Philippe. 1972a. "La découverte des maladies microbiennes: Béchamp et Pasteur (Suite)." *Arch. internat. Claude Bernard* 2:41–131.

Decourt, Philippe. 1972b. "Editorial: L'histoire et la vérité." *Arch. internat. Claude Bernard* 2:5–26.

Decourt, Philippe. 1972c. "Sur une histoire peu connue: La découverte des maladies microbiennes: Béchamp et Pasteur." *Arch. internat. Claude Bernard* 2:27–118.

Decourt, Philippe. 1974a. "A propos de la rage: Une lettre inédite de Pasteur." *Arch. internat. Claude Bernard* 5:161–165.

Decourt, Philippe. 1974b. "Renseignement préliminaires sur 'une histoire méconnue': L'invention des vaccins modernes au XIXe siècle," *Arch. internat. Claude Bernard* 5:165–184.

Decourt, Philippe. 1975. "Deuxième lettre à nos amis." *Arch. internat. Claude Bernard* 8:1–41.

Decourt, Philippe. 1984. "Les zymases ou ferments solubles de Béchamp: À la lumière des connaissances du XXe siècle." *Hist. sci. méd.* *18*:147–151.

Decourt, Philippe. 1986. "A propos d'un centenaire: Erreurs types de raisonnement en médecine." *Hist. sci. méd.* *20*:249–256.

Decourt, Philippe. 1988. "Précisions sur les premiers essais d'application à l'homme du vaccin de Roux-Pasteur contre la rage." *Hist. sci. méd. 22*:29–35.

Decourt, Philippe. 1989. *Les vérités indésirables*. Vol. 1, part 1: *Faut-il réhabiliter Galilée?*; part 2: *Comment on falsifié l'histoire: Le cas Pasteur*. Paris: La Vieille Taupe.

De Kruif, Paul. 1959. *The Microbe Hunters*. New York: Pocket Books.

Delaunay, Albert. 1962. *Pasteur et la microbiologie*. 2d ed. Paris: Presses universitaires de France. 3d ed., 1967.

Diara, Anna. 1984. "Un débat français vu par la presse, 1858–1869: L. Pasteur–F.A. Pouchet et la génération spontanée." *Actes du Muséum de Rouen*, pp. 175–210.

Dierckx, François. 1890. *Une visite à l'Institut Pasteur en 1890: Le traitement de la rage*. Louvain: Charles Peeters, Libraire-éditeur.

Dolman, Claude. 1973. "Robert Koch." In Charles C. Gillispie, ed., *The Dictionary of Scientific Biography*, vol. 7, pp. 420–435. New York: Charles Scribner's Sons.

Dubos, René. 1950. *Louis Pasteur: Free Lance of Science*. Boston: Little, Brown & Co.

Dubos, René. 1955. *Louis Pasteur: Franc-tireur de la science*. Trans. from the English by Elisabeth Dussauze, with a preface by Robert Debré. Paris, 1955.

Dubos, René. 1988. *Pasteur and Modern Science*. New illustrated ed. Ed. T. D. Brock, with foreword by G. L. Geison. Madison, Wis.: Science Tech.

Duclaux, Émile. [1895]. "Le laboratoire de M. Pasteur à l'École normale." *Revue scientifique*, 4th series, *15*:449–454. Reprinted in Ecole Normale, *Le Centenaire de l'Ecole Normale, 1795–1895* (Paris, 1895); and in Institut Pasteur, *Pasteur, 1822–1922* (Paris, 1922), pp. 39–54.

Duclaux, Émile. [1896]. *Pasteur: Histoire d'un ésprit*. Sceaux, France: Charaire. Trans. Erwin F. Smith and Florence Hedges as *Pasteur: The History of a Mind*. Philadelphia: W. B. Saunders, 1920. Reprint of 1920 edition with a new foreword by Rene Dubos. Library of the New York Academy of Medicine, History of Medicine Series, no. 39. Metuchen, N.J.: Scarecrow Press, 1973.

Ecole Normale. 1895. *Le centenaire de l'Ecole Normale, 1795–1895*. Paris.

Enaux, Joseph and François Chaussier. 1785. *Méthode de traiter les morsures des animaux enragé et de la vipère, suivie d'un précis sur la pustule maligne*. Dijon, France: Defay.

Erikson, Erik. 1965. *Childhood and Society*. Harmondsworth, U.K.

Evans, Richard. 1987. *Death in Hamburg: Society and Politics in the Cholera Years 1830–1910*. Oxford: Oxford University Press.

Faivre, E. 1860. "La question des générations spontanées." *Mém. de l'Acad. des Sci., Belles Lettres et Arts, Lyon 10*:172.

Farley, John. 1972. "The Spontaneous Generation Controversy (1700–1860): The Origin of Parasitic Worms." *J. Hist. of Biol. 5*:95–125.

Farley, John. 1974. "The Initial Reactions of French Biologists to Darwin's *Origin of Species*." *J. Hist. Biol. 7*:275–300.

Farley, John. 1977. *The Spontaneous Generation Controversy from Descartes to Oparin*. Baltimore: Johns Hopkins University Press.

Farley, John. 1978. "The Social, Political, and Religious Background to the Work of Louis Pasteur." *Ann. Rev. of Microbiol. 32*:143–154.

Farley, John, and Gerald L. Geison. 1974. "Science, Politics and Spontaneous Generation in Nineteenth-Century France: The Pasteur-Pouchet Debate." *Bull. Hist. of Med. 48*:161–198.

Farley, John, and Gerald L. Geison. 1982. "Le débat entre Pasteur et Pouchet: Science, politique et génération spontanée." In Michel Callon and Bruno Latour, eds., *La science telle qu'elle se fait*, pp. 1–50. [Paris]: Pandore.

Feinstein, Alvan R. 1987. "The Intellectual Crisis in Clinical Medicine: Medaled Models and Muddled Mettle." *Persp. in Biol. and Med. 30*:215–230.

Fishbein, Daniel B., and Laura E. Robinson. 1993. "Rabies." *New England J. of Med.*, 25 November 1993, pp. 1632–1638.

Fleck, Ludwig. [1936]. *Genesis and Development of a Scientific Fact*. Ed. Thaddeus J. Trenn and Robert K. Merton, trans. Fred Bradley and Thaddeus J. Trenn. Foreword by Thomas S. Kuhn. Chicago: University of Chicago Press, 1979.

Fleming, G. 1886. *Pasteur and His Work, from an Agricultural and Veterinary Point of View*. London.

Flourens, M.J.P. 1863. *Comptes rendus 57*:845.

Flourens, M.J.P. 1864. *Examen du livre de M. Darwin sur l'origine des espèces*. Paris.

Forman, Paul. 1971. "Weimar Culture, Causality, and Quantum Theory, 1918–1927: Adaptation by German Physicists and Mathematicians to a Hostile Intellectual Environment." *Hist. Stud. Phys. Sci. 3*:1–115.

Forman, Paul, et al. 1975. "Physics circa 1900: Personnel, Funding, and Productivity of the Academic Establishments." *Hist. Stud. Phy. Sci. 5*:1–185.

Forrester, John. 1984. "The Pasteurization of France." *History of Science 22*:425–427.

Foster, William D. 1970. *A History of Medical Bacteriology and Immunology*. London: William Heinemann Medical Books.

Fox, Robert. 1976. "Scientific Enterprise and the Patronage of Research in France, 1800–1870." In G. L'E. Turner, ed., *The Patronage of Science in the Nineteenth Century*, chap. 1. New York: Science History Publications.

Fox, Robert, and George Weisz, eds. 1980. *The Organization of Science and Technology in France, 1808–1914*. Cambridge, U.K.: Cambridge University Press.

Frank, Robert M., and Denise Wrotnowska, eds. 1968. *Correspondence of Pasteur and Thuillier Concerning Anthrax and Swine Fever Vaccinations*. Tuscaloosa: University of Alabama Press.

Frankland, Percy F., and Grace C. 1898. *Pasteur*. New York: Macmillan.

Frederico, P. J. 1937. "Louis Pasteur's Patents." *Science 86*:327.

Friedlander, W. J. 1967. "Louis Pasteur and His Stroke: A Lesson in Perseverance." *Nebraska Med. J. 52*:112–114.

Fruton, Joseph. 1972. *Molecules and Life: Historical Essays on the Interplay of Chemistry and Biology*. New York: John Wiley & Sons.

Gaffky, Georg. 1881. "Experimentell erzeugte Septicämie, mit Rücksicht auf progressive Virulenz und accomative Züchtung." *Mittheilungen aus dem Kaiserlichen Gesundheitsamte* 1:1–133.

Gajdusek, D. Carleton. 1977. "Unconventional Viruses and the Origin and Disappearance of Kuru." *Science* 197:943–990.

Galérant, G. 1974. "Deux toulousains et un rouennais contre Pasteur au sujet de la génération spontanée." In *Congrès national des Sociétés savantes, 96e, Toulouse, 1971. Comptes Rendus. Histoire des sciences* 1:193–200.

Galtier, Pierre-Victoire. 1879. "Etudes sur la rage." *Comptes rendus de l'Acad. des sci.* 89:444–446.

Galtier, Pierre-Victoire. 1880. *Traité des maladies contagieuses et de la police sanitaire des animaux domestiques*. Lyon: Beau.

Galtier, Pierre-Victoire. 1881a. "Sur la transmissibilité de la rage du chien au lapin." *Bull. Acad. méd.* 2d ser., 10:90–94.

Galtier, Pierre-Victoire. 1881b. "Les injections de virus rabique . . . semblent conférer l'immunité." *Comptes rendus de l'Acad. des sci.* 93:284–285.

Gálvez, Antonio. 1988. "The Role of the French Academy of Sciences in the Clarification of the Issue of Spontaneous Generation in the Mid-Nineteenth Century." *Annals of Science* 45:345–365.

Gamet, André. 1975. *La rage*. Paris: Presses universitaires.

Gascar, Pierre. 1986. *Du côté de chez Monsieur Pasteur*. Paris: Editions O. Jacob; Le Seuil.

Geison, Gerald L. 1971. "Ferdinand Cohn." In Charles C. Gillispie, ed., *The Dictionary of Scientific Biography*, vol. 3, pp. 336–341. New York: Charles Scribner's Sons.

Geison, Gerald L. 1974. "Louis Pasteur." In Charles C. Gillispie, ed., *The Dictionary of Scientific Biography*, vol. 10, pp. 350–416. New York: Charles Scribner's Sons.

Geison, Gerald L. 1977. Review of Louis Nicol, *L'Épopée pastorienne et la médecine vétérinaire*. *J. Hist. Med.* 32:89–91.

Geison, Gerald L. 1978. "Pasteur's Work on Rabies: Reexamining the Ethical Issues." *Hastings Center Report* 8:26–33.

Geison, Gerald L. 1979. "Divided We Stand: Physiologists and Clinicians in the American Context." In Morris Vogel and Charles E. Rosenberg, eds., *The Therapeutic Revolution: Essays in the Social History of American Medicine*, pp. 67–90. Philadelphia: University of Pennsylvania Press.

Geison, Gerald L. 1981 "Pasteur on Vital versus Chemical Ferments: A Previously Unpublished Paper on the Inversion of Sugar." *Isis* 72:425–445.

Geison, Gerald L. 1985. "Pasteur: A Sketch in Bold Strokes." In *World's Debt to Pasteur*, ed. Hilary Koprowski and Stanley A. Plotkin, pp. 5–27. The Wistar Symposium Series, vol. 3. New York: Alan R. Liss, Inc.

Geison, Gerald L. 1990. "Pasteur, Roux, and Rabies: Scientific versus Clinical Mentalities." *J. Hist. Med.* 45:341–365.

Geison, Gerald L. 1991. "Les à-côtés de l'expérience." *Les cahiers de science & vie*, no. 4 (August 1991): 72–78.

Geison, Gerald L. 1993. "Microbes and Men." Review of BBC Video Series. Special Section: The History of Science in Film. *Isis 84*:761–762.

Geison, Gerald L., and James A. Secord. 1988. "Pasteur and the Process of Discovery: The Case of Optical Isomerism." *Isis 79*:6–36.

George, André. 1958. *Pasteur*. New ed., rev. and enlarged. Paris: Albin Michel.

Gillispie, Charles C. 1981. *Science and Polity in France at the End of the Old Regime*. Princeton, N.J.: Princeton University Press.

Glachant, Victor. 1938. "Pasteur disciplinaire: Un incident à l'École normale supérieure (novembre 1864)." *Revue universitaire 47*:97–104.

Gooding, David. 1982. "Empiricism in Practice: Teleology, Economy, and Observation in Faraday's Physics." *Isis 73*:46–67.

Gooding, David. 1989. "History in the Laboratory: Can We Tell What Really Went On?" In Frank A.J.L. James, ed., *The Development of the Laboratory*. London: Macmillan.

Gooding, David. 1990. *Experiment and the Making of Meaning: Human Agency in Scientific Observation and Experiment*. Dordrecht: Kluwer Academic.

Gooding, David. 1992. "Putting Agency Back into Experiment." In A. Pickering, ed., *Science as Practice and Culture*. Chicago: University of Chicago Press.

Gooding, David, Trevor Pinch, and Simon Schaffer, eds., 1989. *The Uses of Experiment: Studies in the Natural Sciences*. Cambridge, U.K.: Cambridge University Press.

Gould, Stephen Jay. 1988. Review of *Simple Curiosity: Letters from George Gaylord Simpson to His Family, 1921–1970*. *New York Times Book Review*, 14 February 1988, p. 14.

Goutina, V. N. 1971. "L. Pasteur et la microbiologie russe du XIXe siècle." In *12th Internat. Congr. Hist. Sci., Paris, 1968, Actes 8*:51–54.

Grancher, Jacques-Joseph. 1895. "Louis Pasteur." *Ann. d'hygiène publique et de méd. légale*, 3d series, *34*:385–400.

Grmek, Mirko D. 1973. *Raisonnement expérimental et recherches toxicologiques chez Claude Bernard*. Geneva: Droz.

Gruber, Howard E. 1974. *Darwin on Man: A Psychological Study of Scientific Creativity*. New York: E. P. Dutton.

Guérard, Albert L. 1920. *French Prophets of Yesterday: A Study of Religious Thought under the Second Empire*. New York: D. Appleton.

Guizot, François. 1862. *L'église et la société chrétiennes en 1861*. 4th ed. Paris: Michel Levy frères.

Hacking, Ian. 1983. *Representing and Intervening*. Cambridge, U.K.: Cambridge University Press.

Hall, T. S. 1969. *Ideas of Life and Matter*. 2 vols. Chicago: University of Chicago Press.

Hankel, Wilhelm Gottlieb. 1843. "Acide tartrique, forme crystalline du tartrique potassique." In J. Berzelins, ed., *Rapport annuel sur les progrès de la chimie*. 3d year. Paris.

Hansen, Bert. 1991. "La réponse américaine à la victoire de Pasteur contre la rage: Quand la médecine fait pour la première fois la 'une'." In Michel Morange, ed., *L'Institut Pasteur*, pp. 88–102. Paris: Editions la Découverte.

Hemphill, Marie-Louise. 1977. "Pasteur at Arbois." *ASM News 43*:298–299.

Heron, J. R. 1982. "Louis Pasteur: The Nature of His Illness." In R. Passmore, ed., *Proc. Roy. Coll. Physicians of Edinburgh Tercentenary Congress, 1981*, pp. 116–127. Edinburgh: Royal College of Physicians of Edinburgh.

Herter, A. 1904. *Influence of Pasteur on Medical Science*. New York: Dodd, Mead & Co.

Hesse, Mary. 1980. *Revolutions and Reconstructions in the Philosophy of Science*. Brighton, Sussex, U.K.: Harvester Press.

Hoenig, L. J. 1986. "Triumph and Controversy: Pasteur's Preventive Treatment of Rabies as Reported in *JAMA*." *Archives of Neurology 43*:397–9.

Holmes, Frederic L. 1974. *Claude Bernard and Animal Chemistry: The Emergence of a Scientist*. Cambridge, Mass.: Harvard University Press.

Holmes, Frederic L. 1984. "Lavoisier and Krebs: The Individual Scientist in the Near and Deeper Past." *Isis 75*:131–142.

Holmes, Frederic L. 1985. *Lavoisier and the Chemistry of Life: An Exploration of Scientific Creativity*. Madison: University of Wisconsin Press.

Holmes, Frederic L. 1986. "Patterns of Scientific Creativity." *Bull. Hist. Med. 60*:19–35.

Holmes, Frederic L. 1987. "Scientific Writing and Scientific Discovery." *Isis 78*:220–235.

Holmes, Frederic L. 1988. "Lavoisier's Conceptual Passage." *Osiris 4*:82–92.

Holmes, Frederic L. 1990. "Laboratory Notebooks: Can the Daily Record Illuminate the Broader Picture?" *Proc. Amer. Phil. Soc. 134*:349–366.

Holmes, Frederic L. 1991a. *Hans Krebs: The Formation of a Scientific Life, 1900–1930*. New York: Oxford University Press.

Holmes, Frederic L. 1991b. "Argument and Narrative in Scientific Writing." In Peter Dear, ed., *The Literary Structure of Scientific Argument: Historical Studies*, pp. 164–181. Philadelphia: University of Pennsylvania Press.

Holmes, Frederic L. 1993. *Hans Krebs*, vol. 2, *Architect of Intermediary Metabolism, 1933–1937*. New York: Oxford University Press.

Holmes, Frederic L. 1993a. "Research Trails and the Creative Spirit." Unpublished typescript of 26 pages.

Holmes, Frederic L. 1993b. "The Investigative Pathway as an Organizing Metaphor for Scientific Biography." Unpublished typescript of 22 pages.

Holton, Gerald. 1978a. "Subelectrons, Presuppositions, and the Millikan-Ehrenhaft Dispute." In Holton, *The Scientific Imagination: Case Studies*, pp. 25–83. Cambridge, U.K.: Cambridge University Press.

Holton, Gerald. 1978b. "Fermi's Group and the Recapture of Italy's Place in Physics." In Holton, *The Scientific Imagination: Case Studies*, pp. 155–198. Cambridge, U.K.: Cambridge University Press.

Huas, Jeanine. 1985. "Le plus célèbre des mordus." *Historama 16*:64–66.

Huber, Dorian. 1969. "Louis Pasteur and Molecular Dissymmetry, 1844–1857." M.A. thesis, Johns Hopkins University, Baltimore.

Hughes, Sally Smith. 1977. *The Virus: A History of the Concept*. New York: Science History Publications.

Hulin, N. 1986. "La rivalité Ecole-Normale-Ecole Polytechnique. Un antécédent: L'action de Pasteur sous le Second Empire. *Hist. de l'educ. 30*:71–81.

Hume, Ethel Douglas. 1923. *Béchamp or Pasteur? A Lost Chapter in the History of Biology*. Chicago: Covici-McGee. New editions under titles: *Pasteur Exposed: Germs, Genes, Vaccines: The False Foundations of Modern Medicine*. Denmark and Australia: Bookreal Publishers, 1989; and Robert Baillie Pearson, *Pasteur: Plagiarist, Imposter!: The Germ Theory Exploded*. Denver: Health, Inc., 1942.

Hume, E. E. 1927. *Max von Pettenkoffer: His Theory on the Etiology of Cholera, Typhoid Fever and Other Intestinal Diseases: A Review of His Arguments and Evidence*. New York: Paul B. Hoeber.

Hutchinson, John. 1985. "Tsarist Russia and the Bacteriological Revolution." *J. Hist. Med. 40*:420–439.

Ihde, Aaron. [1964]. *The Development of Modern Chemistry*. Reprint, New York: Dover, 1983.

Institut national de la propriété industrielle, Paris 1857, 1861, 1865, 1871, 1872, 1873. *Catalogue des brevets d'invention*. 3 February 1857: *21* (brevet 30646); 9 July 1861: *201* (brevet 50359); 11 April 1865: *114* (brevet 67006); 28 June 1871: *48* (brevet 91941); 21 August 1871: *99* (brevet 92505); 13 March 1873: *78* (brevet 98476).

Institut national de la propriété industrielle, Paris. 1886. *Bulletin officiel de la propriété industrielle et commercial* 2:21 (brevet 174611: 8 March 1886); 248 (brevet 176387: 27 May 1886).

Institut Pasteur. 1888. *Inauguration de L'Institut Pasteur le 14 novembre 1888 in présence de M. Le Président de la Republique. Compte rendu*. Sceaux, France: Chraraire et fils, Imprimeurs des Annales de L'Institut Pasteur.

Institut Pasteur. 1922. *Pasteur, 1822–1922*. Paris: Hachette.

Jacob, François. 1988. *The Statue Within: An Autobiography*. Transl. Franklin Philip. New York: Basic Books.

Jacoby, Susan. 1986. Review of Leo Braudy, *The Frenzy of Renown: Fame and Its History*. In *New York Times Book Review*, 7 September 1986, p. 12.

Jacques, Jean. 1986. "Preface." In *Sur la dissymétrie moléculaire: Louis Pasteur, J. H. van't Hoff, et A. Werner*, pp. 7–45. Paris: Christian Bourgois Éditeur.

James, D. G. 1982–84. "The Hunterian Oration on Louis Pasteur's Final Judgement: Host Reaction, Soil or Terrain." *Trans. Med. Soc. London 99–100*:131–147.

Japp, F. R. 1899. "Stereochemistry and Vitalism." *Report of the 68th Meeting of the Brit. Assoc. for the Advancement of Science, London*, pp. 813–828.

Joravsky, David. 1983. "Unholy Science." *New York Review of Books*, 13 October, pp. 3–5.

Jubilée de M. Pasteur. 1893. Paris: Gauthier-Villars.

Kapoor, Satish C. 1971. "Jean-Baptiste Dumas." In Charles C. Gillispie, ed., *The Dictionary of Scientific Biography*, vol. 4, pp. 242–248. New York: Charles Scribner's Sons.

Kapoor, Satish C. 1973. "Auguste Laurent." In Charles C. Gillispie, ed., *The Dictionary of Scientific Biography*, vol. 8, pp. 54–61. New York: Charles Scribner's Sons.

Kete, Katheline. 1988. "*La Rage* and the Bourgeoisie: The Cultural Context of Rabies in the French Nineteenth Century." *Representations* 22:89–107.

Klosterman, Leo J. 1985. "A Research School of Chemistry in the Nineteenth Century: Jean Baptiste Dumas and His Research Students." *Annals of Science* 42:1–80.

Knorr-Cetina, Karin D., and Michael Mulkay, eds. 1983. *Science Observed: Perspectives on the Social Study of Science*. London: Sage.

Koch, Robert. 1882. *Ueber die Milzbrandimpfung. Eine Entgegnung auf den von Pasteur in Genf gehaltenen Vortrag*. Leipzig: Thieme.

Koelbing, H. M. 1969a. "Chance Favours the Prepared Mind Only (Louis Pasteur)." *Agents Actions 1* (July):53–69.

Koelbing, H. M. 1969b. "Pasteur's Fixed Virus of Rabies." *Agents Actions 1* (November):49–52.

Kohler, Robert. 1971. "The Background to Eduard Buchner's Discovery of Cell-Free Fermentation." *J. Hist. Biol.* 4:35–61.

Kopp, Hermann. 1840. "Considerations sur le volume atomique, l'isomorphisme et les poids specifique." *Ann. de chimie*, 2d ser. 75:416.

Koprowski, Hilary, and Stanley A. Plotkin, eds. 1985. *World's Debt to Pasteur: Proceedings of a Centennial Symposium Commemorating the First Rabies Vaccination*. The Wistar Symposium Series 3. New York: Liss.

Kottler, Dorian B. 1978. "Louis Pasteur and Molecular Dissymmetry, 1844–1857." *Stud. Hist. Biol.* 2:57–98.

Kronick, David. 1976. *A History of Scientific and Technical Periodicals: The Origins and Development of the Scientific and Technical Press, 1665–1790*. 2d ed. Metuchen, N.J.: Scarecrow Press.

Kuhn, Thomas S. 1970. *The Structure of Scientific Revolutions*. 2d ed. Chicago: University of Chicago Press.

Lagrange, Émile. 1954. *Monsieur Roux*. Brussels: Ed. Goemaere.

Latour, Bruno. 1983. "Give Me a Laboratory and I Will Move the World." In Karin Knorr-Cetina and Michael Mulkay, eds., *Science Observed*. London: Sage.

Latour, Bruno. 1984. *Les microbes: Guerre et paix; suivi de irréductions*. Paris: Métailié.

Latour, Bruno. 1986. "Visualization and Cognition: Thinking with Eyes and Hands." In H. Kuklick, ed., *Knowledge and Society: Studies in the Sociology of Culture Past and Present*, pp. 1–40. Greenwich, Ct.: Jai Press.

Latour, Bruno. 1987. *Science in Action: How to Follow Scientists and Engineers through Society*. Cambridge, Mass.: Harvard University Press.

Latour, Bruno. 1988. *The Pasteurization of France.* [Translation of *Les microbes*, 1984.] Cambridge, Mass.: Harvard University Press.

Latour, Bruno. 1989. "Pasteur et Pouchet: Hétérogenèse de l'histoire des sciences." In Michel Serres, ed. *Eléments d'histoire des sciences*, pp. 423–445. Paris: Bordas.

Latour, Bruno. 1992. "Pasteur on Lactic Acid Yeast: A Partial Semiotic Analysis." *Configurations 1*:129–145.

Latour, Bruno, and Steven Woolgar. 1986. *Laboratory Life: The Construction of Scientific Facts.* 2d ed. Princeton, N.J.: Princeton University Press.

Laurent, A. 1847. "Action des alcalis chlorés sur la lumière polarisée et sur l'économie animale." *Comptes rendus de l'Acad. des sci. 24*:220.

Lechevalier, Hubert A., and Morris Solotorovsky. 1974. *Three Centuries of Microbiology*. New York: Dover.

Ledoux, E. 1941. *Pasteur et la Franche-Comté: Dole, Arbois, Besançon.* Besançon, France: Chaffanjon.

Leicester, H. M. 1973. "Herman Kopp." In Charles C. Gillispie, ed., *Dictionary of Scientific Biography*, vol. 7, pp. 463–464. New York: Charles Scribner's Sons.

Léonard, Jacques. 1979. "Pasteur, un vétérinaire qui ira loin." *Histoire 12*:78–80.

Léonard, Jacques. 1983. "La pensée médicale au XIXe siècle." *Revue de synthèse 104*:29–52.

Lépine, Pierre. 1948. "Rage—Virus Rabique." In C. Levaditi and Pierre Lépine, eds., *Les ultravirus des maladies humaines*, 2d ed., vol. 1, pp. 301–444. Paris: Librairie Maloine.

Löffler, Friedrich. 1881. "Zur Immunitätsfrage." *Mittheilungen aus dem Kaiserlichen Gesundheitsamte 1*:134–187.

Loir, Adrien. 1937–1938. "A l'ombre de Pasteur." *Mouvement sanitaire 14*:43–47, 84–93, 135–146, 188–192, 269–282, 328–348, 387–399, 438–445, 487–497, 572–573, 619–621, 659–64; *15*:179–181, 370–376, 503–508. [From some citations in French sources, it is clear that this series of articles was gathered together in a single volume (Adrien Loir, *A l'ombre de Pasteur*, Paris, 1938), which, however, I have never seen.]

Lorinser, Fr. W. 1885a. "Gegen den Impfzwang." *Wiener Med. Wochenschrift 35*:1511–1513.

Lorinser, Fr. W. 1885b. "Pasteurs Impfschutz gegen Tollwuth der Hunde und Lyssa der Menschen." *Wiener Med. Wochenschrift 35*:1562–1565.

Lorinser, Fr. W. 1888. "Die Impffage, mit Rücksicht auf 'Die Beiträge des kais. deutschen Gesundheitsamtes zur Beurtheilung des Nutzen der Schutzpockenimpfung." *Wiener Med. Wochenschrift 38*:1493–1496.

Lumet, Louis. 1923. *Pasteur, sa vie, son oeuvre: Ouvrage orné de 121 gravures.* Paris: Hachette.

Lutaud, Auguste J. 1887a. *Hydrophobia in Relation to M. Pasteur's Method and the Report of the English Committee* (brochure). London.

Lutaud, Auguste J. M. 1887b. *Pasteur et la rage.* Paris.

Macfarlane, Gwyn. 1984. *Alexander Fleming: The Man and the Myth*. Cambridge, Mass.: Harvard University Press. (Paperback, Oxford University Press, 1985.)

McKie, D. 1948. "The Scientific Periodical from 1665 to 1798." *Phil. Mag. 39*:122–132.

Mason, S. F. 1991. "From Pasteur to Parity Violation: Cosmic Dissymmetry and the Origins of Biomolecular Handedness." *Ambix 38*:85–99.

Maulitz, Russell C. 1979. "Physician versus Bacteriologist: The Ideology of Science in Clinical Medicine." In Morris Vogel and Charles E. Rosenberg, eds., *The Therapeutic Revolution: Essays in the Social History of American Medicine*, pp. 91–107. Philadelphia: University of Pennsylvania Press.

Maurice-Raynaud, R., and J. Théodoridès. 1986. "Une lettre de Pasteur à propos de la rage." *Hist. sci. méd. 20*:23–30.

Mauskopf, Seymour. 1976. "Crystals and Compounds: Molecular Structure and Composition in Nineteenth-Century French Science." *Trans. Amer. Phil. Soc.* n.s., *66*:1–82.

Medawar, Peter B. 1964. "Is the Scientific Paper Fraudulent?" *Saturday Review*, 1 August, pp. 42–43.

"Medical Research: Statistics and Ethics." 1977. Special issue of *Science 198*:677–705.

Melhado, Evan M. 1980. "Mitscherlich's Discovery of Isomorphism." *Hist. Stud. Phys. Sci. 11*:87–123.

Mercier, P. 1984. "La fondation de l'Institut Pasteur." In A. Pecker, ed., *La Médecine à Paris du XIIIe au XXe siècle*, 173–82. Paris: Fondation Singer-Polignac, Editions Hervas.

Mercier, P., and P. Atanasiu. 1985. "La vaccination antirabique chez l'homme et son évolution." *Bull. Acad. natl. med. 169*(6) (June):779–783.

Merton, Robert K. 1968. *Social Theory and Social Structure*. Enlarged ed. New York: Free Press.

Merton, Robert K. 1973. "Priorities in Scientific Discovery." In Norman W. Storer, ed., *The Sociology of Science*. Chicago: University of Chicago Press.

Merton, Robert K. 1978. "The Sociology of Science: An Episodic Memoir." In Robert K. Merton and Jerry Gaston, eds., *The Sociology of Science*, pp. 3–141. Carbondale, Ill. Southern Illinois University Press.

Merz, John Theodore. [1904–1912]. *A History of European Thought in the Nineteenth Century*. 4 vols. Reprint, New York: Dover, 1965.

Metchnikoff, Élie [Ilya]. 1933. *Trois foundateurs de la médécine moderne: Pasteur, Lister, Koch*. Paris: Librairie Felix Alcan.

Metchnikoff, Élie. [Ilya]. 1939. *The Founders of Modern Medicine: Pasteur, Koch, Lister*. Trans. D. Berger. New York: Walden.

Metchnikoff, Élie [Ilya]. 1965. "Recollections of Pasteur." *Ciba-Symposium 13*:108–111.

Miles, A. 1982. "Reports by Louis Pasteur and Claude Bernard on the Organization of Scientific Teaching and Research." *Notes and Records Roy. Soc. London 37*:101–118.

Miller, Michael B. 1981. *The Bon Marché: Bourgeois Culture and the Department Store, 1869–1920*. Princeton: N.J.: Princeton University Press.

Milne-Edwards, H. 1859. "Remarques sur la valeur des faits qui sont considérées par quelques naturalistes comme étant propers à prouver l'existence de la génération spontanée des animaux." *Comptes Rendus 48*:24.

Mollaret, H. H. 1987. "L'ésprit d'invention." Review of Pierre Gascar, *Du côté de chez Monsieur Pasteur. Méd. et maladies infectieuses 5*:280–283.

Moore, Francis. 1972. "Therapeutic Innovation." In Paul A. Freund, ed., *Experimenting with Human Beings*. London: Allen and Unwin.

Morange, Michel, ed. 1991. *L'Institut Pasteur: Contributions à son histoire*. Paris: Editions la Découverte.

Moreau, Richard. 1986. "La vaccination et le prix de vertu de Jean-Baptiste Jupille (à propos d'un texte de Maxime du Camp)." *Revue de l'Institut Pasteur de Lyon 19*:263–274.

Moreau, Richard. 1989. "Le dernier pli cacheté de Louis Pasteur à l'Académie des sciences." *La vie des sciences 6*:403–434.

Moreau, Richard. 1992. "Les expériences de Pasteur sur les générations spontanées. Le point de vue d'un microbiologiste. Première partie: La fin d'un mythe. Deuxième partie: Les conséquences." *La vie des sciences 9*:231–260, 287–321.

Morson, Gary Saul. 1985. "Two Voices in Every Head." *New York Times Book Review*, 10 February 1985, p. 32. A review of Mikhail Bakhtin, *The Dialogical Principle*.

Morson, Gary Saul, and Caryl Emerson, eds. 1989. *Rethinking Bakhtin: Extensions and Challenges*. Evanston, Ill.: Northwestern University Press.

Morson, Gary Saul, and Caryl Emerson. 1990. *Mikhail Bakhtin: Creation of a Prosaics*. Stanford, Calif.: Stanford University Press.

Moschcowitz, Eli. 1948. "Louis Pasteur's Credo of Science: His Address When He Was Inducted into the French Academy." *Bull. Hist. Med. 22*:451–466.

Moulin, Anne Marie. 1991a. *La dernier langage de la médecine: Histoire de l'immunologie de Pasteur au Sida*. Paris: Presses Universitaires de France.

Moulin, Anne Marie. 1991b. "L'inconscient Pasteurian: L'immunologie de Metchnikoff à Oudin, 1917–1940." In Michele Morange, ed., *L'institute Pasteur: Contributions à son histoire*, pp. 144–164. Paris: Editions la Découverte.

Moulin, Anne Marie. 1992a. "Patriarchal Science: The Network of the Overseas Pasteur Institutes." In Patrick Petitjean, Catherine Jami, and Anne Marie Moulin, eds., *Sciences and Empires: Historical Studies about Scientific Development and European Expansion*, pp. 307–322. Dordrecht: Kluwer.

Moulin, Anne Marie. 1992b. "La métaphore vaccine: De l'inoculation à la vaccinologie." *Hist. Phil. Life Sci. 14*:271–297.

Mulkay, Michael, and Nigel Gilbert. 1984. *Opening Pandora's Box: A Sociological Analysis of Scientists' Discourse*. Cambridge, U.K.: Cambridge University Press.

Mulligan, Hugh A. 1994. "Liberating Moment in Paris." *Trenton Times*, 21 August 1994. Section cc, pp. 1, 12–13.

Myers, Greg. 1990. *Writing Biology: Texts in the Social Construction of Scientific Knowledge*. Madison: University of Wisconsin Press.

Nelson, John S., Allan Megill, and Donald N. McCloskey, eds. 1987. *The Rhetoric of the Human Sciences: Language and Argument in Scholarship and Public Affairs*. Madison: University of Wisconsin Press.

Nickles, Thomas. 1978. "Introductory Essay: Scientific Discovery and the Future of Philosophy of Science." In Thomas Nickles, ed., *Scientific Discovery, Logic, and Rationality*, pp. 1–60. Boston Studies in Philosophy of Science 56. Dordrecht: D. Reidel.

Nickles, Thomas. 1985. "Beyond Divorce: Current Status of the Discovery Debate." *Philosophy of Science* 52:177–207.

Nicol, Louis. 1974. *L'épopée pastorienne et la médecine vétérinaire*. Garches: Auxs dépenses de l'auteur.

Nicolle, Charles. 1932. *Biologie de l'invention*. Paris: Librairie Felix Alcan.

Nicolle, Jacques. 1953. *Un maître de l'enquête scientifique, Louis Pasteur*. Paris: La Colombe. Trans. as *Louis Pasteur: a Master of Scientific Enquiry*. London: Hutchinson, 1961.

Nicolle, Jacques. 1966. *Louis Pasteur: The Story of His Major Discoveries*. New York: Fawcett.

Nonclercq, Marie. 1982. *Antoine Béchamp, 1816–1908: L'homme et le savant originalité et fécondité de son oeuvre*. Paris: Maloine.

Notter, A. 1989. "Pasteur et la peinture." In *Conférences d'histoire de la médecine, cycle 1988–1989*, pp. 139–51. Lyon: Institut d'histoire de la médecine.

Nye, Mary Jo. 1986. *Science in the Provinces: Scientific Communities and Provincial Leadership in France, 1860–1930*. Berkeley: University of California Press.

Outram, Dorinda. 1984. *Georges Cuvier: Vocation, Science, and Authority in Post-Revolutionary France*. Manchester, U.K.: Manchester University Press.

Owen, Richard. 1868. *On the Anatomy of Vertebrates*. Vol. 3. London: Longmans, Green.

Paget, Stephen. 1910. "Louis Pasteur." *The Spectator* (London) *105*:509–510.

Paine, T. F. 1985. "Louis Pasteur Commemorated on French Five Franc Note." *J. Tenn. Med. Assoc. 78*(10):622–624.

Palladino, Paolo. 1990. "Stereochemistry and the Nature of Life." *Isis 81*:44–67.

Partington, J. R. 1961–1970. *A History of Chemistry*. 4 vols. London: Macmillan.

Pasteur, Louis. See "Author's Note on Notes and Sources" for information about the following references: Pasteur, *Cahiers*; Pasteur, *Correspondance*; Pasteur, "Notes divers"; Pasteur, *Oeuvres*; and Papiers Pasteur.

Paul, Harry W. 1972a. "The Issue of Decline in Nineteenth-Century French Science." *French Historical Studies* 7:416–450.

Paul, Harry W. 1972b. *The Sorcerer's Apprentice: The French Scientist's Image of German Science*. Gainesville: University of Florida Press.

Paul, Harry W. 1985. *From Knowledge to Power: The Rise of the Science Empire in France, 1860–1939*. Cambridge, U.K.: Cambridge University Press.

Pauly, Philip J. 1987. *Controlling Life: Jacques Loeb and the Engineering Ideal in Biology*. New York: Oxford University Press.

Pennetier, Georges. 1907. *Un débat scientifique, Pouchet et Pasteur (1858–68)*. Rouen: J. Girieud.

Perett, D. B. 1977. "Ethics and Error: The Dispute between Ricord and Auzias-Turenne over Syphilization, 1845–1870." Ph.D. diss., History of Science Dept., Stanford University, Stanford, Calif.

Pfizer and Company, Inc. 1958. *The Pasteur Fermentation Centennial, 1857–1957. A Scientific Symposium on the occasion of the one hundredth anniversary of the publication of Louis Pasteur's Mémoire sur la fermentation appelée lactique*. With contributions by René Dubos and Pasteur Vallery-Radot. New York: Pfizer.

Pickering, Andrew, ed. 1992. *Science as Practice and Culture*. Chicago: University of Chicago Press.

Pictet, F.-J. 1853. *Traité de paléontologie ou histoire naturelle des animaux fossiles*. 2d ed. Paris.

Piel, Gerard. 1986. "The Social Process of Science." *Science* 231 (January 17):201.

Polanyi, Michael. 1958. *Personal Knowledge: Towards a Post-Critical Philosophy*. Chicago: University of Chicago.

Porter, J. R. 1964. "The Scientific Journal—300th Anniversary." *Bacteriological Reviews* 28:211–230.

Pouchet, Félix. 1847. *Théorie positive de l'ovulation spontanée et de la fécondation des mammifers et de l'espèce humaine*. Paris.

Pouchet, Félix. 1848. "Note sur les organes digestifs et circulatoires des animaux infusoires." *Comptes rendus* 28:516–518.

Pouchet, Félix. 1853. *Histoire des sciences naturelles au moyen âge, ou Albert le Grand et son époque considérés comme point de départ de l'école expérimentale*. Paris.

Pouchet, Félix. 1858. "Note sur des proto-organismes végétaux et animaux, nés spontanément dans l'air artificiel et dans le gaz oxygène." *Comptes rendus* 47:979–984.

Pouchet, Félix. 1859a. "Remarques sur les objections relatives aux proto-organismes recontrés dans l'oxygène et l'air artificiel." *Comptes rendus* 48:149.

Pouchet, Félix. 1859b. *Hétérogenie, ou Traité de la génération spontanée*. Paris: Baillière.

Pouchet, Félix. 1862. *Les créations successives et les soulevèments du Globe. Lettres à M. Jules Desnoyers*. Paris.

Pouchet, Félix. 1864. *Nouvelles expériences sur la génération spontanée et la résistance vitale*. Paris, 1864.

Pouchet, Félix, N. Joly, and Ch. Musset. 1863. "Expériences sur l'hétérogenie exécutées dans l'intérieur des glaciers de la Maladetta." *Comptes rendus* 57:558–561.

Prévost, Marie-Laure. 1977. "Manuscrits et correspondance de Pasteur à la Bibliothèque nationale." *Bull. Bibliothèque nationale*, 2d year, 3:99-107.

Priestley, Eliza. 1897. "The French and the English Treatment of Research." *Nineteenth Century* 42:113–123.

Priestley, Eliza. [Lady Eliza]. 1908. *The Story of a Lifetime*. London: Kegan Paul.

Provostaye, Frédéric de la. 1841. "Recherches cristallographiques." *Ann. de chimie*, 3d series, 3:136–137.

Ramon, G. 1962. "Contribution des vétérinaires à la recherche scientifique en biologie et en médecine expérimentale et comparée." *Rev. méd. vét.* 113:512–526.

Raspail, Xavier. 1916. *Raspail et Pasteur; trente ans de critiques médicales et scientifiques, 1884–1914*. Paris: Vigot fréres.

Reid, Robert. 1975. *Microbes and Men*. New York.

Richet, Charles Robert. 1923. *L'oeuvre de Pasteur*. Paris: Alcan.

Ritvo, Harriet. 1987. *The Animal Estate: The English and Other Creatures in the Victorian Age*. Cambridge, Mass.: Harvard University Press.

Roll-Hansen, N. 1972. "Louis Pasteur—A Case against Reductionist Historiography." *Brit. J. Phil. Sci.* 23:347–61.

Roll-Hansen, N. 1979. "Experimental Method and Spontaneous Generation: The Controversy between Pasteur and Pouchet, 1859–64." *J. Hist. Med. Allied Sci.* 34: 273–292.

Roll-Hansen, Nils. 1983. "The Death of Spontaneous Generation and the Birth of the Gene: Two Case Studies of Relativism." *Social Studies of Science* 13:481–519.

Root-Bernstein, Robert. 1979. "Pasteur's Cosmic Asymmetric Force: The Public Image and the Private Mind." Unpublished typescript of about 55 pages.

Root-Bernstein, Robert. 1989. *Discovering*. Cambridge, Mass.: Harvard University Press.

Roscoe, Henry. 1889. "Louis Pasteur." Originally published in *Nature*, 10 October 1889. Reprinted in Bessie Zaban Jones, ed., *The Golden Age of Science: Thirty Portraits of the Giants of 19th-Century Science by Their Scientific Contemporaries*, pp. 448–465. New York: Simon and Schuster, 1966.

Rosenblatt, Roger. 1993. "Lewis Thomas." *New York Times Magazine*. 21 November 1993, pp. 50–53, quote on pp. 50–51.

Rosset, R., ed. 1985. *Pasteur et la rage*. Paris: Informations techniques des services vétérinaires.

Rossi, Darius C. 1870. *Le Darwinisme et les generations spontanees*. Paris: C. Reinwald.

Rossignol, H. 1881. *De la possibilité de rendre les bestiaux réfractoires au charbon par la méthode des inoculations preventives: Rapport sur les expériences de Pouilly-le-Fort faites sous la direction de M. Pasteur avec la collaboration de Mm. Chamberland et Roux*. Melun: Societé d'Agriculture de Melun. Brochure of 95 pages.

Roussillat, Jacques. 1964. "La vie et l'oeuvre du professeur Jacques-Joseph Grancher." Doctoral thesis, Faculté de Médecine de Paris. Guéret, France: Les Presses du Massif Central.

Roux, Emile. 1883. "Nouvelles acquisitions sur la rage." *Thèse de Paris.*

Roux, Émile. 1890a. "Les inoculations preventives." *Proc. Roy. Soc. London 46*:154–172.

Roux, Émile. 1890b. "Bactéridie charbonneuse asporogène." *Ann. de l'Institut Pasteur 4*:25–34.

Roux, Émile. [1896]. "L'oeuvre médicale de Pasteur." In *Agenda du chimiste.* Paris. Reprinted in Institut Pasteur, *Pasteur 1822–1922.* Trans. by Erwin F. Smith as "The Medical Work of Pasteur," *Scientific Monthly 21* (1925):365–389.

Roux, Emile. [1910]. "Madame Pasteur." In Institut Pasteur, *Pasteur, 1822–1922* (see above), pp. 102–104.

Roux, Emile. [1911]. "L'oeuvre agricole de Pasteur." In Institut Pasteur, *Pasteur, 1822–1922* (see above), pp. 89–101.

Royer, Clemence. 1862. *De l'origine des espèces par sélection naturelle.* Paris.

Rudwick, Martin J. S. 1972. *The Meaning of Fossils.* London: MacDonald.

Rudwick, Martin J. S. 1976. "The Emergence of a Visual Language for Geological Science, 1760–1840." *History of Science 14*:149–195.

Rudwick, Martin J. S. 1985. *The Great Devonian Controversy: The Shaping of Scientific Knowledge among Gentlemanly Specialists.* Chicago: University of Chicago Press.

Saint Romain, A. de. 1984. "Pasteur: Les vraies raisons d'une gloire." *Histoire 74*: 40–41.

Salomon-Bayet, Claire, ed. 1986a. *Pasteur et la révolution pastorienne.* Paris: Payot.

Salomon-Bayet, Claire. 1986b. "Postface." In *Sur la dissymétrie moléculaire: Louis Pasteur, J. H. van't Hoff, et A. Werner*, pp. 255–281. Paris: Christian Bourgois Éditeur.

Sapp, Jan. 1987. "What Counts As Evidence or Who Was Franz Moewus and Why Was Everybody Saying Such Terrible Things about Him?" *Hist. Phil. Life Sci. 9*:277–308.

Sapp, Jan. 1990. *Where the Truth Lies: Franz Moewus and the Origins of Molecular Biology.* Cambridge, U.K.: Cambridge University Press.

Schaffer, Simon. 1986. "Scientific Discoveries and the End of Natural Philosophy." *Soc. Stud. Sci. 16*:387–420.

Schaffer, Simon. 1991. "The Eighteenth Brumaire of Bruno Latour." *Stud. Hist. Phil. Sci. 22*:174–192.

Secord, James A. 1986. *Controversy in Victorian Geology: The Cambrian-Silurian Dispute.* Princeton, N.J.: Princeton University Press.

Secord, James A. 1989. "Extraordinary Experiment: Electricity and the Creation of Life in Victorian England." In Gooding et al. 1989, pp. 337–383.

Shapin, Steven. 1982. "The History of Science and Its Sociological Reconstructions." *History of Science 10*:157–211.

Shapin, Steven. 1984, "Experiments Are the Key." *Isis 75*:125–130.

Shapin, Steven, and Simon Schaffer. 1985. *Leviathan and the Air-Pump: Hobbes, Boyle, and the Experimental Life.* Princeton, N.J.: Princeton University Press.

Simonnet, Henri. 1947. *L'oeuvre de Louis Pasteur.* Paris: Masson.

Sturdy, S. 1991. "The Germs of a New Enlightenment." *Stud. Hist. Phil. Sci.* 22:163–73.

Suzor, Renaud. 1887. *Hydrophobia: An Account of M. Pasteur's System Containing a Translation of All His Communications on the Subject, The Techniques of His Method, and The Latest Statistical Results*. London: Chatto & Windus.

Szabadvàry, F. 1974. "Eilhard Mitscherlich." In Charles C. Gillispie, ed., *The Dictionary of Scientific Biography*, vol. 9, pp. 423–426. New York: Charles Scribner's Sons.

Taylor, Kenneth L. 1978. "Gabriel Delafosse." In Charles C. Gillispie, ed., *Dictionary of Scientific Biography*, vol. 15, pp. 114–115. New York: Charles Scribner's Sons.

Temple, D. 1986. "Pasteur's Theory of Fermentation: A Virtual Tautology?" *Stud. Hist. Phil. Sci.* *17*:487–503.

Théodoridès, Jean. 1968. *Un grand médecin et biologiste: Casimir-Joseph Davaine (1812–1882)*. Oxford: Pergamon Press.

Théodoridès, Jean. 1972. "Un precurseur de Pasteur: Pierre-Victoire Galtier." *Arch. internat. Claude Bernard* 2:167–171.

Théodoridès, Jean. 1973. "Quelques grands précurseurs de Pasteur." *Hist. sci. méd.* 7:336–343.

Théodoridès, Jean. 1974. "Considerations historiques sur la rage." *Arch. internat. Claude Bernard* 5:151–160.

Théodoridès, Jean. 1977. "A propos de Henri Toussaint (1847–1890) et de son oeuvre microbiologique." *Hist. sci. méd.* *11*:201–202.

Théodoridès, Jean. 1982. "Paris enragé." *Rev. Hist. pharm. (Paris)* *29*(252):193–198.

Théodoridès, Jean. 1986. *Histoire de la rage: Cave canem*. Paris: Masson.

Théodoridès, Jean. 1989. "Pasteur and Rabies: The British Connection." *J. Roy. Soc. Med.* *82*(8):488–490.

Toussaint, H. 1880a. "De l'immunité pour le charbon, acquise à la suite d'inoculations préventives." *Comptes rendus de l'Acad. des sci.* *91*:135–137.

Toussaint, H. 1880b. "Procédé pour la vaccination du mouton et du jeune chien." *Comptes rendus de l'Acad. des sci.* *91*:303–304.

Toussaint, H. 1881. "Vaccinations charbonneuses." *Association française pour l'avancement des sciences, Compte rendu de la 9 session*, Reims, 1880 (Paris, 1881), pp. 1021–1025,

Tyndall, John. 1885. "Introduction." In *Louis Pasteur: His Life and Labours*, by his son-in-law, René Vallery-Radot, trans. Lady Claud Hamilton, pp. xi–xlii. New York: Appleton.

U.S. Center for Disease Control. 1977. *Morbidity and Mortality Weekly Report* *26*:275.

Vallery-Radot, Maurice. 1982. "Les options politiques de Pasteur." *Rev. des sci. morales & politiques* *137*:461–475.

Vallery-Radot, Maurice. 1985. *Pasteur, un génie au service de l'homme*. Lausanne, Switzerland: P.-M. Favre.

Vallery-Radot, Pasteur. 1954. *Pasteur inconnu*. Paris: Flammarion.

Vallery-Radot, Pasteur. 1956. *Images de la vie et de l'oeuvre de Pasteur: Documents photographiques, la plupart inédits provenant de la collection Pasteur Vallery-Radot.* Paris: Flammarion.

Vallery-Radot, Pasteur. 1958. *Louis Pasteur: A Great Life in Brief.* Trans. Alfred Joseph. New York: Knopf.

Vallery-Radot, Pasteur. 1962. *Pasteur.* Encylopédie par l'image. Paris: Hachette.

Vallery-Radot, Pasteur. 1968. *Pages illustres de Pasteur.* Paris: Hachette.

Vallery-Radot, René. 1883. *Pasteur, histoire d'un savant par un ignorant.* Paris: J. Hetzel et cie. [Strictly speaking, this book was published anonymously, but everyone knew that its author was Pasteur's son-in-law, René Vallery-Radot.] 5th ed., 1884.

Vallery-Radot, René. 1900. *La vie de Pasteur.* 2 vols. Paris: Flammarion.

Vallery-Radot, René. 1911. *The Life of Pasteur.* Trans. by Mrs. R. L. Devonshire. 2 vols. London: Constable.

Vallery-Radot, René. 1914. *Madame Pasteur.* Paris: Emile Paul.

Vallery-Radot, René. 1926. *The Life of Pasteur.* Trans. by Mrs. R. L. Devonshire, with an introduction by Sir William Osler. Abbreviated one-volume ed. Garden City, N.Y.: Doubleday.

Vandervliet, Glenn. 1971. *Microbiology and the Spontaneous Generation Debate during the 1870's.* Lawrence, Kan.: Coronado Press.

Van Ness, G. B. 1969. *Louis Pasteur and the Ecology of Anthrax* (brochure). Hyattsville, Md.: U.S. Agricultural Research Service, Animal Health Division.

Von Frisch, Anton. 1887. *Die Behandlung der Wuthkrankheit.* Vienna.

Wachhorst, Wyn. 1981. *Thomas Alva Edison: An American Myth.* Cambridge, Mass.: MIT Press.

Walton, John K. 1979. "Mad Dogs and Englishmen: The Conflict over Rabies in Late Victorian England." *J. Soc. Hist. 13*:219–239.

Ward, Lorraine. 1994. "The Cult of Relics: Pasteur Materials at the Science Museum." *Medical History 38*:52–72.

Webster, Leslie T. 1942. *Rabies.* New York: Macmillan.

Weindling, Paul. 1992. "Scientific Elites and Laboratory Organisation in *fin de siècle* Paris and Berlin: The Pasteur Institute and Robert Koch's Institute for Infectious Diseases Compared." In Andrew Cunningham and Perry Williams, eds., *The Laboratory Revolution in Medicine*, pp. 170–188. Cambridge, U.K.: Cambridge University Press.

Weisberg, Richard. 1995. "The Representation of Doctors at Work in Salon Art of the Early Third Republic in France." Ph.D. diss., New York University.

Weiss, René. 1924. *La commémoration du centenaire de Pasteur par la Ville de Paris.* Paris: Imprimèrie nationale.

Weisz, George. 1983. *The Emergence of Modern Universities in France, 1863–1914.* Princeton, N.J.: Princeton University Press.

Weisz, George. 1987. "The Posthumous Laennec: Creating a Modern Medical Hero, 1826–1870." *Bull. Hist. Med. 61*:541–562.

Weisz, George. 1988. "The Self-Made Mandarin: The *Éloges* of the French Academy of Medicine, 1824–47." *History of Science 26*:13–40.

Williams, L. Pearce. 1953. "Science, education and the French Revolution." *Isis 44*:311–330.

Winslow, C.-E. A. 1944. *The Conquest of Epidemic Disease*. Princeton, N.J.: Princeton University Press.

Wright, G. 1966. *France in Modern Times*. Chicago: Rand-McNally.

Wrotnowska, Denise. 1958. "Candidatures de Pasteur à l'Académie des sciences." *Hist. de la méd.*, special ed., pp. 1–23.

Wrotnowska, Denise. 1959. "Pasteur précurseur des laboratoires auprès des musées." *Bull. du lab. du Musée du Louvre 4* (Sept.):46–61.

Wrotnowska, Denise. 1967a. "Pasteur, professeur à Strasbourg (1849–1854)." In *92ème Congrés national des sociétés savantes I*, pp. 134–144. Strasbourg-Colmar.

Wrotnowska, Denise. 1967b. "Pasteur et Lacaze-Duthiers, professeur d'histoire naturelle à la Faculté des sciences de Lille." *Hist. sci. méd. 1*:1–13.

Wrotnowska, Denise. 1972. "Disease of the Silk Worms: Pasteur in the Cevennes. His Relations with Adrien Jeanjean, Mayor of Saint-Hippolyte-Du-Fort." *Ala. J. Med. Sci. 9*:205–211.

Wrotnowska, Denise. 1973. Le rouget du porc: Pasteur et Achille Maucuer d'après une correspondance en partie inédite." *Rev. hist. sci* (Paris) *26*:339–364.

Wrotnowska, Denise. 1974a. "Pasteur et la 'Royal Society of London.'" In *Proceedings of the 23d International Congress of the History of Medicine*, pp. 647–653. London.

Wrotnowska, Denise. 1974b. "A propos de la rage: Une lettre inédite de Pasteur." *Arch. internat. Claude Bernard 5*:161–165.

Wrotnowska, Denise. 1975. *Pasteur, professeur et doyen de la Faculté des Sciences de Lille (1854–1857)*. Mémoires de la Section des sciences, 4. Paris: Bibliothèque Nationale.

Wrotnowska, Denise. 1975–1976. "Pasteur et Davaine d'après des documents inédits." *Hist. sci. méd. 9*:213–230, 261–90.

Wrotnowska, Denise. 1976a. "Pasteur, membre du Comice agricole de l'arrondissement de Lille." In Congrès National des Sociétés Savantes, 101e, Lille, *Comptes Rendus*, Section des sciences, fasc. 3, pp. 133–144. Paris, Bibliothèque national, 1976.

Wrotnowska, Denise. 1976b. "Un ami de Pasteur: Charles Chappuis." In Congrès National des Sociétés Savantes, 99e, Besançon. *Comptes Rendus*, Section des sciences, fasc. 5, pp. 171–184. Paris: Bibliothèque National, 1976.

Wrotnowska, Denise. 1978. "Le vaccin anti-charbonneux Pasteur et Toussaint, d'après des documents inédits." *Hist. sci. med. 12*(1):12–14.

Wrotnowska, Denise. 1979. "Claude Bernard et Pasteur: Commission du choléra, en 1865." *Hist. sci. méd. 13*(1):25–32.

Wrotnowska, Denise. 1981. "Pasteur: Premières recherches sur les fermentations," *Clio Medica 15*:191–199.

Wrotnowska, Denise. 1982. "Maladie des vers à soie: Pasteur s'intéresse à la race perpignanaise." In Congrés National des Sociétés Savantes, 106e, Perpignan, *Comptes rendus*, Section des sciences, 1982. fasc. 4, pp. 197–208. Paris: Bibliothèque National, 1982.

Zeldin, Theodore. 1977. *France, 1848–1945*, vol. 2, *Intellect, Taste and Anxiety*. Oxford: Clarendon Press.

Zuckerman, Harriet. 1977. "Deviant Behavior and Social Control in Science." In Edward Sagarin, ed., *Deviance and Social Change*, pp. 87–138. London.

Index

Académie de médecine: debates over Pasteur's rabies research, 220–29, 253–55, 337n.26; fermentation research discussed at, 269; role in Toussaint-Pasteur rivalry, 160–69, 173–76
Académie des sciences: optical isomer research presented by Pasteur at, 85; Pasteur manuscripts at, 3; Pasteur's membership in, 26–28; Pouilly-le-Fort anthrax trial recounted to, 146–47, 325n.9; rabies commission, 192–93; rabies research presented by Pasteur to, 190–95, 201–202, 212–18, 220; spontaneous generation debate and, 112–21, 125–29; theories on origin of life at, 321–22; Toussaint-Pasteur rivalry and, 160–69, 173–76
Académie française, Pasteur's membership in, 261, 270
agrégé-préparateurs, Pasteur's encouragement of, 29, 48, 237
Agricultural Society of Melun: as official sponsor of Pouilly-le-Fort trial, 147, 151, 156, 158, 171, 325n.7; publishes account of Pouilly-le-Fort trial, 325n.3; text of protocol for Pouilly-le-Fort trial, 281–85
AIDS, ethical dilemmas regarding, 230–31
alcoholic fermentation, Pasteur's research on, 41, 107–109
Alexandre the Third (Prince), 265
alkalinity, spontaneous generation theory and, 119, 131
alternative medicine, Pasteur's research and, 275
Amoroso, Dr., 223–24
amyl alcohol, Pasteur's fermentation research and, 95–103, 134–35
amylates, Pasteur's research concerning, 99
Annales de chimie, 62
Annales scientifiques de l'Ecole Normale Supérieure, 29
anthrax vaccine: antiseptic version, development of, 20–21, 148–59, 326nn.18; commercialization of, 41–42; criticism of, 327n.27; Enaux-Chaussier book on, 181; ethics of research, in Pasteur's work on, 17–18, 145–76; germ theory of disease and, 33; laboratory notebooks on Pouilly-le-Fort trial, 11, 151–59; *modus fasciendi* as described by Pasteur, 148–59, 325n.13, 326n.16, 331n.84; Pasteur's research methods, 33, 181–82, 237–38; Pasteur's monopoly over, 174–76; Pouilly-le-Fort trial, 145–76, 192; protocol for Pouilly-le-Fort trial, 147, 151–56, 170–71, 281–85, 325n.7; public account of Pouilly-le-Fort trial, 146–50; rabies research influenced by, 183–85, 186–88, 192, 228; revisionist accounts of Pasteur's research in, 276–77; Roux's role in, 150, 158, 164–65, 168–69, 172–76, 237–39; Toussaint-Pasteur competition for development of, 158–69, 172–76, 286–98, 328n.30; trials with least virulent to most virulent strains, 245
anti-vaccination movement, 218
anti-vivisectionists, 218, 265
antiseptic techniques, Lister's campaign for, 33, 262
antiseptic vaccines: conversion to, for rabies, 231; Pasteur's use of, in anthrax trials, 20–21, 149–50, 167, 172–76, 326nn.18, 331n.84; Toussaint-Bouley trials with, 70, 164–69, 172–76, 330nn.62–65
Appel, Toby, 122
Appert, 118
Archives Nationales, Pasteur manuscripts at, 3
arsenious acid, Pasteur's thesis on, 60, 315n.17
asparagine, Pasteur's research concerning, 98–99
aspartates, Pasteur's research concerning, 98–99, 102–103, 135–36
asymmetry: "law of hemihedral correlation" and, 97–103; organic compounds and, Pasteur's view of, 103–105, 135–38; Pasteur's experiments concerning, 138–42, 272; Ruhmkorff apparatus for research on, 139–40

atheism, spontaneous generation debate and, 123–25, 127
atmospheric dust, role of, in spontaneous generation research, 114–19
"atomic volume" concept, 62–64
attenuation techniques: anthrax vaccine development with, 148–50, 326n.16, 327n.26, 328n.30; caustic potash as agent, 213–15; Pasteur's work with, 159–60, 166–69, 170–71, 212–13, 329n.59, 330n.70; rabies vaccine development with, 184–88, 191, 203–205, 232; theoretical basis for, 245; trials with, recounted in laboratory notes, 151–59, 242–43; virulence of spinal cord emulsification, Pasteur's trials with, 243–50. *See also* intracranial injection; oxygen attenuation; serial passage of viruses; spinal cord emulsification
Auzias-Turenne, Joseph-Alexandre, 276

Bacon, Francis, 12
bacteriology, spontaneous generation theory and, 128. *See also* microbes
Bakhtin, Mikhail, 5, 309n.7
Balard, Antoine Jérôme, 53, 59–60, 116, 126
Barbet, M., 24
Bardsley, Samuel Argent, 301–304
Barnes, S. B., 310n.24
Bastian, Henry Charlton, 128–29
Béchamp, Antoine, 275–76
Béchamp or Pasteur?, 275
Beclard, Dr., 337n.26
beer fermentation, brewers' support of Pasteur's work in, 41
Bensaude-Vincent, Bernadette, 323n.41, 337n.23, 341n.32
Bernal, J. D., 47, 58, 93–94, 314n.10, 317nn.44–45, 319n.12
Bernard, Claude, 125; biographies of, 11–13; "germ theory" of fermentation challenged by, 18–21; Pasteur's attack on, 134
Berthelot, Marcellin, "chemical" theory of fermentation, 19
Berzelius, Jacob, fermentation research of, 106
Bigo, M., 27, 92–95, 100
biological theory of immunity: Pasteur's commitment to, 158, 163–66, 329n.46; Pasteur's switch to chemical theory, 245–50
Biot, Jean-Baptiste, 38; amyl alcohol research, 96; optical isomer research of, 56–57, 84–85, 101, 318n.54; Pasteur influenced by, 59–61, 71–72, 87–89, 135, 139
bitartrates, Pasteur's research on, 66–68
Bon Marché department store, 41, 49
Bonaparte, Napoleon, 44, 122, 236, 339n.7
Bonaparte, Princess Mathilde, 110, 132
Boucicaut, Madame, 41
Bouley, Henri, 36, 160–69, 172, 174, 192, 290–93, 328nn.30, 330n.67
Boussingault, Joseph, 129
Broad, William, 309n.16, 332n.96
Brogniart, Adolphe-Théodore, 125
Brouardel, Dr., 30, 35, 227, 337nn.25–26, 338n.51
Brown-Séquard, Charles, 46
Buchner, 108
Bulloch, William, 323n.45
Burke, Donald, 276, 341n.48
Burndy Library, 273, 314n.7

Cabanis, Georges, 121–22
Cadeddu, Antonio, 21, 40, 151, 276–77, 326n.18, 338n.2, 341n.51
carbolic acid: development of antiseptic anthrax vaccine with, 156, 158, 164–65, 168–69, 172–73, 331nn.84–85; inactivated rabies vaccine with, 231
Carnot, Sadi, 259, 261–62
Catholic Church: spontaneous generation debate and, 123–25
caustic potash, attenuation with, 213–15, 234–35
Chamberland, Charles, 33, 48–49, 129; pledges vaccine proceeds to Institut Pasteur, 261; rabies vaccine research with Pasteur, 182, 246; role in anthrax vaccine development, 26, 70, 151, 156, 158, 164, 168–69, 171–76, 299–300, 327n.21, 330n.66
Chamberland-Pasteur filter, 41
Chappius, Charles, 138–39
Charcot, 221
Charlton, D. G., 323n.32
Chaussier, François, 181
Chauveau, Auguste, 59, 164, 184, 247, 329n.46
"chemical" theory of fermentation, 19, 106–109, 163–68, 329n.59

chemical theory of immunity: Pasteur's conversion to, 245–50
Chevreul, Michel Eugène, 54, 116
chicken cholera: parallel research by Pasteur and Toussaint on, 166–69, 329n.55; Pasteur's vaccine against, 33, 39–40, 182, 311n.46; rabies research influenced by, 182–85, 187–88, 228; revisionist account of Pasteur's research in, 276–77; trials with least virulent to most virulent strains, 245; vaccine linked with anthrax vaccine research, 148–50, 159, 162
clinical issues in rabies research, critiques of Pasteur's methods, 220–29
Cohn, Ferdinand, 131–32, 323n.60
Colin, Gabriel, 160–61, 328n.38
Collège d'Arbois, 23
Collège Rollin, 33
Collège Royale de Besançon, 23
Collins, Harry, 15, 310n.27
Communard uprising, 31–32, 140
Constantine of Russia (Grand Duke), 259
corneal transplants, rabies transmission through, 202
correspondence: "private science" and, 6; sources of, 305–306
Coste, Jacques, 125
Cressac, Mary, 339n.5
crystalline structure: "law of hemihedral correlation" and, 97–103; Pasteur's optical isomer discovery and, 73–79, 81–85, 317n.44
crystallization, waters of, Pasteur's tartrate research and, 66–72
crystallography, Pasteur's early research on, 25–26
cult of Pasteur, international dimensions of, 264–67
Cuny, Hilaire, 312n.58, 313n.63
Cuvier, Georges, 121–24, 127, 132

Dagognet, François, 318n.51
Dansette, A., 323nn.32–33
Darwin, Charles: Huxley's defense of, 219; prestige of, 267; "repression" of private thoughts, 5–6, 309n.8; spontaneous generation debate and, 124–25, 127
Davaine, Casimir-Joseph, 168–69, 181–82, 247, 276, 326n.18, 330n.65, 332n.85, 333n.15
Dear, Peter, 341n.31
Decourt, Philippe, 48, 55, 276, 326n.18, 338n.50, 341n.42
De Kruif, Paul, 18, 93, 266–67, 310n.29, 340n.25
Delafosse, Gabriel, 59, 61, 84, 87, 315n.15
De la Provostaye, 75, 87
de Latour, Charles Cagniard, 106
Delaunay, Albert, 340n.13
de Monet, Jean-Baptiste, 121
Dessaignes, Victor, 101–102, 135–36
diamagnetism, empirical research in, 318n.50
Diara, Anna, 49, 127, 323n.41
Dibner, Bern, 314n.7
Dictionary of Scientific Biography, 276
dimorphism: Laurent's research on, 59, 62; Pasteur's tartrates research and, 61–62, 316n.26
dog massacres, myths surrounding rabies and, 179
Dubos, René, 311n.36, 312n.47; biography of Pasteur, 275; on fermentation research, 94; on hay bacillus in Pasteur's anthrax vaccine, 327n.26; on rabies vaccine research, 332n.1, 334n.23; on Pasteur's religious beliefs, 312n.59; on spontaneous generation research, 138
Du Camp, Maxime, 270, 341n.33
Duclaux, Emile, 45, 311n.14, 320n.29, 323n.45; biography of Pasteur, 272, 341n.41; collaboration with Pasteur on chicken cholera vaccine, 40; on fermentation research by Pasteur, 100–101, 107; on Pasteur's obsession with secrecy, 309n.10; on Pasteur's working habits, 47–48, 313nn.80–81; on spontaneous generation research by Pasteur, 54, 59, 69, 116–17, 129, 131, 133, 135, 322, 323n.45, 324n.67; relations with Pasteur, 48–49, 236–37
Dugès, Antoine, 121
Duhem, Pierre, 48
Dujardin-Beaumetz, Dr. Georges, 195, 197–98, 200, 203, 337n.26
Dumas, Alexandre, 110
Dumas, Jean-Baptiste, 38, 44, 71–73, 88–89, 319nn.61–62; spontaneous generation debate and, 126, 128–29
Duruy, Victor, 110

Ecole Centrale des Arts et Manufactures, 232
Ecole des Beaux-Arts, Pasteur's lectures at, 28–29
Ecole Normale Supérieure, 237, 271; Pasteur admitted to, 23–24; Pasteur appointed teaching assistant to Balard, 53; Pasteur named director of scientific studies, 27–32, 92; student disturbances at, 30–31
Ecole Polytechnique, 24, 29
Edelfelt, Albert, 341n.37
Edison, Thomas A., 278
Einstein, Albert, 267
empirical novelty: Pasteur's tartrate research and, 79–85; rabies vaccine and, 227–29, 245–46
Empress Eugénie, 38, 41
Enaux, Joseph, 181
England, cult of Pasteur in, 265–66
English Commission on Rabies, Report of, 225–26, 338n.42
Erikson, Erik, 277, 341n.53
ethics in research: assessment of Pasteur's research, 16–18, 35–45; competition and deception in Pasteur's anthrax vaccine trials, 145–76; human trials with rabies vaccine and, 178, 193, 197, 229–33, 253–55; infectious disease, ethics of preventive measures for, 229–30; Meister vaccination as issue of, 17, 181, 193, 205–209, 215–17, 231, 233, 238–39, 335n.1, 336nn.2–3; Pasteur's comments on, 97, 232–33, 320n.17; process of discovery and, 57–58, 78–85, 88–89, 314n.12; role of laboratory notebooks in, 11–16, 195–205, 229–33, 253–55; Pasteur-Pouchet debate and, 114–21; "private patients" in Pasteur's rabies experiments and, 195–205; rabies vaccine, issues surrounding, 220–31; Roux's tension with Pasteur regarding, 236–39, 253–55; Toussaint's sealed note (pli cacheté) and, 160–62, 286. *See also* fraud in scientific research; human trials of rabies vaccine
evolutionary theory, spontaneous generation debate and, 121–29, 323n.37. *See also* Darwin, Charles
Examen du livre de M. Darwin sur l'origine des espèces, 127, 323n.47
"exhaustion" theory of immunity, 246
experimental techniques: critiques of Pasteur's methods in rabies research, 220–29; details of, in Pasteur's laboratory notebooks, 37, 240–43; human experiments, Pasteur's reservations regarding, 178, 193, 232–33, 320n.17; Pasteur-Pouchet debate on, 114–21; "post-exposure" experimental trials, 238–42. *See also* specific techniques, e.g., attenuation technique; intracranial injection

Faculty of Sciences (Lille), 26–27, 90–93
Faculty of Sciences (Paris), 29
Faculty of Sciences (Strasbourg), 25–26
Faivre, Ernst, 123–24, 323n.35
Falloux Law, 123
Faraday, Michael, 12
Farley, John, 50–53, 128, 319n.63, 321–22, 323n.48
Faure, François Félix, 259
Feinstein, Alvan R., 339n.6
fermentation: "chemical" theory of, 19, 106–109; "germ theory" of, 18–21, 90–91, 106–109, 163, 219–20; law of hemihedral correlation and, 95–103; Pasteur's research on, 26–27, 32–33, 36, 134–35, 90–109, 268–69; spontaneous generation theory and, 113–21; theoretical *vs.* industrial motivations for research in, 91–95; U.S. patents awarded to Pasteur for work in, 266
Fermi, Enrico, 313n.86
Flourens, Pierre, 47, 125–27, 323n.43
"force plastique": spontaneous generation theory and, 113
Fox, Robert, 319n.62, 341n.28
Franco-Prussian War, Pasteur's reaction to, 31–32, 45
Frankland, Percy F., 341n.35
fraud in scientific research: construction of scientific knowledge and, 11–16; Pasteur experiments and, 10–11, 175–76, 309n.16, 332nn.95–96
Frederico, P. J., 340n.23
French Association for the Advancement of Science, 165–66, 329n.52
French Rabies Commission, 240
Freud, Sigmund, 267
Friedlander, W. J., 313n.74
Friendly Association of Former Students, 232
Fruton, Joseph, 320n.45

Galérant, 323n.41
Galileo, 139–40, 267
Galtier, Pierre-Victor, 32, 184–85, 189, 276, 334nn.27–29
Galvez, Antonio, 321–22
Gamaleia, Dr., 340n.19
Gascar, Pierre, 7, 20, 339n.5, 340n.13, 341n.36
Gay-Lussac, Joseph Louis, 118
Geison, Gerald L., 49–50, 319n.63, 320n.30, 338n.45, 339n.6, 340nn.5–6, 341nn.28–29
Geoffroy Saint-Hilaire, Etienne, 121–25, 127, 132, 322n.30
geology, spontaneous generation debate and, 122–23, 323n.31
George, André, 340n.1; on Pasteur's religious beliefs, 312n.59
Gerhardt, M., 70–71
Germany, Pasteur's attitude toward, 45; reaction to Pasteur in, 264–65
germ theory of disease: emergence of, 32–33, 90–91, 170; Pasteur's rabies vaccine and, 193, 227
"germ theory" of fermentation: Bernard's challenge of, 18; Pasteur's research on, 36, 90–91, 106–109, 163, 219–20; spontaneous generation and, 113–21
Gillispie, Charles C., 309n.5
Girard, M., private patient in Pasteur's rabies experiments, 195–205, 231, 250, 335n.56
Glachant, Victor, 311n.20
Gooding, David, 12, 318n.50
Gould, Stephen Jay, 6
government grants for research, Pasteur's use of, 40–42, 312n.50
Grancher, Jacques-Joseph (Dr.), 206–207, 215–18, 238–39, 251, 254–55, 272, 336n.2, 337n.26, 341n.39, 341n.40
Grand Cross of the Legion of Honor, Pasteur's receipt of, 260
Grmek, M. D., 11–12, 15, 309n.17
Gruber, Howard E., 309n.8
Guérard, Albert, 123, 323n.34
Guizot, François, 124, 323n.36

Hankel, Wilhelm Gottlieb, 71, 75–76
Hansen, Bert, 340n.24
Haüy, Abbé Rene-Just, 59, 72, 79, 84, 97
hay bacillus: addition of, in anthrax vaccine, 327n.26
hemihedrism, Delafosse's research on, 59; fermentation research by Pasteur, 95–103; optical isomers and, 81–85, 87–89; paratartrate structure and, 80–85, 318n.51; and tartrate research, 55, 71–72, 75–78, 85–86, 314n.6, 317n.46. *See also* law of hemihedral correlation
Hemphill, Marie-Louise, 341n.43
Heron, J. R., 313n.74
Herter, Christian (Dr.), 274, 340n.22
Herter, Richard, 341n.44
heterogenesis, spontaneous generation theory and, 112–13, 120–21
Heterogenesis: A Treatise on Spontaneous Generation, 112–13, 125
Histoire d'un savant par un ignorant, 273–74
history of science: Pasteur's interest in, 20–21
HIV virus, rabies virus compared with, 231
Holmes, F. L., 11–13, 15–16, 19, 25, 309n.17, 310n.23
Holton, Gerald, 12–13, 15–16, 309n.18, 310n.22, 313n.86
"horse typhoid," Pasteur's research on, 182
Hugo, Victor, 262
human trials of rabies vaccine: ethical concerns over, 178, 193, 197, 232–33, 320n.17; Pasteur's "private patients" as, 195–205; Roux's reluctance regarding, 238–39. *See also* ethics in research
Hume, Ethel Douglas, 275, 341nn.45–46
Hutchinson, John, 340n.20
Huxley, T. H., 218–20

immunity: biological theory of, Pasteur's commitment to, 158, 163; Pasteur's switch from biological to chemical theory, 245–50; relativity of, 185–86
industrialists: Pasteur's research and, 41, 92–95
infectious disease, ethics of preventive measures for, 229–30; Pasteur's research on, 32–33, 237
inorganic compounds: Pasteur's research concerning, 104–105
Institut de France, 160–61
Institut Pasteur, 178; burial of Pasteur at, 260; employee relations in, 49, 313n.87; founding of, 35; inauguration of, 218, 261, 270–71; international reputation of, 263; Pasteur in residence at, 259

International Congress of Medical Sciences, 261; Pasteur's speech on rabies vaccine at, 178, 192–93, 232–33
"In the Shadow of Pasteur," 7, 149
intracranial injection, rabies vaccine research with, 189–95, 213–18, 224. *See also* serial passage of viruses
Introduction to the Study of Experimental Medicine, 19
isomorphism: Laurent's research on, 59, 62; Pasteur's tartrates research and, 62–73, 316n.26

Jacob, François, 263–64, 339n.7, 340nn.14–15
Jacoby, Susan, 341n.26
Jacques, Jean, 318n.51
Jenner, Edward, 172, 228
Joly, Nicolas, 125–26, 321–22
Joravsky, David, 309n.16
Joubert, Jules, 129
Jungfleisch, Emile, 136–38
Jupille, Jean-Baptiste: Pasteur's rabies vaccine and, 177–78, 193, 207, 210–12, 216–18, 242; receives Académie des sciences prize for virtue, 217, 336n.16; chemical theory of immunity and trials on, 248; commemorative statue of, 263

Kapoor, Satish C., 314n.13, 319n.63
Kete, Katheline, 12, 333n.7
Klosterman, Leo J., 319n.62
Koch, Robert, 33, 36, 39, 264–65; criticism of Pasteur's anthrax vaccine, 158, 176; on virulence through serial passage, 184; Pasteur's dispute with, 45, 50; state-funded research support for, 41
Kopp, Hermann, 62–64
Krebs, Hans, 12–13

laboratory notebooks (generally): Bernard's notebooks posthumously published, 18–21; scientific fraud and construction of scientific research, 11–16
laboratory notebooks of Pasteur: attenuation technique described in, 151–59, 242–43; description of, 3–4, 7–11, 306; dimorphism/isomorphism research in, 62, 68–72, 316n.26; ethical issues raised by, 11–16, 195–205, 229–33, 253–55; experimental techniques recounted in, 37, 240–43; Jupille's treatment recorded in, 207–12, 336n.17; Loir's accounts as different from, 274–77, 338n.2; "Meister method" (from least virulent to most virulent) recorded in, 250–52; Meister's treatment recorded in, 205–209, 336n.3; missing first notebook, entries reproduced from, 57–58, 63, 65, 67–68, 74–77, 306, 314n.10, 317nn.44–47; optical isomer research reconstructed from, 11, 57–58, 66–73; Pasteur's conversion from biological to chemical theory recounted in, 246–50; Pasteur's insistence on total control of, 8–10, 226; Pouilly-le-Fort anthrax trial recorded in, 151–59, 170–71; post-exposure rabies trials noted in, 233–42; "private patients" in rabies experiments noted in, 195–203, 334n.52, 335n.54, 335nn.54–55; rabies vaccine research as documented in, 8–9, 11, 181–88, 239–43; restricted access to, 3; spontaneous generation debate and, 135–36; tartrates research, 62, 68–73, 316n.26
lactic acid fermentation: Pasteur's research on, 90–90, 106–107, 113–14, 319n.2
Lagrange, Emile, 8–10, 20, 255–56, 326n.18, 339n.5
Lamarck, 121–22
Lannelongue, Dr., 182
Latour, Bruno, 15, 132–33, 223, 310n.27, 319n.2, 324nn.65–66, 337n.33, 341n.54
Laurent, Auguste: influence on Pasteur of, 17, 44, 58–62, 64–65, 70, 76–77, 79–85, 135, 268, 314n.13, 315nn.14, 317nn.42, 318n.47; optical isomer research of, 101; Pasteur's disavowal of influence of, 86–89, 318n.58
Lavoisier, Antoine, 12–13, 21
L'Avenir national, 127
"Law of hemihedral correlation," 97–103, 134–35
Ledoux, E., 310nn.2–3, 332n.1
Leicester, H. M., 316n.27
Le mouvement sanitaire, 275–77
Lépine, Pierre, 333n.9, 338n.50
Lewis, David, 327n.28
Lewis, Sinclair, 266
Life of Pasteur (*La vie de Pasteur*), 53, 56, 273–75, 277
life, origins of: asymmetric forces and, Pasteur's research into, 138–42, 271–72;

fermentation research and, 101–103; nineteenth-century theories on, 321–22; organic compounds and, 103–105, 135–38
Lindsay, John, 180, 301–304
Lister, Joseph, 33, 36, 164, 262
Littré, Emile, 43
Loir, Adrien, 7–8, 46, 80–83, 313nn.76–77; account of anthrax trials by, 21, 26, 78–79, 149–50, 155–56, 158–59, 171, 326nn.17–18, 327n.24, 328n.30, 330n.71; deconstruction of Pasteur myth by, 274–77; on Pasteur's "private" rabies patients, 195, 198; on Pasteur's monopoly of anthrax vaccine, 174–75, 332n.90; on Pasteur's rabies research, 334n.39, 337n.27; on tension between Pasteur and Roux, 11, 234–36, 254–55, 338nn.1–2, 339nn.3, 4, 11; schoolboy account of Pasteur by, 273–74
Louis Napoleon (Emperor): abdication, 44; assassination attempt on, 30; dissolves Constituent Assembly, 44; Dumas' links with, 88, 126; reforms of, 123; support of Pasteur, 27, 31–32, 38, 41, 44, 89, 312n.49
Louis Pasteur: Free Lance of Science, 275
Louis Philippe, 53
Lumet, Louis, 16, 340nn.1–5, 341n.41
Lutaud, Dr. Auguste, 58, 220, 228, 312n.54, 332n.91, 333n.5, 337n.24
Lycée Saint-Louis, 24

malates, Pasteur's research on, 98–99, 102–103, 135–36
materialism: asymmetry and origins of life and, 141; Pasteur's attitude toward, 43; spontaneous generation debate and, 121–25, 127, 322n.30
Maulitz, Russell C., 339n.6
Maurice-Raynaud, R., 340n.1
Mauskopf, Seymour, 58, 60–61, 63, 314n.10, 315nn.14–15, 316n.28, 317nn.44–46, 318n.54, 319nn.61–62, 320n.30
McKie, D., 310n.24
Medawar, Peter B., 13–15, 310n.26
Meister, Joseph: chemical theory of immunity reflected in trials on, 248–50; ethical issues surrounding vaccination of, 17, 181, 193, 205–209, 215–17, 231, 233, 238–39, 335n.1, 336nn.2–3; treatment of, as reversal of Pasteur's previous research, 245–50
"Meister method" of rabies vaccination, 250–52
Mémoire sur les corpuscles organisés qui existent dans l'atmosphère, 114, 118–19
mercury, spontaneous generation theory and, 119–20, 130–32
Merton, Robert K., 12, 24, 310nn.20, 21, 332n.95
Metchnikoff, Ilya, 265
Method of Treating the Bites of Rabid Animals, 181
Microbe Hunters, The, 18, 93, 266–67
Microbes and Men, 276, 326n.18, 341n.50
microbe theory, Pasteur's fermentation research and, 103, 131–32, 163; Pasteur's research on, 93–94; rabies vaccine research and, 182–85, 191, 227
Millikan, Robert, 12–13, 309n.18
Milne-Edwards, Henri, 125–26, 129, 324n.44
Mitscherlich, Eilhard, 54, 59, 70–71, 81, 84–85, 87
molecular asymmetry, Pasteur's tartrate research and, 80–85, 318n.51
Montyon prize, 112
Morange, 340nn.12–13
Moreau, Richard, 270, 341n.33
Morson, Gary Saul, 309n.7
Muni, Paul, 267
Muséum d'histoire naturelle, 127
Musset, Charles, 125–26, 321–22
myth of Pasteur: deconstruction of, by Loir's manuscript, 274–77; historical context of, 277–78; limited access to family documents and, 273–74; Pasteur's health as factor in, 269–72; Pasteur's role in construction of, 267–69; Pastorians' posthumous perpetuation of, 272–74

National Assembly, award to Pasteur from, 260
National Guard, Pasteur's service in, 24–25, 53
National Library of Medicine, Pasteur manuscripts at, 3
nationalism: as catalyst in Pasteur's research activities, 219–20; cult of Pasteur and, 265; as spur to fermentation research, 108–109

Naturphilosophen: spontaneous generation debate and, 121–23, 322n.30, 323n.31
Newark Dog Scare, 266
Newton, Isaac, 139–40, 267
Nicol, Louis: on Pouilly-le-Fort trial, 324n.3, 331n.73; on Toussaint-Pasteur rivalry, 36, 49, 55, 59, 70, 328n.30, 329n.46, 330n.66
Nicolas of Greece (Prince), 259
Nicolle, Charles, 255, 327n.23, 339n.5
nutritional needs of pathogenic organisms, Pasteur's theories on, 245–46

"Olympian silence" of Pasteur, 6, 309n.10
"On the Asymmetry of Naturally Occurring Organic Compounds," 140–41
On the Origin of Species, 124
optical isomers: chronology of Pasteur's research on, 12, 57–58, 314n.10; fermentation research linked with, 95–103; hemihedrism and, 81–85; laboratory notebooks as key to, 11, 57–58, 66–73; Laurent's research influences Pasteur, 58–61, 315n.17; origins of life and, 87–89, 271–72; Pasteur's research on, 25–26, 36, 39, 56–58; tartrate research and, 53–55, 73–78; spontaneous generation debate and, 135–38. *See also* hemihedrism; specific tartrates
organic compounds: asymmetry of, 103–105, 139–42; Pasteur's research on, 35–36
"Organized Corpuscles that Exist in the Atmosphere," 108–109
Osler, William, 265–66
Outram, Dorinda, 319n.62
Owen, Richard, 124–25, 132, 323n.38, 324n.64
oxygen attenuation: anthrax vaccine development with, 148–50, 155–56, 166–69, 170–71, 325n.13, 326n.16, 327n.26, 328n.30, 329n.59, 330n.70; rabies vaccine development with, 184–88

Paget, Stephen, 265–66
papers and manuscripts of Pasteur, limited access to, 3, 47, 135, 274, 341n.38, 341n.44. *See also* laboratory notebooks of Pasteur
Papiers Pasteur: account of rabies vaccine research in, 178, 332n.2; description of, 3–4, 7–11, 306; Jupille's treatment recorded in, 336n.4. *See also* laboratory notebooks of Pasteur
paratartrates, Pasteur's research on, 39, 70–72
Pasteur and Modern Science, 94
Pasteur Exposed: The False Foundations of Modern Medicine, 275
Pasteur: Free Lance of Science, 94
Pasteur, Jean-Baptiste, 47, 261–62, 271
Pasteur, Louis: Bernard critiqued by, 18–21; Bouley's correspondence with, 160–62, 292–93; boyhood memories of rabies, 177–78; death of, 35, 259–60; early life, 22–23; financial resources and income, 40–42, 312n.58; government research support for, 40–41, 312n.50; health of, 45–46, 269–72, 313n.74; heroic images of, 260–63; honors and prizes awarded to, 34, 260–61; international reputation of, 264–67; marriage and family life of, 25, 46–47; nearsightedness of, 39, 45–46; outline of career of, 28; outline of major research interests, 37; patriotism and chauvinism of, 45; personality traits of, 38–39, 45–50; political activities of, 43–45, 88–89; post-optical isomer research of, 85–86, 318n.57; "private science" of, 5–7, 175–76; proprietary attitudes toward research, 5–7, 175–76; published correspondence described, 305–306; published works described, 305; religious beliefs of, 42–43, 133–38, 259, 312n.59; reputation of, 10–11, 23–35, 37, 262–63, 269–72; research activities of, 32–45, 85–86; Roux's relations with, 48–49, 234–39, 253–55, 268, 272–73, 338n.2, 339nn.5–8; scientific research assessed, 35–45; secrecy as obsession of, 6–7, 20–21, 175–76, 223, 226–29, 309n.10; self-promotion by, 267–69; seventieth birthday celebrated, 261–62; spontaneous generation debate with Pouchet, 110–42; suppression of laboratory notes on anthrax vaccine development, 151–59; teaching reputation of, 29–31, 48–49; Toussaint's rivalry with, 160–69, 172–76; working habits of, 8–9, 47–49, 187–88. *See also* specific achievements, e.g. anthrax and rabies vaccines
Pasteur, Marie-Louise, 47

Pasteur, Mrs. Louis, 22, 46–48, 138–39
Pasteur Plagiarist, Imposter!, 275
pasteurization technique, emergence of, 91
patents: Pasteur income derived from, 41–42, 58, 312n.53; procedures for, 160; U.S. patents awarded to Pasteur, 266
paternalism of Pasteur's research group, 48–49, 313n.87
Paul, Harry, 94
Pennetier, Georges, 45–46, 125, 323nn.39–40
peripneumonia, Pasteur's research on, 182
Perrot (Mayor of Villers-Farlay), 177–78, 207, 212, 332n.2, 336n.16
Peter, Dr. Michel, 50, 220–29, 240, 242, 251, 253, 337n.26, 338nn.47–48
physics, Pasteur's thesis in, 60–61
Poincaré, Raymond, 259–60
polarimetry: Pasteur's fermentation research and, 102–103; Pasteur's optical isomer research and, 56–57, 61, 81–85
polio vaccine, 229
politics: Pasteur's work and, 43–45, 88–89; spontaneous generation debate and, 112–13, 121–25, 131, 323n.32
Pope Pius IX, 123
Popper, Karl, 14
popular press, treatment of Pasteur in, 220, 337n.23
Porter, J. R., 310n.24
positivism, Pasteur's rejection of, 43
"post-exposure" rabies trials, account of, in laboratory notebooks, 238–42
potassium bichromate: antiseptic action of, in anthrax vaccine, 21, 149–150, 173–76, 326nn.18, 331n.85; results of Chamberland's experiments with, 299–300; trials with, recounted in laboratory notes, 151–59, 171, 327n.26
Pouchet, Felix-Archimède: Pasteur's debate with, 110–42, 225, 321–22; spontaneous generation research of, 112–13, 125–26
Pouchet, George, 127
Poughon, Julie-Antoinette, 198–205, 231
Pouilly-le-Fort, anthrax vaccine trial at, 145–76; background to, 170–71; diagram of trial, 157; laboratory notes, suppression of, 151–59; public account of, 146–50; revisionist accounts of Pasteur's role in, 276–77; Roux's role in, 237–39; text of protocol for, 281–85
Pravaz syringe injections, rabies vaccine research with, 195–205, 215–18
"preconceived ideas," Pasteur's embrace of, 16, 324n.68; Pasteur's fermentation research and, 96–103; Pasteur's optical isomer research and, 87–89; spontaneous generation theory and, 132–38, 324n.68; Tyndall's skepticism regarding, 5, 309n.6
Preminger, Alex, 341n.42
Prévost, Marie-Laure, 310n.32
Priestly, Eliza, 240n.21
private science: as public knowledge, 18–21; construction of scientific knowledge and, 12–16; defined, 5–7. *See also* sociology of scientific research
Prussian Ordre Pour le Mérit, Pasteur's refusal of, 45
pyroelectricity, tartrate research and, 71, 85–86

quartz, hemihedrism in, 84–85, 99

"Rabies," 335n.1
rabies vaccine: alternative treatments of rabies and, 230–31; corneal transplant transmission of rabies and, 202; critiques of Pasteur's research on, 206–33; "dead" vaccine, development of, 248–49; deaths of patients following treatment with, 222–29; diagrams of Pasteur's path to, 194, 244; ethics of research, in Pasteur's work on, 16–18, 177–81, 193–205, 220–31, 253–55; incubation period of disease and, 179–80, 184–88, 190–91, 230–31, 247, 250; "intensive method," ethics of, 253–55; laboratory notebooks as key to, 8–9, 11, 181–88; "Meister method" (from least virulent to most virulent), 250–52; morbidity and mortality rates with disease, 10, 178–81, 229–31, 333n.5, 338n.57; myths surrounding disease and, 179;on nonfatal cases of, 179, 333n.10; "paralytic" vs. "furious" form of rabies and, 189, 228; Pasteur's research on, 33, 35, 212–18; "post-exposure" experimental trials, 238–42; post-Pasteur developments of, 231, 338n.60; pre-vaccine pathogenesis and epidemiology of rabies, 177–81, 301–304, 333n.9; "private patients" and human experimentation, 177–81, 193–205; publications by Pasteur on, 188–95;

rabies vaccine (*cont.*)
revisionist accounts of Pasteur's research on, 276–77; Roux's tension with Pasteur regarding, 238–39; "spontaneous" rabies and, 179, 186–88; uncertainty of rabies diagnosis ("false rabies"), 200–205, 220, 335n.56 "vaccinal substance" theory, 247–48
racemic acid: chemical formula for, 54; fermentation research and, 102–103; Pasteur's early research on, 25–26
Ramon, G., 326n.18
Recueil de médecine vétérinaire, 328n.30
Reid, Robert, 326n.18
religion: asymmetric forces and origin of life and, 138–42, 272; Pasteur's lack of interest in, 42–43, 312n.58; spontaneous generation theory and, 112–13, 121–25, 323n.32
Renan, Ernst, 124, 333n.12
republicanism: Pasteur's early embrace of, 43–44; political reaction against, 122–25
"Researches on Dimorphism," 62
rhetoric, role of, in Pasteur's research, 268–69, 341n.32
Rigal, Dr., 195, 197–98, 200, 203
Ritvo, Harriet, 333nn.6–7, 338n.45
Roll-Hansen, Nils, 321–22, 341n.29
Root-Bernstein, Robert, 324n.80
Roqui, Jeanne-Etiennette (Madame Pasteur), 22, 46–48, 138–39
Rosenblatt, Roger, 333n.11
Rossignol, Hippolyte, 145–47, 170–71, 326n.16
Rous, Peyton, 274
Roussillat, Jacques, 36, 339n.12, 340n.39
Roux, Emile, 6–8, 33; anthrax vaccine trials, role in, 66–67, 70, 150, 158, 164–65, 168–69, 172–76, 237–39, 287–89, 294–96, 326n.20, 329n.49, 330n.62; chicken cholera vaccine research, 40; clinical mentality of, 255–56; Croonian Lecture of 1890, 326n.20; intracranial injection technique, 189–95, 213–18, 224; on "private" rabies patients, 195, 197; on Pasteur's marriage and family, 47, 313nn.77–78; pledges vaccine proceeds to Institut Pasteur, 261; Pouilly-le-Fort vaccine trial and, 146, 325n.6; public reticence about Pasteur, 255–56; rabies vaccine research with Pasteur, 182, 184–85, 187, 189–95, 235, 246, 250, 252–55, 334n.39; relations with Pasteur, 48–49, 234–39, 253–55, 268, 272–73, 338n.2, 339nn.5–8; reservations regarding rabies vaccine, 252–55; warns Pasteur about his obsession with secrecy, 226
Royer, Clémence, 124, 323n.37
Rudwick, Martin, J., 323n.31
Ruhmkorff apparatus, 139–40
Russia, cult of Pasteur in, 265

"saliva microbe": Pasteur's rabies research and, 182–83
Salomon-Bayet, Claire, 318n.51, 341n.54
Sand, Georges, 110
Sapp, Jan, 309n.16
Schwann, Theodore, 106, 115
Science and Industry in the Nineteenth Century, 93
scientific creativity, philosophical and sociological analysis of, 14–16
scientific knowledge, construction of, fraud and, 11–16
Scientific Method: critiques of Pasteur's rabies research and, 226–29; influence of, on medical research, 219–20; role of, in scientific research, 14–16; scientific fraud and, 10; spontaneous generation debate and, 129–33; theories on origin of life and, 321–22
scientific papers: Pasteur's publications on rabies research, 188–95, 212–18; role of, in research, 13–14, 310n.24
scientism, cult of Pasteur and, 264–65
Second Republic, formation of, 24–25
Seignette salt, Pasteur's research on, 68–69, 317n.45
serial passage of viruses: post-exposure experimental trials and, 242–43; rabies vaccine research and, 185–88, 190–91, 212–18, 334n.33. *See also* antiseptic vaccines; attenuation techniques; intracranial injection
Serres, Antoine, 125
Shapin, Steven, 341n.31
silkworm blight, Pasteur's research on, 11, 28–29, 32, 276
Simmonet, Henri, 340n.13
smallpox vaccine, 172, 228–29

Société chimique de Paris, 56, 103–105, 118–19
Société de Pharmacie, Pasteur receives prize from, 26
sociology of scientific knowledge: critiques of Pasteur's rabies research and, 222–29; private science concept and, 5–6, 14–16, 132–33; rhetoric as factor in, 269, 341n.32
sodium ammonium tartrate: hemihedral crystals in, 54–55, 76–78, 317nn.45–46; Mitscherlich's research on, 54–55; Pasteur's notebook entries on, 64, 70–72, 74–85, 316n.29, 317n.44
sodium ammonium paratartrate: Mitscherlich's research on, 54–55; Pasteur's research on, 76–85, 318n.47
"Some Reflections on Science in France," 45
Sorbonne: Pasteur's appointment to, 31–32; Pasteur's spontaneous generation lecture at, 110, 119–21, 129–30
Souvenirs littéraires, 270, 341n.33
"sowing" of microbes: lactic fermentation and, 91
Spectator, 266
spinal cord emulsification: attenuation with caustic potash, 213–15; first trial (from most virulent to least virulent), 243–50; "Meister method" (from least virulent to most virulent), 250–52; rabies vaccine and, 203, 227
spiritualism, 43
spontaneous generation: Académie des sciences and, 125–29; Pasteur's research on, 32–33, 268, 337n.23; Pasteur-Pouchet debate over, 110–42; religious and philosophical aspects of, 43, 110–13, 121–25; Scientific Method and, 129–33
"spontaneous ovulation" theory, 112–13
statistics: in Pasteur's rabies vaccine research, 220, 224–31; success rate of Pasteur's dog trials with, 240–43
Statue Within, The, 263–64
Story of Louis Pasteur, The, 267
"Story of a Scientist by a Layman, The," 56, 314n.7
Strick, James, 323n.50, 324n.80
"swan-necked" flasks: role of, in spontaneous generation research, 116–18
swine fever, Pasteur's research on, 182, 191, 228, 234, 245

Taine, Hippolyte, 123
tartrates: anomalies of, Pasteur's research on, 66–73, 79, 318n.47; "atomic volume" concept and, 63–64; crystalline structure of, 54–55, 66–73; empirical research and, 78–85; optical isomer research with, 53–55, 73–78, 136–37; Pasteur's notebook entries on, 39, 64–73, 316n.29; Pasteur's physics research on, 60–61; "privileged material" concept of, 80, 318n.51;
Temple, D., 341n.29
Théodoridès, Jean, 14, 29, 31–32, 326n.18, 329n.55, 331n.85, 333n.9, 334n.27; revisionist assessment of Pasteur and, 276, 341n.48
Third Republic: formation of, 32; Pasteur's rejection of, 44; scientism in, 265; state-supported science research during, 41
Thomas, Lewis, 180–81, 333n.11
Toussaint, Jean-Joseph Henri: Bouley and, 160–65, 328n.36; chemical theory of immunity and, 247; chicken cholera research parallels Pasteur's, 159, 166–69, 329n.66; mental health of, 173–74, 332n.86; Pasteur's competition with, in vaccine development, 158–76, 182, 276, 286–98, 327n.21, 328n.30, 329n.59, 330n.70, 331n.85; publicity surrounding vaccine research of, 160–61; sealed note (*pli-cacheté*) of, 160–65, 286, 328n.36; speech to French Association for the Advancement of Science, 165–66, 329n.52
transformism, spontaneous generation debate and, 121–22, 127, 322n.30
Treatment of Rabies: An Experimental Critique of the Pastorian Techniques, The, 224
Trousseau, Dr., 200–201
Tyndall, John: skepticism regarding "preconceived ideas," 5, 309n.6; spontaneous generation research, 131

U.S. Center for Disease Control, 229, 338n.57
United States, cult of Pasteur in, 265–66
University of Strasbourg, 139

vaccines. *See* specific vaccines
Valléry-Radot, Louis Pasteur: as guardian of Pasteur's papers, 3, 44, 47, 135, 274, 306, 341n.38; on Pasteur's political views,

Valléry-Radot, Louis Pasteur (*cont.*) 66–67, 313n.63, 319n.63; on Pasteur's religious beliefs, 312n.59; on Pouchet correspondence, 321–22; publication of selected Pasteur papers, 4, 40, 311nn.21–22
Vallery-Radot, Maurice: on myth of Pasteur, 32, 341n.28; on Pasteur's political views, 313n.63, 319n.63; on Pasteur's religious beliefs, 312n.59
Valléry-Radot, René, 46, 311nn.13, 33–34, 313n.78; "Eureka" account of Pasteur's optical isomer discovery, 53, 56, 314n.7; on fermentation research by Pasteur, 92–95, 319n.3; marriage to Marie-Louise Pasteur, 47; Meister experiment recounted by, 335n.1; myth of Pasteur, role in creation of, 22, 273–75, 277–78, 340n.1, 341nn.42–43; on Pouilly-le-Fort trial, 324n.3; on preconceived ideas, as embraced by Pasteur, 324n.68; on rabies vaccine research, 332nn.1–2
van Leeuwenhoek, Antonie, 39
Vandervliet, Glenn, 323n.60
Viala, Eugène, 313n.87
Vie de Jesus, 124
virulence of rabies virus, research on: "Meister method" (from least virulent to most virulent), 250–52; most virulent to least virulent strains, Pasteur's trials with, 243–50; rabies vaccine development and, 185–88, 190–91, 213–18, 228–29, 334n.33; relativity of, 185–86; *vs.* quantities injected, Pasteur's theories on, 246–47. *See also* immunity; serial passage of viruses
viruses: rabies vaccine research and, 185–90, 334n.33
Von Frisch, Anton, 38–40, 223–24, 337n.36, 338n.59
Von Liebig, Justus, 42, 100, 264; antagonism toward Pasteur, 50, 108; fermentation research of, 106–109
Vone, Thédore, 206
Vulpian, Dr. E.F.A., 206–207, 215–18, 222, 238–39, 336n.2, 337n.26

Wachhorst, Wyn, 341n.56
Wade, Nicholas, 309n.16, 332n.96
Walton, John K., 338n.45
Webster, Leslie T., 338nn.59–60
Weis, René, 319n.62
Weisberg, Richard, 341n.37
Wellcome Institute for the History of Medicine, 3
Wright, G., 323n.32
Wrotnowska, Denise, 311nn.15–16; on Toussaint-Pasteur rivalry, 92, 328n.30, 329n.55, 332n.86

yellow fever, Pasteur's research on, 182

Zeldin, Theodore, 340n.11
Zuckerman, Harriet, 309n.16